Introductory Astronomy

Introductory Astronomy

Keith Holliday
Department of Physics and Applied Physics
University of Strathclyde

JOHN WILEY & SONS
Chichester · New York · Weinheim · Brisbane · Singapore · Toronto

Baffins Lane, Chichester,
West Sussex PO19 1UD, England

National 01243 779777
International (+44) 1243 779777
e-mail (for orders and customer service enquiries): cs-books@wiley.co.uk
Visit our Home Page on http://www.wiley.co.uk
or http://www.wiley.com

Other Wiley Editorial Offices

John Wiley & Sons, Inc., 605 Third Avenue,
New York, NY 10158-0012, USA

WILEY-VCH Verlag GmbH, Pappelallee 3,
D-69469 Weinheim, Germany

Jacaranda Wiley Ltd, 33 Park Road, Milton
Queensland 4064, Australia

John Wiley & Sons (Asia) Pte Ltd, Clementi Loop #02-01,
Jin Xing Distripark, Singapore 129809

John Wiley & Sons (Canada) Ltd, 22 Worcester Road,
Rexdale, Ontario M9W 1L1, Canada

Library of Congress Cataloging-in-Publication Data

Holliday, Keith.
Introductory astronomy / Keith Holliday.
p. cm.
Includes index.
ISBN 0-471-98331-4 (hardcover : alk. paper). — ISBN 0-471-98332-2 (pbk. : alk. paper)
1. Astronomy. I. Title.
QB43.2.H65 1998
520—dc21 98-18192
CIP

British Library Cataloguing in Publication Data
A catalogue record for this book is available from the British Library

ISBN 0 471 98331 4 (cloth)
ISBN 0 471 98332 2 (pbk)

Typeset in Great Britain by Alden Group, Oxford.
Printed and bound in Great Britain by Biddles Ltd, Guildford and King's Lynn.
This book is printed on acid-free paper responsibly manufactured from sustainable forestry, in which at least two trees are planted for each one used for paper production.

Contents

Preface

The principal objective of this book is to provide an affordable introduction to astronomy. The presentation is scientific rather than sensational and it is intended that the reader will emerge with an understanding of the workings of the universe rather than a scattering of facts and figures. The book covers all aspects of astronomy, starting with observation from Earth and moving on to the Solar System, the workings of stars, the structure of galaxies, and theories of how it all came to be and where it will all end. A fuller description of the content of the book can be found in the Introduction so what follows are a few personal words that talk round the book rather than about it.

The book was inspired by the course for non-astronomers that I teach at the University of Strathclyde. I hope that it will help my students to understand the concepts associated with astronomy and also enhance their enjoyment of the subject, and that students elsewhere and, in particular, the general reader will find the book both inspiring and useful.

Introductory Astronomy has benefited greatly from help received in the course of writing it. In particular from four scientists who reviewed the first draft of the text and who supplied me with lists of mistakes, mis-spellings and mishaps. Their comments were intelligent, enormously useful, and helped to transform the draft into the book you now see. Special thanks are due to Dr Andy Agousti of Kingston University, Dr Phil Willmott of the University of Zürich, Dr Bill Barnes of the University of Exeter, and Professor Christopher Kitchin, director of Hertfordshire University Observatory. Another vital contribution came from my mum, not only for years of encouragement but, more practically, for typing the text. This allowed me to spend my evenings writing at home rather than in my cold office at the university. Kelly McNamara, my fiancée, showed devotion beyond the call of duty in helping me index in that very same cold office (in July!). I would also like to thank my dad and many friends around the world for putting up with being alternately ignored and then moaned at! Thanks for sticking with me.

The work that is presented here is very definitely not mine. I merely present the accumulated knowledge and understanding of centuries of work by astronomers and physicists. I'm not about to start thanking Newton and Einstein but the most relevant acknowledgement is to the many scientists and support workers in various divisions of NASA whose efforts have allowed me to reproduce photos of everything from the Earth to the edges of the universe (if you'll pardon the expression). If you want to see more of their work I can recommend a couple of sites on the Internet. They are *http://nssdc.gsfc.nasa.gov* and *http://oposite.stsci.edu/pubinfo/pictures.html*. The starting point for all NASA sites is at *http://www.nasa.gov*.

As this is being written before the book has been published, it is not easy to know who to thank (or blame!) at Wiley. For now, I'll just thank Andy Slade for his calming words and maybe get to everyone else when the second edition appears!

K.H.

Introduction

An introduction to *Introductory Astronomy*! The whole of the universe in a couple of pages? Well no, just a few words to describe what is to come, how it is arranged and how this book might best be used and enjoyed.

The book is written in narrative style and it is recommended that the reader starts at the beginning and moves forwards until reaching the end! For the more informed reader, the book can function as a reference work to be dipped into randomly but it is really aimed at the beginner who should therefore begin at the beginning! There is a liberal sprinkling of worked examples throughout the book. They are placed with care and should ideally be read as they appear though the non-mathematically minded should rest assured that these examples can be skipped without losing the main drive of the book. The vast majority of worked examples serve to illustrate simple ideas and many do not use mathematics at all.

Though the book, by definition, is concerned with things extra-terrestrial, it starts with its feet firmly planted on the Earth. The first two chapters are concerned with the light that showers down upon us from the heavens. The nature of light is explained and a few glimpses as to what it will tell us later in the book are afforded before the collection and imaging of light with the use of telescopes is discussed.

Knowing where to look in the sky for objects of interest and making a telescope stay with a chosen body as the night goes by is of obvious importance. Such problems are dealt with in Chapters 3 and 4 and the discussion leads on to related matters such as why the Earth's weather is seasonal, why the stars of the night sky vary throughout the year and why astronomers must use a different time system from that of 'normal' people.

In Chapter 5 the narrative blasts off for outer space with a discussion of the orbiting bodies of the Solar System; the planets and their moons, asteroids, comets and meteoroids. These bodies are seen only by the light that they reflect and this introductory treatment focuses on what we see both from Earth and at closer quarters.

The Solar System has often been likened to a clockwork toy that never needs winding up, apparently repeating its motions endlessly. Chapter 6 investigates and elucidates the force that drives the motions of the Solar System: gravity. The approach given here is similar to that taken more than three centuries ago by Newton, based on the rules of planetary motion developed by Kepler. That the Solar System is far from being a simple clockwork toy is shown by a series of examples that include tidal forces and their actions, synchronous orbits and resonant orbital interactions such as the production of Cassini's division in Saturn's rings. The theory of gravitation developed here is applied to stellar and galactic phenomena later in the book.

With a knowledge of the way that the Solar System looks and how gravity operates it becomes possible to discuss how the Solar System came to be in the first place (about 5 billion years ago) and why it is how it is now. A paradigm for the formation and evolution of the Solar System comprises Chapter 7. The following two chapters take a much closer look at the bodies of the Solar System, including their internal structure and processes that are taking place both in their bodies and in their atmospheres.

The master of the Solar System, the Sun itself, is detailed in Chapter 10. The Sun is just an average star that happens to be very close to us on our little planet. However, the information that has been gleaned from studying the Sun is of critical importance to understanding stars of all types. From the conversion of mass to energy via nuclear reactions in the core of the Sun to the transport of the energy to the surface and its interaction with the solar atmosphere, all is salient to general stellar studies and so is described in detail at this stage.

A start on describing the universe of stars beyond the Solar System is made in Chapter 11. This is a sort of mechanics' guide to stars; how various stellar parameters relate to each other, how these parameters can be measured and what combinations of parameters appear to be allowed. The important technique of spectroscopy is introduced here and the use of the Hertzsprung–Russell diagram introduced.

Chapters 12 and 13 build on the information collected in the previous three chapters (and a little from Chapter 7) to describe the life history of stars of all sizes. During the early years not too much varies but as stars get older, all the exotic terminology of astronomy—red giants, white dwarfs, supergiants, supernovae, neutron stars and black holes—appear. The story is presented as a physical model before the observational evidence for these fascinating objects and events is described.

How stars are organised throughout space is discussed in Chapter 14. The Milky Way is described, including the techniques that can be used for mapping our galaxy while sitting at one edge, apparently unable to obtain perspective. Local galaxies are mentioned but so too are extraordinary active galaxies such as quasars that eject enormous quantities of energy and matter from their centres.

An attempt to give a history of all that has ever been and will ever be is given in Chapter 15! Cosmology is perhaps the most controversial area of astronomy but also one of the most exciting. A collection of evidence is presented that includes the recession of the galaxies (described by Hubble's law) and studies of the cosmic microwave blackbody radiation. The implications of these findings is discussed in terms of the Big Bang theory that suggests the current form of the universe sprang into action about 15 billion years ago.

At the end of every chapter there are ten questions. The problems are intended to provide practice with the mathematical aspects of the subject. The exercises are suggestions for revision tasks or sometimes essay topics that mainly rely on recycling the text of the book but which are also useful self-tests of whether the main ideas have been digested and understood. Most chapters also contain a few teasers that aim to make the reader think a little beyond what is presented in this text and truly let the mind wander among the stars!

Finally, it is worth commenting on a few things that are not here. The principal objective of this book is to provide an affordable insight into astronomy, to understand the universe on a budget! For this reason it is not packed with colour, either literally in terms of palette or metaphorically in terms of tales of the great astronomers. This book presents all that is important in black and white, just as a Yorkshireman should tell it!

1 Light

Astronomy is the science that more than any other relies on observation. The physicist, the chemist and the biologist can all devise experiments that will fit neatly into a laboratory. The size of the apparatus may sometimes be very large, maybe a particle accelerator that is several kilometres long or a jungle that is many thousands of square kilometres in area, but it is always on a scale that is accessible. The astronomer's interest doesn't really awaken until you get to the Moon and that is the furthest a human has ever travelled. For the most part, the astronomer's realm is all those places that humankind has never been. To understand what is happening in the rest of the universe we rely on what *it* sends *us*. Astronomers collect light by pointing telescopes at the sky and try to work out what the patterns of brightness and darkness are telling them. To understand the message it is first necessary to understand the messenger.

The messenger is light. Stars, including the Sun, are generators of light and from the colours and intensities of the light it is possible to diagnose the processes that produce it. Other bodies, such as planets, reflect light. At the simplest level this allows us to see them but further analysis provides information on details such as the constitution of planetary atmospheres and the type of material that composes their surfaces. The sparse gas and dust that lies between the condensed bodies of the universe can also be detected and identified as it absorbs and scatters light. Extremely massive objects such as galaxies can even cause detectable changes in the path of light that is otherwise bound to travel in straight lines. To arrive at a sensible interpretation of the information that is raining down upon us, the behaviour of light must be understood. Fortunately, the astronomer can rely on scientific colleagues who have been busy answering just such questions in their laboratories on Earth.

Light as a Wave

There is more than one way of describing the nature of light depending on which aspect of its behaviour is under examination. To start with, light will be considered to be a wave. Like waves on the ocean, light waves move in a certain direction, with a characteristic distance between peaks and troughs and at a definite speed, all of which depend on the prevailing conditions. Unlike mechanical waves, light does not depend upon the presence of a medium to propagate. In fact, matter tends to get in the way of light, causing it to be variously reflected, scattered, slowed or deflected. This is because the waves that compose light are oscillating electric and magnetic fields that interact with the electrically charged electrons and nuclei that make up the majority of the universe's known matter. It is reasonable to think of such an electromagnetic wave as being like an ocean wave, so long as it is remembered that electromagnetic radiation is not made of matter and that the light that we see is composed of waves with peaks only around 5×10^{-7} m (half a micron, or micrometre) apart.

To understand the idea of a light wave consider Figure 1.1. The wave is travelling in the z-direction. The magnitude of the electric field varies sinusoidally in the y-direction and the magnetic field oscillates with the same waveform but in the x-direction. The electric and magnetic fields always have the same waveforms as each other and their magnitudes are always in the same proportion. It is the distance between peaks, known as the wavelength and denoted by λ, that determines the nature of the light. Daylight, provided

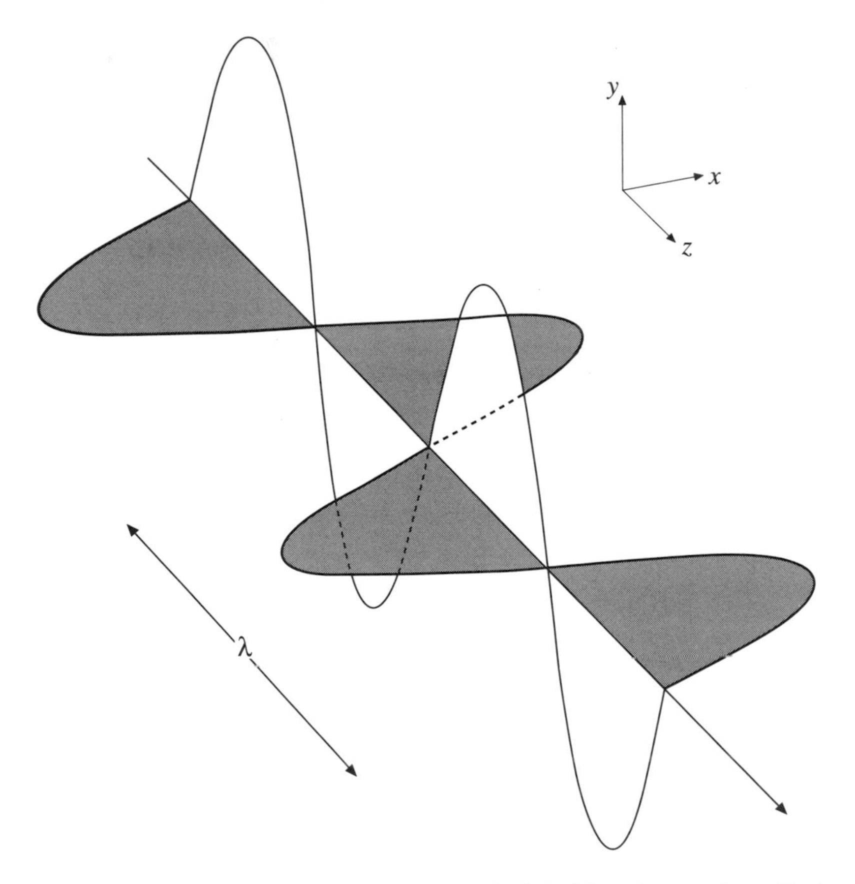

Figure 1.1 The form of an electromagnetic wave. Electric (in bold) and magnetic oscillating waves, with sinusoidal form, are in phase with each other though in perpendicular planes. Any part of the wave (for instance, a particular peak) travels through space at the speed of light, c. The wavelength, λ, is indicated as being the distance between repetitions of the waveform. The period of the wave is given by the time, at any point in space, taken by the wave to repeat itself, that is, complete an oscillation. The frequency, f, is the reciprocal of the period, that is, the number of oscillations per second of the electric and magnetic field intensities at any point in space

by the Sun, is composed of waves of many different wavelengths. Together they add up to white light but the waves can be sorted according to constituent colours, by a rain storm to create a rainbow or by a prism, for instance. Then the light is spread according to its wavelength. The human eye can distinguish the visible spectrum into a few hundred shades[1] (or traditionally only the seven colours of the rainbow) but actually the wavelength of light varies continuously. At the red end of the visible spectrum the wavelength is about twice that of the violet end but in both cases it is still less than one thousandth of a millimetre

The speed of the wave in a vacuum is a fundamental constant of nature and is usually denoted by c. The value of c is about 300 million metres per second. It is slowed by less than 1% when propagating through a gas, such as air, and by a few tens of a per cent by condensed matter, like water or glass. For now, it is only important to consider a ray of white light. The white light is composed of contributions at all wavelengths but each component is travelling at the same speed. It is clear that peaks will arrive at different rates depending on the wavelength of the wave in question. The frequency, f, at which peaks (or troughs) arrive is simply the ratio of the speed of the wave divided by its wavelength. Rearranging this relationship leads to the expression

$$c = f\lambda \tag{1.1}$$

Substituting values for c and λ for visible light shows that the frequency of oscillation of visible light is more than 10^{14} hertz (cycles per second). The eye is fooled into believing that television pictures are continuous when in fact they are only updated 50 times per second so that the frequency of oscillation of light is enormously beyond human range. We see light in continuous rays that are either off or on. Frequency and wavelength characteristics are only apparent to us because our brains are able to interpret in terms of colour the messages that chemical receptors in the eye send it.

WORKED EXAMPLE 1.1

Q. The eye can see light at wavelengths between about 350 nm and 800 nm. To what frequency range does this correspond?

A. The solution can be obtained by rearranging equation (1.1), knowing that $c = 3.0 \times 10^8\ \mathrm{m\ s^{-1}}$:

$$\text{For } \lambda = 350\,\text{nm},\quad f = \frac{c}{\lambda} = \frac{3.0 \times 10^8}{350 \times 10^{-9}} = 8.6 \times 10^{14}\,\text{Hz}$$

$$\text{For } \lambda = 800\,\text{nm},\quad f = \frac{c}{\lambda} = \frac{3.0 \times 10^8}{800 \times 10^{-9}} = 3.8 \times 10^{14}\,\text{Hz}$$

So the frequency of visible light varies from about 400 THz to about 900 THz.

---

[1] The human eye can distinguish between literally millions of different colours but most are multichromatic, that is, composed of mixtures of different wavelengths. Only a few hundred monochromatic colours can be distinguished.

Light and Time

It is clear that light exists on a scale that is remote from human experience. Consider sitting at the back of an auditorium and observing a lecturer speak. Because sound propagates more slowly than light, the speaker's lips appear to move about 0.1 s before the sound arrives. In this case the travel time of the light is effectively zero. For light to take a period of 0.1 s to arrive the source would have to be 30 000 km away, a distance greater than the diameter of the Earth. This is beginning to enter the realm of the astronomer. The Moon is about 400 000 km away and so light takes more than 1 s to reach us from there (hence the reason that there were always short pauses in conversations between the men that went to the Moon and their support team on Earth). Light from the Sun takes over 8 minutes to reach the Earth. Light from the next nearest star travels for more than four years and from nearby galaxies hundreds of thousands of years. For light to cross a typical galaxy it would take about 100 000 years. The distance that light travels in a year, about 10^{13} km,[2] is a convenient unit for describing stellar and galactic distances and is called a light year.

The implication of the enormity of astronomical distances is that the finite nature of the speed of light can no longer be ignored. The further away one looks, the longer ago it was that the events we observe now occurred. For instance, detecting light from a galaxy that is ten million light years away means that the events of 10 million years ago are being observed, before humans had evolved. This is, however, a short period of time on the scale of the universe. The universe as we know it is thought to have been created by the Big Bang about 10–20 billion years ago. Some objects (quasars) have been detected that are thought to be 12 billion light years away. In observing such objects we are observing the universe as it was when it was less than half its current age. A large percentage of the universe's life can be seen by looking in the right place and that implies that we can see a large proportion of the universe.

To return to the comparison between the magnitude of human comprehension and the size of the universe it is instructive to scale down the history of the universe to fit into one day, beginning at midnight. During this day, the Earth formed during mid-afternoon and plants began to produce oxygen in the early evening but humans didn't evolve until two seconds before midnight. Ferdinand Magellan first circumnavigated the globe in what we now consider to be an era barely beyond medieval times but on the reduced scale used here it was only three thousandths of a second before midnight. On this scale, we can each expect to live for less than one thousandth of a second.

The Electromagnetic Spectrum

Human frailty again becomes apparent when the spectrum of light is considered. The light that we can see has a wavelength that is restricted to the region between about 350 and 800 nm but electromagnetic radiation is not limited to this range. As the wavelength gets longer, beyond the red limit of our eyes, it is called infrared radiation. Heat detectors in the skin can sense strong infrared sources, for instance the glow of a raging bonfire. If

[2] The speed of light is 3.0×10^8 m s^{-1}. In one year there are $(60 \times 60 \times 24 \times 365) = 3.15 \times 10^7$ seconds. In one year light can therefore travel $(3.15 \times 10^7 \times 3.0 \times 10^8) = 9.5 \times 10^{15}$ m or 9.5×10^{12} km, just under 10^{13} km.

we could see infrared radiation then a beam of light would be visible every time we point a remote control at a television. The tiny lasers that 'read' the music on a compact disc also operate in this region of the spectrum. Once the wavelength of light has increased beyond about one millimetre it becomes known as microwave radiation. If our eyes could see in this region then domestic microwave ovens would glow brightly every time they are turned on (if we could also see through the protective door!). Beyond this region, as the wavelength increases beyond about one metre, a new name is used: radio. If radio waves were visible then we would never be left in darkness. Radio transmitters allow us to listen to music or hear the latest news almost anywhere on the planet. Fortunately, a package of electronics is required to detect and decode the radio signals before translating them into audible waves (that are mechanical rather than electromagnetic in nature). The wavelength of a typical radio broadcast is around one billion times longer than the limit of human sight.

As the wavelength of light gets shorter, beyond the violet limit of our eyes another name is adopted. Ultraviolet radiation is the part of the electromagnetic spectrum that tans human skin upon exposure. As the wavelength gets still shorter to about 10 nanometres, light becomes known as X-rays, used in medicine to diagnose broken bones among other things. Finally, very short wavelength electromagnetic radiation, below about 10 picometres, is known as gamma radiation. This is emitted by radioactive nuclei and is used to sterilise food so that it can be transported over long distances without rotting.

A summary of the electromagnetic spectrum is given in Figure 1.2. It is important to remember that the constitution of all parts of the spectrum is the same. It is only the

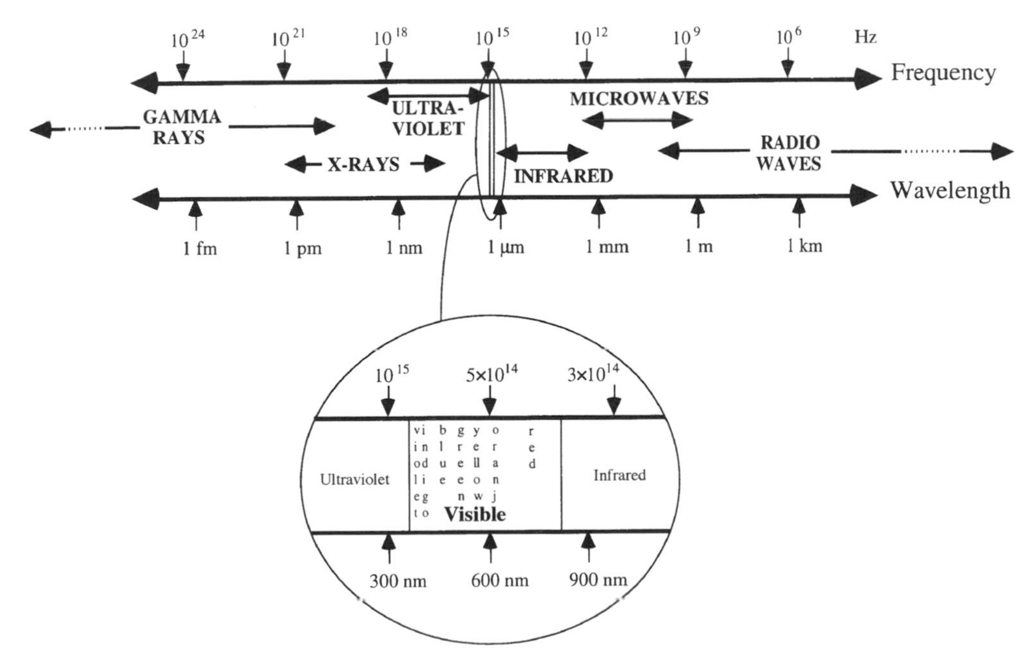

Figure 1.2 The electromagnetic spectrum. Note that the product of wavelength and frequency at all points in the spectrum is equal to 3×10^8 m s^{-1}, the speed of light in free space. The named regions of the spectrum do not have well defined cross-overs. The visible part of the spectrum is very small and is expanded below to indicate the wavelengths and frequencies of the colours of the rainbow

magnitude of the wavelength and frequency that vary. The different names given to parts of the spectrum are merely for convenience. Apparent differences in the behaviour of the different regions are due to the behaviour of the materials with which they interact and not intrinsic properties of the radiation.

The whole of the electromagnetic spectrum is important to the astronomer because observation is not limited to the eye. Light can be imaged and detected in all parts of the spectrum and the bodies that compose the universe send us light in all these regions. By understanding the circumstances under which light of different wavelengths is created or interacts with matter, a much bigger and better picture of the universe can be produced than that obtained by observation in just the visible part of the spectrum.

WORKED EXAMPLE 1.2

Q. A galaxy at a distance of 2×10^{22} km emits radio waves that are detected on Earth by a radio receiver that is tuned to the same frequency as 1FM at 99 MHz. What is the wavelength of the detected signal. How long ago did the events take place that are being detected?

A. Radio waves are electromagnetic radiation and therefore conform to the same rules as visible light. Rearranging equation (1.1) again:

$$\text{For } f = 99\,\text{MHz}, \quad \lambda = \frac{c}{f} = \frac{3.0 \times 10^8}{99 \times 10^6} = 3.03\,\text{m}$$

So the wavelength of the radio waves is a little over 3 m.

The radio wave travels at the velocity c and so, by definition, travels one light year per year. One light year is about 10^{13} km and so the galaxy is $(2 \times 10^{22}/10^{13}) = 2 \times 10^9$ light years away. This means that the detected radiation set off 2×10^9 years ago and so the observed events occurred 2 billion years ago. The same solution can be obtained by dividing the distance travelled (in metres) by the speed of light (m s^{-1}) to give the time of flight in seconds. This is 6.7×10^{16} s or 2 billion years.

Blackbody Radiation

Of greatest importance to the astronomer is a description and understanding of the light that is radiated by the energy sources, the stars. At the turn of the century Max Planck derived equations to explain the wavelength and frequency dependence of the radiation emitted by a condensed body of arbitrary temperature in thermal equilibrium. He showed that a definite spectrum[3] can be assigned to any body by simply measuring its absolute temperature[4], T, so long as it absorbs and radiates electromagnetic energy in an 'ideal' manner. Such an object is given the name 'blackbody' to indicate that it reflects no light of any wavelength (although light is emitted via radiation). Nothing behaves in a perfectly ideal way but condensed matter, including stars, approximate the results of Planck's

[3] A spectrum is a plot of the wavelength (or frequency) dependence of a process. The process in this case is the radiation of electromagnetic energy.

[4] The absolute temperature scale uses the kelvin (K) as a unit. The magnitude of the kelvin is the same as a degree Celsius but the zero point for the Kelvin scale is taken as absolute zero rather than the nominal zero (the temperature of solidification of pure water at one atmosphere of pressure) used for the Celsius scale. Absolute zero is defined as being the lowest possible temperature, that is, the temperature at which bodies have the minimum possible thermal energy. Absolute zero is $-273\,^\circ\text{C}$ and so a simple relationship between the two temperature scales is given by $T(C) = T(K) - 273$.

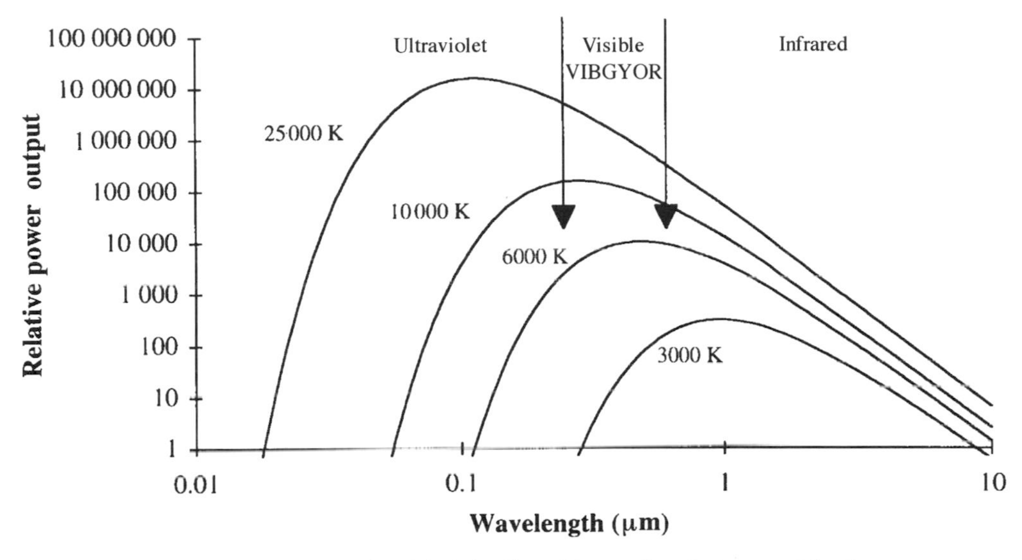

Figure 1.3 Energy emitted by perfect black bodies with equal surface areas. The four spectra are chosen to represent typical surface temperatures of stars but the shapes of the curves apply equally well to any black body. Note that both axes are drawn on exponential scales so that the energy per unit area emitted by the body at 25 000 K is very much greater than that at 3000 K

calculations closely. Figure 1.3 shows the wavelength characteristics for bodies at temperatures of 3000, 6000, 10 000 and 25 000 K, typical temperatures for stars. There are two obvious differences in the wavelength dependence of the electromagnetic energy emitted.

The first difference between the spectra in Figure 1.3 is that there are tremendous differences in the quantities of energy emitted by bodies of different temperatures, so much so that a logarithmic scale is required to be able to plot the curves for the selected temperatures on the same graph. The area under the graphs is proportional to the total power per unit area, P_A (in W m^{-2}) being emitted by the body and this is given quantitatively by the Stefan–Boltzmann law:

$$P_A = \sigma T^4 \tag{1.2}$$

where σ is Stefan's constant (5.67×10^{-8} W m^{-2} K^{-4}). The power radiated by two objects of equal surface area is therefore proportional to the fourth power of the body's absolute temperature so that, for instance, doubling the temperature will lead to a sixteenfold increase in power output.

The second clear difference between the spectra shown in Figure 1.3 is the shift in the wavelength, λ_p, at which peak energy is emitted. This can also be quantified and is given by Wien's displacement law:

$$\lambda_p T = 2.9 \times 10^{-3}\ \text{m K} \tag{1.3}$$

An important consequence of Wien's displacement law is the change in colour of a body as it is heated. At normal temperatures on Earth, objects emit energy in the infrared (at a peak wavelength of around 10 μm) but very little visible light. We see such objects due to the light they reflect. Materials can be heated until they become 'red hot', however. At temperatures of a few thousand kelvin the peak output wavelength is still in the infrared (at around 1 μm) but there is now also significant emission in the long wavelength part of the visible spectrum and so the object is seen by the eye to glow red. The coolest stars have surface temperatures of around 3000 K and so also appear red. For a temperature of 6000 K, the peak output wavelength is at 480 nm, in the blue part of the visible spectrum. However, as can be seen from Figure 1.3, there is significant radiation emitted across the visible spectrum with a slight bias towards the red end of the spectrum. To the eye, the object appears yellow like the colour of the Sun. This simple observation indicates that the temperature of the Sun is in the region of 6000 K. At 10 000 K the peak wavelength is on the boundary between violet and ultraviolet and the spread of light across the visible part of the spectrum is such that the eye sees the object as being white. In other words, the combination of the wavelength dependence of the light emitted and the sensitivity of the eye makes it seem as though the light is spread evenly across the visible part of the spectrum. When the temperature increases beyond 10 000 K the peak emission wavelength moves further into the ultraviolet and the bias towards the violet part of the visible spectrum increases making the object appear blue. The hottest stars have this appearance and have a surface temperature of a few tens of thousand kelvin.

WORKED EXAMPLE 1.3

Q. Use the fact that the Sun's peak output wavelength is at 500 nm to calculate the power it radiates per unit area.
A. The surface temperature of the Sun can be calculated using Wien's displacement law (equation (1.3)),

$$T = \frac{2.9 \times 10^{-3}}{\lambda_{\max}} = \frac{2.9 \times 10^{-3}}{5 \times 10^{-7}} = 5800\,\mathrm{K}$$

Using the absolute temperature of the Sun's surface it is possible to calculate the power radiated per unit area using the Stefan–Boltzmann Law (equation (1.2)),

$$P_A = \sigma T^4 = (5.7 \times 10^{-8}) \times 5800^4 = 6.5 \times 10^7\,\mathrm{W\,m^{-2}} = 65\,\mathrm{MW\,m^{-2}}$$

So the Sun radiates 65 MW of power from every square metre of its surface, enough to simultaneously power 10 000 electric showers if the energy could be converted efficiently. Only 1 mm^2 would be required to power an average light bulb.

WORKED EXAMPLE 1.4

Q. Calculate the Solar constant (the light power from the Sun that is incident on the Earth per unit area) using the fact that the Sun's diameter is 1.4 million km and is 150 million km from Earth.
A. From Worked Example 1.3, the Sun radiates 65 MW m^{-2} and so the total power radiated is this figure multiplied by the total surface area. The surface area of a sphere is given by πd^2, where d is the diameter of the sphere and so the total power output, P, is given by,

$$P = P_A \pi d^2 = (6.5 \times 10^7)\pi(1.4 \times 10^9)^2 = 4 \times 10^{26}\,\mathrm{W}$$

This power is radiated equally in all directions and so, at the Earth's orbit, it is spread over a sphere with a diameter of 300 million km (3×10^{11} m) and a surface area of $\pi(3 \times 10^{11})^2$ so that

$$\text{Power per unit area} = \frac{\text{Total power}}{\text{Total area}} = \frac{4.0 \times 10^{26}}{\pi(3.0 \times 10^{11})^2} = 1400\,\text{W m}^{-2}$$

The Earth receives about 1.4 kW of radiant power per square metre. Not all of this reaches the surface due to cloud reflection and atmospheric absorption.

Reflection of Light

It is important to draw the distinction between the observed colours of cool and hot objects. Cool objects can appear to have any colour due to variation in the wavelengths of light that they reflect. For instance, Mars is an example of a cool, red astronomical body for essentially the same reasons that a British post box is red. Their temperatures are only a few hundred kelvin, rendering their visible radiation too weak to see, but they reflect the light of the Sun more strongly in the red part of the spectrum than elsewhere. The difference between cool and hot objects on Earth is immediately obvious due to the much greater luminosity of hot objects. However, care must be taken when comparing the observed brightness of astronomical objects as nearby reflective, cool bodies often appear brighter than much more distant hotter objects. For instance, Venus, a nearby highly reflective planet, is often the brightest point source of light in the sky to the unaided eye despite the fact that it is visible solely due to the reflection of sunlight. Stars, on the other hand, may appear dimmer despite radiating enormous amounts of radiation, because they are much further away.

The Doppler Effect

A further complication in the appearance of objects such as stars can be caused by the relative motion of the light source and the observer. When a source of radiation is moving radially with respect to an observer, the separation between the source and the observer is varying. As the light always travels at the speed c then the time separation of the arrival of the wave peaks at the observer will vary, causing the frequency and therefore wavelength of the light to be shifted. This effect is known as the Doppler effect and is illustrated in Figure 1.4. It can be shown that the wavelength shift, $\Delta\lambda$, is almost independent of whether it is the source or the observer that is moving and is approximately proportional to their relative speed, v:

$$\Delta\lambda \approx \frac{v}{c}\lambda \tag{1.4}$$

Here, v can be positive, in the case of source and observer separating, which causes an increase in the wavelength, known as a red-shift, or v can be negative when the source and observer are approaching, causing a shortening of the wavelength, known as a blue-shift. Clearly, for the wavelength shift to be large the relative radial speed of the observer and

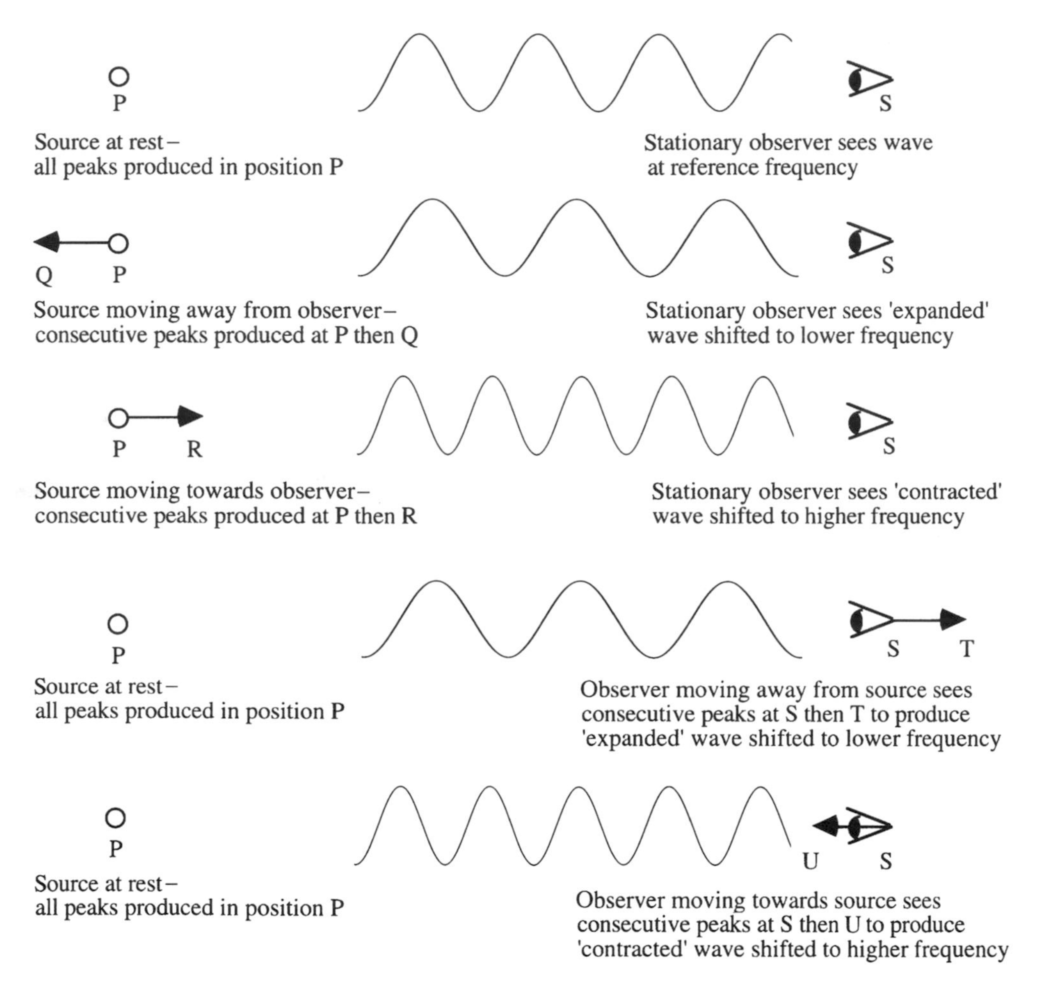

Figure 1.4 The Doppler effect. How the observed wavelength of light (or sound) is influenced by the motion of the source or observer

source must be very high so that the ratio v/c is significant. Small Doppler shifts can be measured using spectroscopic equipment rendering the effect of great importance in astronomy for determining the relative motion of planets, stars and galaxies. Such effects cannot be observed in everyday life on Earth but the sound analogue is clearly audible because the velocity of mechanical waves is much smaller than that of electromagnetic waves. An everyday example of this is the sound of a siren. As an emergency vehicle rushes towards the listener a certain tone, blue-shifted from the sound the driver experiences, is heard. After it passes, the sound becomes lower as it is now red-shifted compared to the sound heard when the vehicle is not moving. In terms of equation (1.4), v changes sign from negative to positive (approaching to receding) as the vehicle passes and so a proportional shift in tone of $2v/c_s$, where c_s is the speed of sound (about 1200 km h^{-1}), is heard.

WORKED EXAMPLE 1.5

Q. How fast would a spacecraft have to be moving away from the Sun so that its peak radiation wavelength appears to be 530 nm? What colour would the Sun appear to be?
A. From Worked Example 1.3, the Sun's peak emission wavelength is 500 nm when measured from the Earth and so the red-shift as observed from the spaceship, $\Delta\lambda$, is +30 nm. Rearranging equation (1.4) gives

$$v \approx \frac{\Delta\lambda}{\lambda} c - \frac{30}{500} 3 \times 10^8 = 1.8 \times 10^7 \text{ m s}^{-1}$$

To cause such a shift in the wavelength requires a very high recessional speed (65 million km h^{-1} in this case), a speed at which more complicated relativistic effects start to become relevant. The diagram shows the observed spectrum of the Sun in the visible region for a stationary observer (in bold) and for an observer in the spacecraft. Though the difference between light at 500 nm and 530 nm could be detected by eye, the difference in the overall appearance of the broad spectra shown in the figure is quite small. The Sun would appear to be slightly more orange due to this red–shift. Very large radial speeds are required before broad spectra are noticeably shifted.

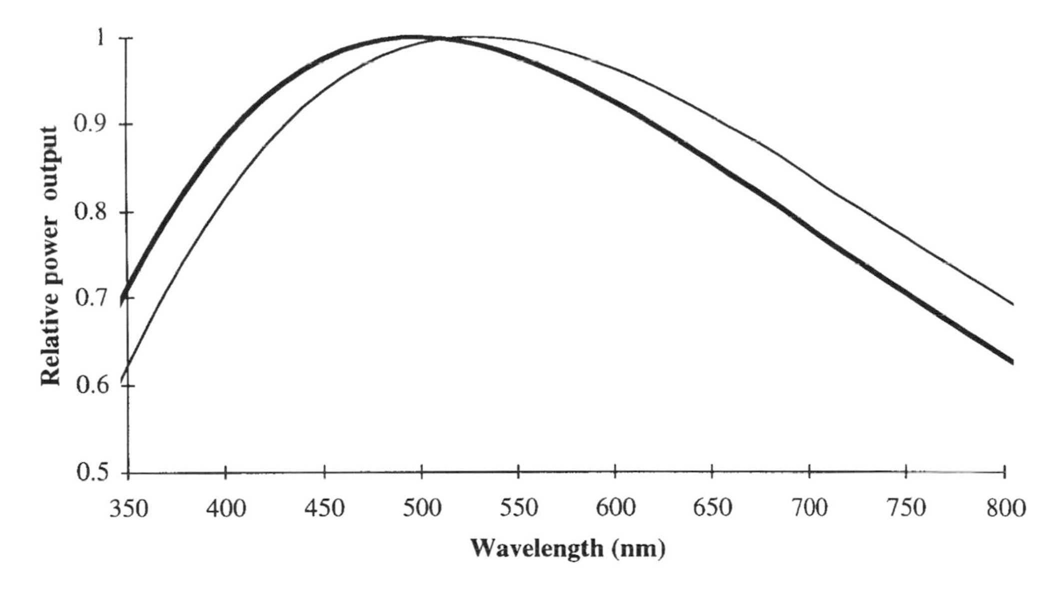

Light as a Particle

The final piece of information about light that is required before embarking on a course on astronomy is perhaps the most unworldly of all. The two words 'quantum' and 'physics', when used together, often strike fear into the hearts of the casual scientific browser. Fortunately, the implications of quantum physics for light are easy to understand even if the fundamental theory that underlies it is not. The central fact of the quantum theory of light is that energy may only be added or removed from a beam of light in definite sized packages or quanta, called photons. The energy, E_p, of a single photon is inversely proportional to the wavelength (and therefore, invoking equation (1.1), proportional to the frequency) of the light field of which it forms a part, thus:

$$E_p = hf = \frac{hc}{\lambda} \tag{1.5}$$

In this expression the constant, h, is known as Planck's constant and is equal to 6.6×10^{-34} J s. The very small size of this constant shows that a photon of visible light has an energy of the order of a few billionths of a billionth of a joule. On the scale of human perception this is a very small quantity and so, once again, there is no contradiction between the nature of light and persistence of vision. Under normal viewing conditions a multitude of photons rain down upon the eye's retina so that individual contributions cannot be recognised. It is only at the microscopic level, for instance when considering the interaction between light and an individual retinal cell, that the photon needs to be considered. Such issues may seem irrelevant to the astronomer who is interested mainly in vast bodies at enormous distances but the photon will prove to be a very handy diagnostic tool later in the book, often in conjunction with the Doppler effect. Both the wave and quantum approaches to light are required but it is always possible to pick the model that suits best for any particular circumstance.

WORKED EXAMPLE 1.6

Q. Over what range does the photon energy of visible light vary?
A. The solution can be obtained using the information in Worked Example 1.1 and from equation (1.5), knowing that $c = 3.0 \times 10^8$ m s^{-1} and that $h = 6.6 \times 10^{-34}$ J s.

$$\text{For } \lambda = 350\,\text{nm}, \quad E_\text{p} = \frac{hc}{\lambda} = \frac{(3.0 \times 10^8)(6.6 \times 10^{-34})}{350 \times 10^{-9}} = 5.7 \times 10^{-19}\,\text{J}$$

$$\text{For } \lambda = 800\,\text{nm}, \quad E_\text{p} = \frac{hc}{\lambda} = \frac{(3.0 \times 10^8)(6.6 \times 10^{-34})}{800 \times 10^{-9}} = 2.5 \times 10^{-19}\,\text{J}$$

So the energy of a photon of visible light can vary from about 0.2 aJ to about 0.6 aJ (1 aJ $= 10^{-18}$ J).

The astronomical messenger, a beam of light, may have set out billions of years ago from its unknowing source. Upon arrival at its destination an appropriate reception is required. The light must be collected, imaged and detected. That is the subject of the next chapter.

Questions

Problems

1 Calculate the approximate energy of photons at the borders of the main electromagnetic radiation regions.

2 (a) A spectrometer is able to detect wavelength differences of 0.01 nm. If the spectrometer is observing a laser beam of monochromatic light at 500 nm being sent from a spacecraft accelerating away from Earth, how fast would the spacecraft have to be moving before the spectrometer detects a shift in the wavelength of the light?

(b) If the eye can detect wavelength changes of 1 nm how fast would the spacecraft have to move to make the laser appear to change colour?

3 How long ago did the light that we observe now set off from the following objects?
(a) Venus (at its closest approach of 41 million km).
(b) Proxima Centauri (the closest star to the Sun at 4.0×10^{13} km).
(c) Betelgeuse (at 1.3×10^{16} km).
(d) Vega (at 2.4×10^{17} m).
(e) Stars at the opposite edge of our galaxy, the Milky Way (at 10^{18} km).
(f) Pollux (at 35 light years).
(g) The Large Magellanic Cloud (at 2×10^{18} km).
(h) The Andromeda Galaxy (the most distant object that can be seen with the naked eye at 2.1×10^{19} km).
(i) A quasar at a distance of 10^{23} km.

4 From Figure 1.2, approximate the centre wavelengths and frequencies of the seven colours of the rainbow.

5 What is the peak emission wavelength of the following stars (surface temperatures given in parentheses)?
(a) Spica (20 000 K)
(b) Sirius (10 000 K)
(c) Betelgeuse (3600 K)

6 The star Aldebaran has a surface temperature of 3800 K and emits 700 times more energy than the Sun (surface temperature 5800 K). How large is its radius compared to the Sun's?

Teasers

7 (a) In looking at the Sun for 1 s (**DON'T DO THIS. IT WILL DAMAGE YOUR EYES**), approximate how many photons would enter your eye. Use the approximation that all visible photons have the same energy. The Sun is 150 million km from the Earth.
(b) In looking at an identical star to the Sun, at a distance of 10^{14} km, for 1 s, approximate how many photons would enter your eye if the light is gathered by a telescope with an aperture (collecting area) of 1 m^2.

8 Why might gamma-rays be referred to as being high-energy radiation and radio waves as low-energy radiation?

9 Consider two stars of equal size; star A has a surface temperature of 3000 K and star B a surface temperature of 30 000 K.
(a) Which star emits the most radiation at 700 nm (red)?
(b) Which star, to the eye, appears redder?

Exercise

10 From Figure 1.3, approximate the relative amount of power per unit area emitted by a star that has a surface temperature of 6000 K at 700 nm (red), 530 nm (green) and 400 nm (violet). Repeat the same exercise for stars that have surface temperatures of 3000 K, 10 000 K and 25 000 K. What would the overall colours of the four stars appear to be?

2 Seeing into Space

Ancient astronomers looked at the night sky and saw almost exactly what we see today, a mass of points of light with an occasional extended object such as the Moon. At that time the sizes of even relatively nearby bodies such as the other planets of the Solar System could not be determined and they appeared just like bright stars. Though the planets could be distinguished as they slowly moved across the fixed patterns of stars (hence the word 'planet' from the Greek for wanderer), there was no obvious difference in their appearance to the eye. The clearest way to discover the difference between planets and stars is to try to produce magnified images of both. The telescope is the instrument that makes this possible in the visible part of the electromagnetic spectrum.

Imaging Light from Space

The light rays that arrive on Earth from what appears to be a point source of light are parallel to each other. Light travels through space in straight lines and so there is only one path from a point source to an observer (under normal circumstances). This simplifies the design of a basic optical telescope as the elements that make up the instrument need only to image light that is travelling into it from one direction.

There are two fundamental optical elements that are used in telescopes. These are lenses and mirrors. Plane (flat) mirrors do not affect the imaging of light. Looking in a bathroom mirror allows you to see yourself but it presents the same image that a nearby observer would see without the mirror. In order to obtain a different image the mirror needs to be curved. For instance, car wing mirrors are often curved in order to improve the field of view and this affects the size of the image observed. Fun fairs often contain mirrors that are curved in complicated ways to make the observer appear fat, thin, short, tall or a combination of these distortions at the same time. Such effects are due to rays of light being reflected in different directions depending on where the ray hits the mirror. Lenses also influence the path of a ray of light but do so not by reflection but by a phenomenon known as refraction. An ideal lens does not reflect any light at all but causes the path of light to bend as it passes through its body. An everyday application of a lens is in a pair of spectacles. Such a pair of lenses changes the path of light as it passes through the glass so that the ray direction is altered to enable a defective eye to image correctly.

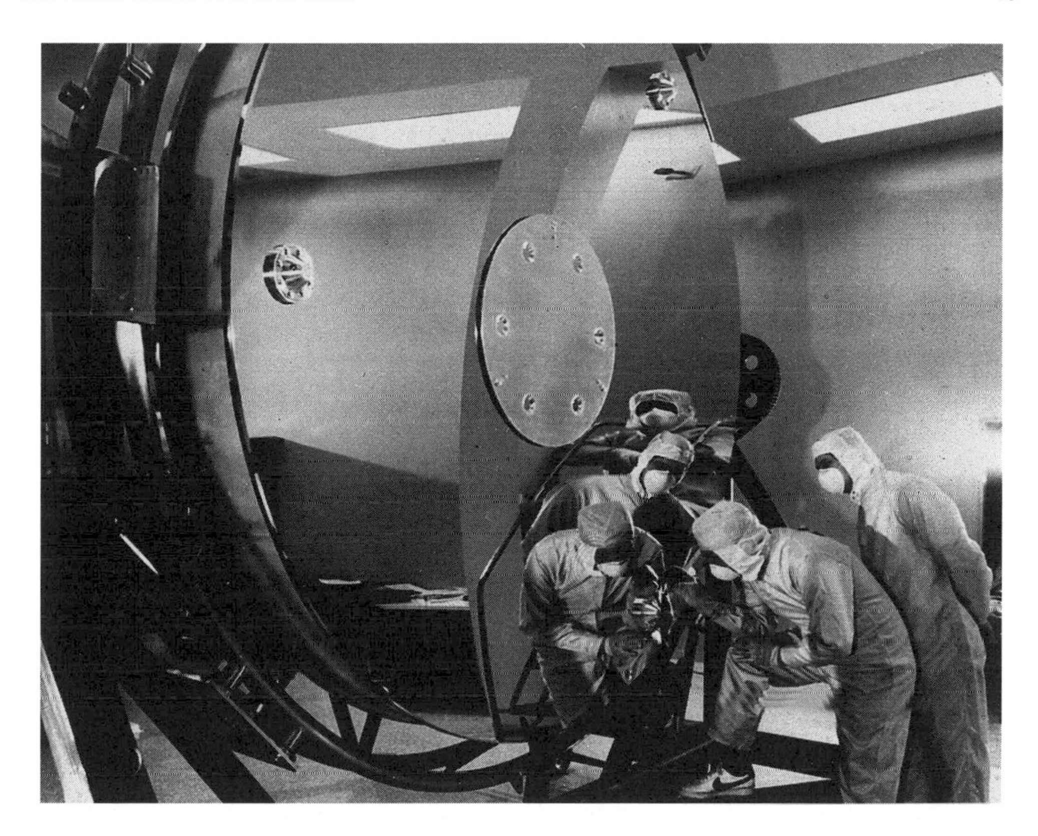

Figure 2.1 Technicians put the finishing touches to the 2.4 m primary mirror of the Hubble Space Telescope. The technicians whose distorted images can be seen in the mirror are standing out of shot (courtesy of NASA, Jet Propulsion Laboratory and National Space Science Data Center)

The astronomer is mainly interested in parallel rays of incoming light from very far-distant sources and so the influence on such rays of standard mirrors and lenses is the main information required to be able to design a useful telescope. Figure 2.2 illustrates this for convex lenses and concave mirrors, both of which cause parallel rays to converge. Where the rays converge a real image appears. This means that if a screen is placed at this position an image of the far distance will be seen. As the screen is moved away from or towards the optical element the image loses sharpness. An example of this is a simple camera. A piece of light-sensitive film is placed at the focal length of a focusing lens when pictures of the far distance are being taken. The distance from the optical element to the image position is known as the focal length.

An important feature of optical elements such as lenses and mirrors is that light will always follow the same path through them in either direction. A light source in the focal plane (at the focal length) will therefore produce a parallel beam of light after it has interacted with a lens or mirror.

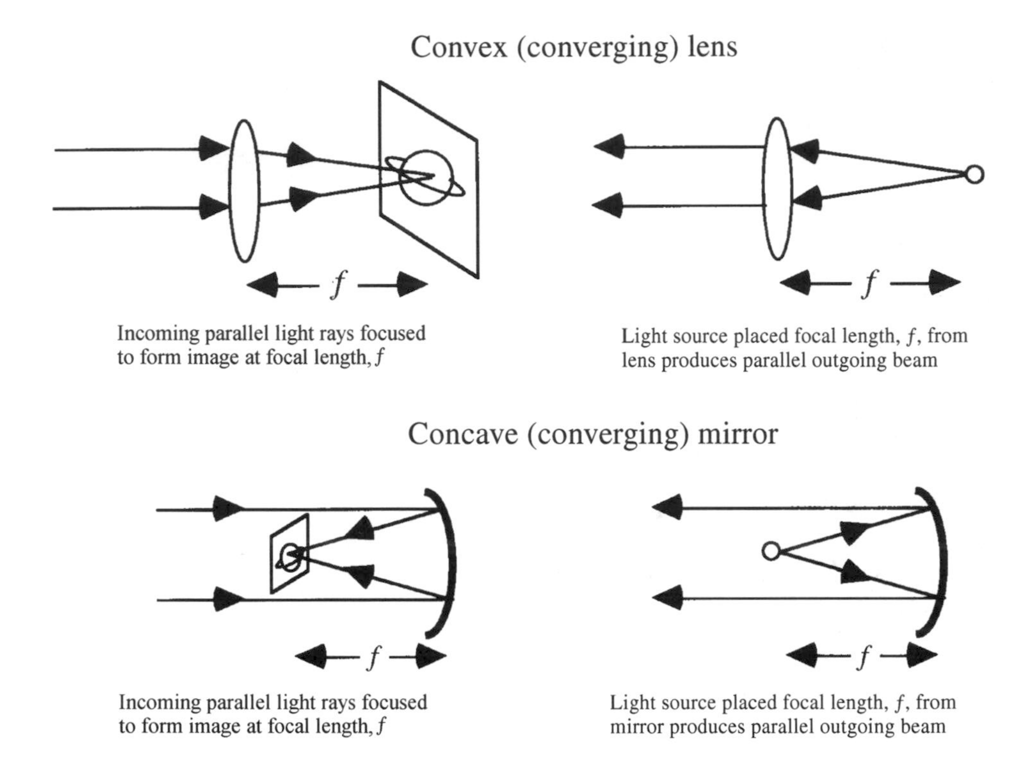

Figure 2.2 The influence of lenses and curved mirrors on parallel beams of light

Objectives of Telescope Design

There are three main objectives for a telescope. The first aim is to maximise image brightness thus allowing faint objects to be observed. The second is to resolve objects that are separated by small angles. Third, it can be useful to magnify the image so that, for instance, the object (or parts of it) can be seen to have size rather than appear as a point source.

Image brightness maximisation is the most important task for a telescope and also the simplest to consider. It is a function of a telescope to gather light and so the larger its aperture, the more light it can gather. The amount of light a telescope can gather is known as its light grasp and is proportional to the aperture cross-section for simple designs. Telescopes almost always have mirrors and lenses with circular cross-sections and so the discussion here assumes this to be the case. The cross-sectional area of a circle is given by $\pi D^2/4$ where D is the diameter and so the following relationship holds:

$$\text{light grasp} \propto D^2 \qquad (2.1)$$

In other words, the image becomes brighter with the square of the aperture diameter. Increasing the light grasp allows fainter objects to be seen, of great importance as astronomers look deeper and deeper into the universe. For instance, about 6000 stars can

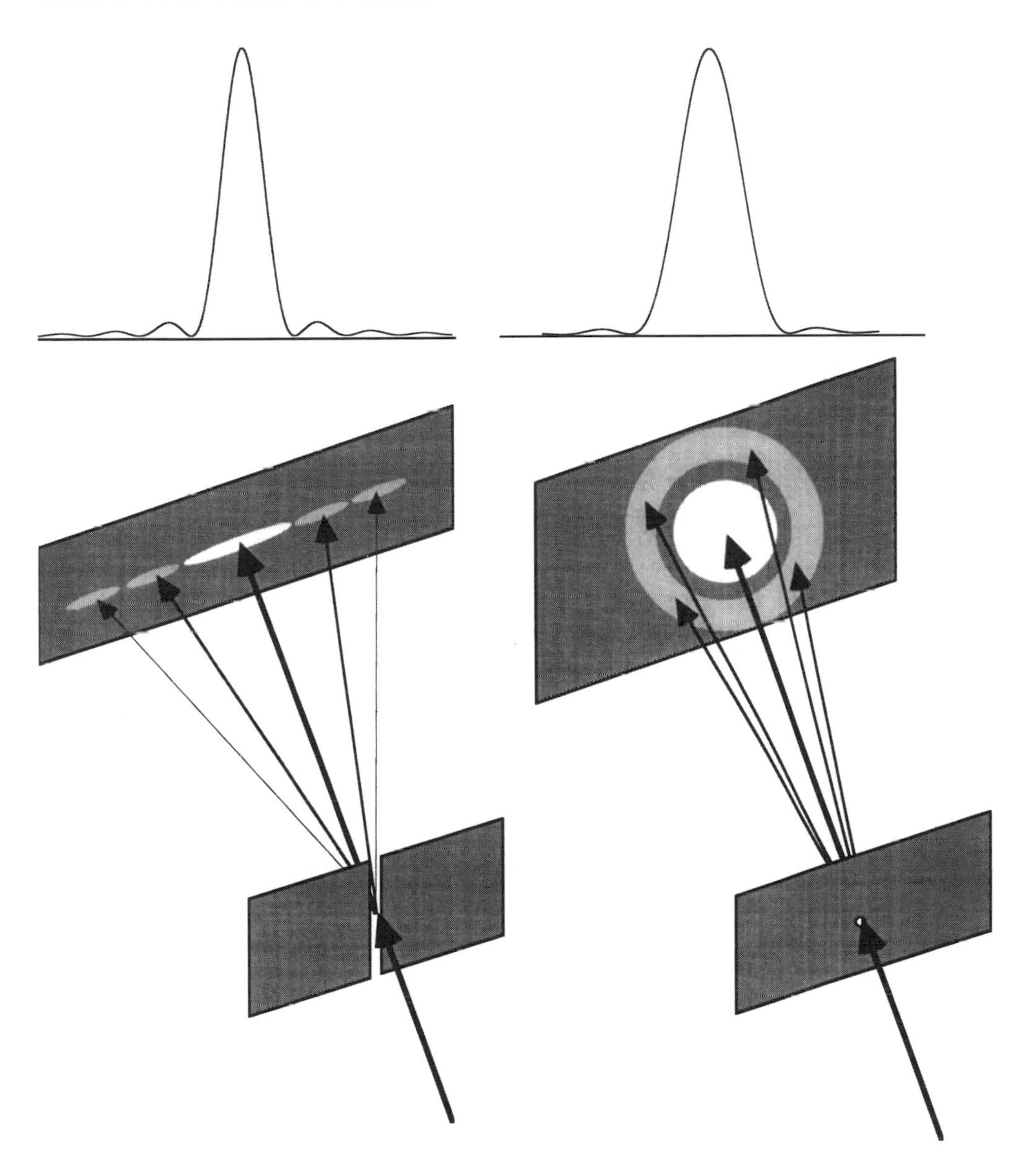

Figure 2.3 The effect on a beam of light of passing through an aperture. The beam's intensity is perturbed as shown by the patterns drawn on screens behind the aperture. The graphs above show the intensity along a central cross-section. Generally, the central maximum decreases in size with increasing aperture size

be seen with the naked eye whereas half a million can be seen using a telescope with an aperture of just the size of an average hand.

The aperture size also determines the resolving power of a telescope. When light enters an aperture there is an effect that can be considered to be an interaction between the electromagnetic wave and the aperture edges. This phenomenon is known as diffraction and is best visualised by considering ocean waves approaching a harbour entrance. The part of the wave that goes through the gap in the sea walls does not continue in a straight line but spreads out into the harbour. The same thing happens for light as illustrated in Figure 2.3. When the aperture is two-dimensional rather than linear, more complicated diffraction patterns consisting of light and dark fringes appear. Bright parts of the pattern occur where diffraction results in a superposition of wave peaks (or troughs) in a certain region and this is called constructive interference. Destructive interference occurs when peaks and troughs occur at the same position so cancelling out to produce dark regions. Figure 2.3 shows the diffraction pattern for a circular aperture. The bright central region is known as the Airy disk and contains most of the light intensity. This is used to define resolving power. Two point sources of light are said to be resolved when their Airy disks can be distinguished from each other. There is more than one definition of when this occurs, and the precise nature of the objects under observation determine which is most appropriate, but there are only small differences between them. Commonly used is the Rayleigh criterion which states that images are resolvable when the Airy disk peak from one image coincides with the first minimum of the other. This leads to the following definition of maximum achievable resolving power, R, for an aperture of diameter, D:

$$R = \frac{1.22\lambda}{D} \tag{2.2}$$

Here, the resolving power is defined as being the angular separation in radians[1] of two point sources of light that can just be resolved and so a small value for R indicates good resolution. As the resolving power is limited by an effect related to the interference between waves, it is not surprisingly a function of wavelength (λ). Good resolution is achieved when the wavelength of the light is much smaller than the aperture so that the light enters the telescope little affected by diffraction. In order to achieve good resolution it is therefore necessary to observe at short wavelengths or build a telescope with a large aperture. It is important to realise that equation (2.2) does not only apply to closely separated stars. For instance, features on a planet can be considered to be the objects that need to be resolved and the same principles apply.

WORKED EXAMPLE 2.1

Q. What size of telescope aperture would be required to resolve the bands that run around Jupiter?
A. The Earth and Jupiter revolve about the Sun at mean distances of about 150 million km and 780 million km respectively and so at their position of closest approach they are separated by about 630 million km. Jupiter has a diameter of 140 000 km. For convenience, consider there to be about 14 stripes with thicknesses of about 10 000 km (= 140 000/14). The telescope must therefore be able to resolve details 10 000 km apart from a distance of 630 million km.

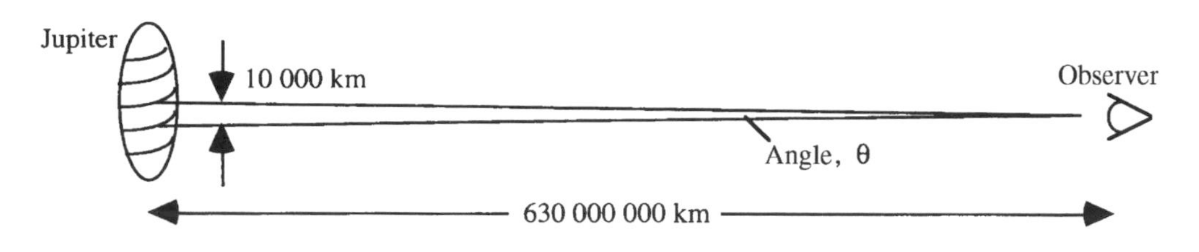

[1] A radian is $360°/2\pi$ which is about 57°.

As the angle is very small, θ (expressed in radians) is given by the ratio of the stripe size to the distance apart, that is, $\theta = 10^4/(630 \times 10^6) = 1.6 \times 10^{-5}$ radians. Take the average wavelength of visible radiation to be 500 nm (also the peak wavelength of reflected light as Jupiter is seen through the reflection of the Sun's rays (see worked example 1.3)) and rearranging equation (2.2):

$$D = \frac{1.22\lambda}{R} = \frac{1.22 \times (500 \times 10^{-9})}{1.6 \times 10^{-5}} = 0.038\,\text{m}$$

And so an optical instrument with an aperture of just 4 cm (for instance, good binoculars) is all that is required to make out surface features on Jupiter.

WORKED EXAMPLE 2.2

Q. What size of telescope aperture would be required to resolve the size of a Sun-like star at a distance of 4 light years?
A. The separation of the observer and object is 4×10^{13} km and the star is assumed to have a diameter similar to that of the Sun at 1.4×10^6 km. The angular size, θ, is again given by the ratio of the size to be resolved to the distance apart, in this case, $(1.4 \times 10^6/4 \times 10^{13}) = 3.5 \times 10^{-8}$ radians. Again take the average wavelength of visible radiation to be 500 nm and rearrange equation (2.2);

$$D = \frac{1.22\lambda}{R} = \frac{1.22 \times (500 \times 10^{-9})}{3.5 \times 10^{-8}} = 17.4\,\text{m}$$

No optical telescope with an aperture greater than 10 m has ever been built and so such a star's size cannot be directly measured in this way (though other techniques can be used). There are no stars closer to the Sun than 4 light years.

Figure 2.4 The first ever image of a star (other than the Sun) showing spatial resolution. Betelgeuse, a star with a diameter more than 1000 times greater than the Sun, shows signs of temperature variation (bright spot at centre) across its light-emitting surface. Betelgeuse is a bright star in Orion (reproduced by permission of AURA/STSCI)

In 1995 the Hubble Space Telescope produced the first image of a star that revealed surface features. Though this telescope only has a 2.4 m aperture it resolved surface detail on the star by observing in the ultraviolet part of the spectrum, where resolution is improved according to equation (2.2), and by choosing a very large star, Betelgeuse, that has a radius 1200 times greater than the Sun's (see Figure 2.4).

A telescope's resolving power can only be utilised if the optical design allows sufficient magnification of the object. For instance, if a telescope has no optics other than a flat mirror to reflect light into the observer's eye then the size of the aperture is almost irrelevant as the aperture of the eye will limit resolving power.

Before discussing optical design, it is necessary to define angular magnification. The concept of a point source of light becomes redundant as a zero-dimensional object cannot be magnified. Even though an object such as a planet appears to be a point source of light when observed by eye from the Earth, it subtends a definite angle in the sky. The aim of the telescope is to make the planet appear to subtend a larger angle. The ratio between the apparent and actual angles subtended is called the angular magnification. Figure 2.5 compares the angular sizes of Venus at its closest approach, the Moon and the Sun. By coincidence, the Moon and the Sun both subtend angles of about half a degree, the Sun having a diameter 400 times greater than that of the Moon but being at a distance 400 times greater. At its closest approach, Venus is more than 100 times further from the Earth than the Moon but has a diameter less than four times that of the Moon. It therefore subtends an angle of less than 0.017° at the Earth and appears like a point source of light to the naked eye. If a telescope could be designed with an angular magnification of (0.5/0.017 =) 30 then using it to observe Venus at its position of closest approach would result in Venus appearing to be the same size as the Moon or the Sun viewed with the unaided eye.

The Astronomical Telescope

Building a telescope with an angular magnification of 30 is a trivial task by the standards of modern optical engineering. In order to understand the way in which a telescope magnifies, the simplest design, the astronomical telescope, is discussed. Astronomical objects generally have a very small angular size, α, and are very far distant so that rays

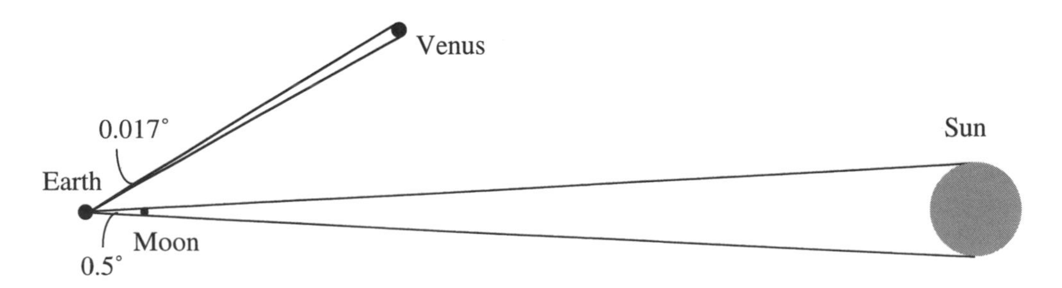

Figure 2.5 Angular size. The (exaggerated) angular sizes of the Moon, the Sun and Venus. Note that the ratio of the Moon's diameter to its distance from Earth is almost the same as that for the Sun so that the Moon and the Sun have very similar angular sizes as observed from Earth

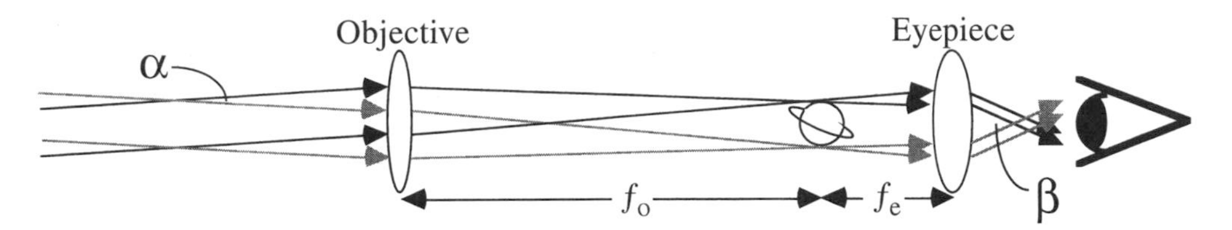

Figure 2.6 The astronomical (refracting) telescope. Rays entering the telescope from opposite sides of the object (indicated by different strength lines) are separated by an angle, α (exaggerated). An image is formed at the primary focus as indicated by the planet. The light then passes through the eyepiece and emerges so that the angle between the rays entering the eye are separated by an angle β (exaggerated). The angular magnification is thus given by β/α

arriving at the telescope are almost parallel. In the limit of α being zero, the light-gathering stage of the telescope, a converging lens known as the objective, forms an image at its focal length. Figure 2.6 shows two rays emanating from each limb of the body being viewed, exaggerating α for clarity. The first image could only be viewed by placing a screen into the telescope but the idea of having a primary image provides a convenient way of describing the light rays inside the telescope. The second lens can be thought of as acting to magnify the first image. The second lens is positioned so that it is separated from the first image by a distance equal to its own focal length. This scenario is the opposite of the image-forming process of the first lens. The new object, the first image, is now at the focus of the second lens and, as light travels through simple optical elements along the same path in either direction, the light will emerge from the telescope as (almost) parallel rays, as illustrated. However, the outgoing rays of light are now at a much greater angle than α. By simple geometry, the outgoing angle, β, must be greater than α by the ratio of the focal length of the objective lens, f_o, to that of the eyepiece, f_e, thus the angular magnification, M, is given by

$$M = \frac{\beta}{\alpha} = \frac{f_o}{f_e} \tag{2.3}$$

It can also be seen from the ray diagram that the final image is inverted. It may seem counter-intuitive that the intermediate image, that is never seen, can be drawn on the diagram but the final image which is seen by the eye is not. The reason for this is that the rays of light that emerge from the telescope are focused by the eye's optics onto the retina. If a screen were to be placed in the eye's position, nothing but a beam of almost collimated light would be seen. If the limit of α being very small is relaxed, equation (2.3) is no longer strictly valid and a perfect image is not formed. However, a small adjustment of the lens separation brings the image back into focus and the geometry is little changed, meaning that equation (2.3) still represents a good approximation.

WORKED EXAMPLE 2.3

Q. If the eyepiece of a refracting telescope has a focal length of 50 mm, what would the minimum focal length of the objective have to be in order for the eye to resolve the bands that run around Jupiter?

A. From Worked Example 2.1, the angle subtended at the Earth by the bands of Jupiter is 1.6×10^{-5} radians. By considering the eye to be a perfect optical instrument operating at 500 nm with an aperture (pupil diameter) of 5 mm, its resolving power can be determined from equation (2.2):

$$R = \frac{1.22\,\lambda}{D} = \frac{1.22 \times (500 \times 10^{-9})}{5 \times 10^{-3}} = 1.2 \times 10^{-4} \text{ radians}$$

For the eye to resolve the bands, it requires the image to be magnified by a factor of $(1.2 \times 10^{-4})/(1.6 \times 10^{-5}) = 7.5$. Rearranging equation (2.3):

$$f_e = M f_o = 7.5 \times 0.05 = 0.38 \text{ m}$$

The focal length of the objective must be about 0.4 m for a perfect eye to resolve the bands. The total length of the telescope therefore needs to be only around half a metre with an aperture of just 4 cm (see Worked Example 2.1) and such a telescope is easy to build. Of course, the eye is not a perfect optical instrument and so the telescope may need to be as much as ten times longer than theoretically predicted.

WORKED EXAMPLE 2.4

Q. If the eyepiece of a refracting telescope has a focal length of 50 mm, what would the minimum focal length of the objective have to be in order for the eye to resolve the size of a Sun-like star at a distance of 4 light years?

A. The calculation follows in exactly the same way as for Worked Example 2.3 but using the angular size calculated in Worked Example 2.2 of 3.5×10^{-8} radians. The angular magnification required is now $(1.2 \times 10^{-4})/(3.5 \times 10^{-8}) = 3400$. Rearranging equation (2.3):

$$f_e = M f_o = 3400 \times 0.05 = 170 \text{ m}$$

The focal length of the objective must therefore be almost 200 m for the eye to see the star as having size. In Worked Example 2.2 it was shown that the telescope would require an aperture greater than 10 m. The dimensions of such a telescope are beyond current engineering capabilities. It should also be noted that a solution cannot be found by reducing the focal length of the eyepiece very much below 50 mm due to the limiting physical properties of optical materials, though more sophisticated designs (see below) can reduce telescopic length considerably (see note on Betelgeuse in Worked Example 2.2).

An important realisation that arises from Worked Examples 2.1–2.4 is that the bodies of the Solar System are at an enormously smaller distance than even the closest stars. Small telescopes are useful to observe details of the Solar System but spatial detail on stars can rarely be resolved using an optical telescope (the star used as an example lies at very close range compared to most stars). Stars can be seen but must generally be regarded as being point sources of light. In discussing the capabilities of telescopes it is therefore important to consider their effect on both point sources and extended objects. The eye has been used as the detector for these examples but similar conclusions are drawn regardless of the instrument that collects the light after it has passed through the telescope.

Difficulties with Lenses

It is a simple matter to build a small telescope by placing an objective lens at one end of a tube and an eyepiece at the other. To achieve good angular magnification the ratio of the focal lengths should be large and the separation of the two lenses should be equal to the sum of their focal lengths with a control to enable the separation to be varied slightly in order to achieve correct focusing. The problem arises when much larger telescopes are required. When looking into the far reaches of the universe, very small and faint objects must be detected. To do this a large aperture is required and this presents major engineering difficulties. The production of large, high-quality lenses is very difficult. Even if such lenses could be produced there remain problems with supporting the lens within the telescope in such a way that light can still pass through it but that it is prevented from slowly sagging out of shape. Finally, there exists a fundamental optical difficulty which is that there is a slight variation in the focal length of a lens as a function of wavelength regardless of the material from which it is made. Consequently, an object that is multichromatic, such as a blackbody radiator like a star, cannot be properly imaged using a lens. The lens focuses light of different wavelengths in slightly different positions. This problem is known as chromatic aberration and is illustrated in Figure 2.7. For the small thin lenses in portable telescopes the effect is negligible but for large telescopes it becomes unacceptable. To keep the thickness of glass small the only option is to use a very long focal length objective lens. This necessitates a very long telescope which leads back to mechanical stability problems. Chromatic aberration can be overcome by building a compound lens from a series of different glasses of different focal lengths that act together as a single lens with no focal length wavelength variation. This solution is impractical for use as a large objective lens and does not solve other difficulties associated with lenses such as absorption of light and construction imperfections such as internal bubbling and inexact surface formation. The best solution is to replace the objective lens with a mirror.

The Reflecting Telescope

The reflecting telescope uses exactly the same principle as the astronomical telescope but replaces the refractive lens with a concave mirror. The concave mirror performs the same

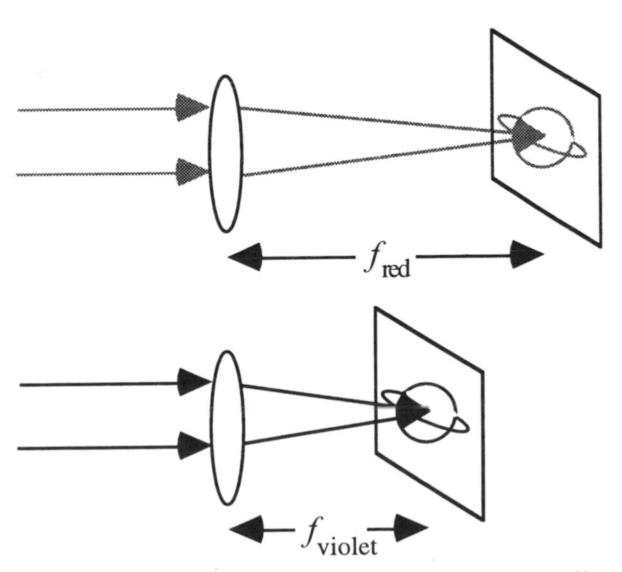

Figure 2.7 Chromatic aberration. The focal length of a lens varies depending on the wavelength of the light that it is focusing. The effect is exaggerated here for red and violet light

functions as the lens, that is, to gather light and form the first image. An eyepiece then produces the angular magnification required. Expression (2.3) still holds in exactly the same way but the design of the telescope becomes more complicated. As the objective is a mirror the light is reflected back down the telescope towards the object being viewed. A small mirror must therefore be added into the body of the telescope to deflect it to a position where it can be viewed, as shown in Figure 2.8. If the mirror is flat then it does not affect the imaging of the telescope though the light it blocks will cause the final image to be slightly dimmer[2]. Mirrors reflect light in the same direction regardless of wavelength and so the problem of chromatic aberration is automatically overcome. The separation of the mirror and lens must still be equal to the sum of the focal lengths but, as the light is being reflected back upon itself and then out of the instrument, the total length of the telescope can be significantly reduced from that of an astronomical telescope, thus improving its

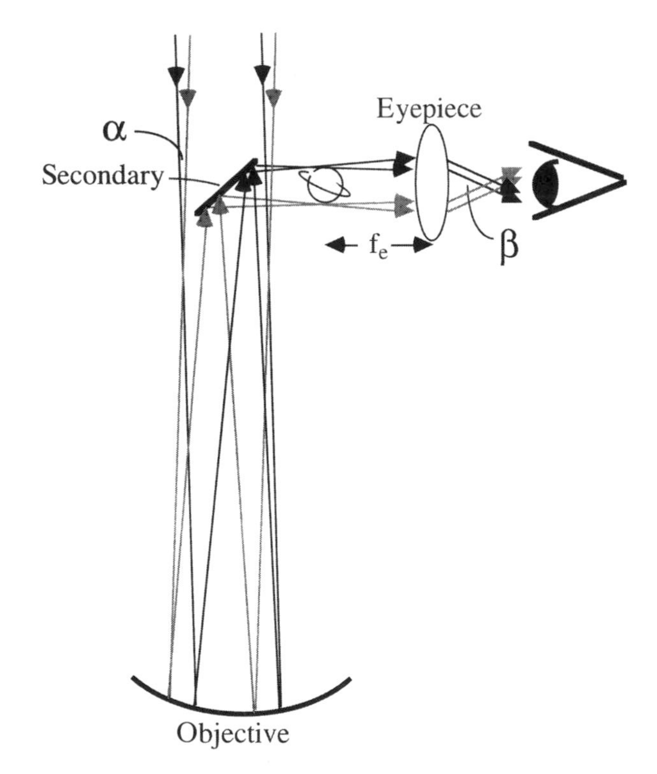

Figure 2.8 The Newtonian (reflecting) telescope. Rays entering the telescope from opposite sides of the object (indicated by different strength lines) are separated by an angle, α (exaggerated). An image is formed at the primary focus as indicated by the planet. The light then passes through the eyepiece and emerges so that the angle between the rays entering the eye are separated by an angle β (exaggerated). The angular magnification is thus given by β/α. The secondary mirror does not affect the focusing of the system but serves to deflect the final image from the barrel of the telescope

[2] The insertion of an object into the telescope does not cause a part of the final image to be obscured if it is inserted at a place where the light is not imaged. The geometry of the design is such that if, for instance, most of the concave mirror were to be painted black the part that remained reflecting would produce an identical image suffering only from reduced brightness.

mechanical stability. Mirrors are also much easier to hold in position than lenses, as they can be firmly supported from behind, and can be produced on a large scale with great accuracy. A final advantage is that when a mirror surface, which is usually just a thin layer of aluminium, becomes degraded it can be removed using nitric acid and replaced without rebuilding the mirror structure itself.

Solving Problems with Telescope Design

It is clear that the reflecting telescope has many advantages over lens-based systems. However, there are difficulties with the reflecting design. The first involves choosing the shape of the mirror. Different mirror shapes produce different problems depending on the task that the telescope is attempting to perform. The details of the analysis of the behaviour of curved mirrors is complicated and only the main points are discussed here. If a spherical shape is chosen the telescope suffers from spherical aberration as illustrated in Figure 2.9. Rays that hit the mirror at different distances from the central axis are focused at slightly different positions, thus blurring the final image. This is clearly a problem when the collecting mirror is large and it can be solved simply by choosing the mirror to be parabolic as also shown in Figure 2.9.

A large number of telescope designs are used in modern observatories in order to perform many different tasks. Probably the most common is the Cassegrain telescope (Figures 2.10 and 2.11) which incorporates two curved mirrors, a parabolic concave mirror to collect the light and a hyperbolic convex secondary mirror. The mirrors may be placed relatively close together, so that the length of the telescope can be minimised to

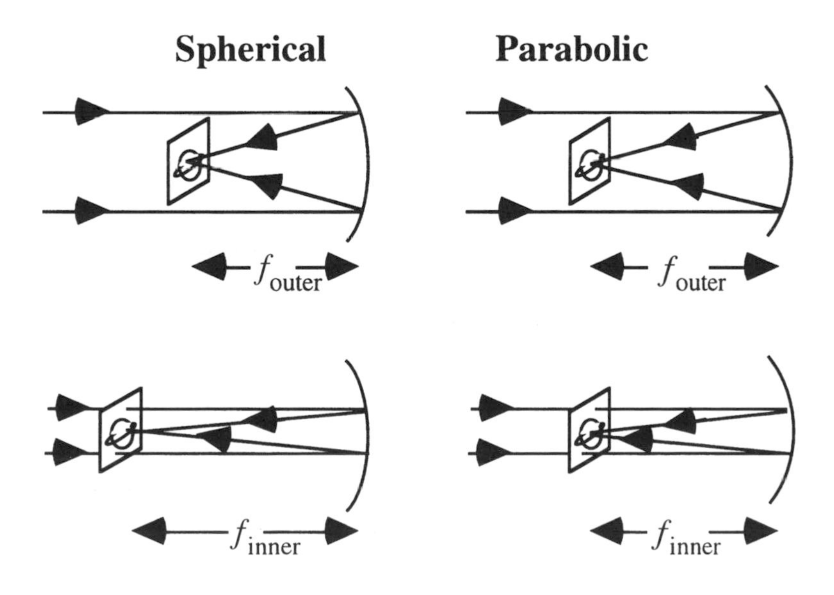

Figure 2.9 Spherical aberration. The focal length of a spherical mirror varies depending on the distance of the rays from the centre of the mirror. The effect is exaggerated here

maintain mechanical stability, but act together as a long focal length objective thus maximising angular magnification. The pair of mirrors cause the light to be reflected from the instrument in the same direction that it enters. A hole is cut in the collecting mirror to allow the light to exit.

It is rare to physically look through large telescopes and so the optics are usually arranged so that the light is focused onto detection instruments such as those discussed in the preceding chapter. Some pieces of detection or analytical apparatus are too heavy or delicate to be moved as the telescope is tracked across the sky. It is then necessary to deflect the image to a common position, independent of the orientation of the telescope. This can be done by placing a third mirror into the telescope. This mirror is flat and performs a similar role to the flat mirror in the simple reflecting telescope discussed above in that it merely deflects the rays of light without influencing the focusing. The mirror can be positioned so that the image produced outside the telescope is always at the same position, known as the Coudé focus.

The parabolic mirror has been identified as a solution to the problem of spherical aberration. Telescopes are often required to image a small portion of the sky but parabolic focusing narrows the field of view substantially so that only axial rays are focused correctly. A telescope's inability to focus axial and non-axial rays simultaneously is known as comatic aberration (or coma) and its effect has already been touched upon for the astronomical telescope. When observing two or more point objects simultaneously it is inevitable that rays enter the telescope at different angles to the axis. Under these circumstances, for at least one of the rays, the entrance angle, α, deviates from zero and the focusing for each set of rays is different, meaning that it is not possible to produce a clear image of all objects simultaneously (see Figure 2.12). The problems of spherical and comatic aberration can only be solved simultaneously for individual tasks. The

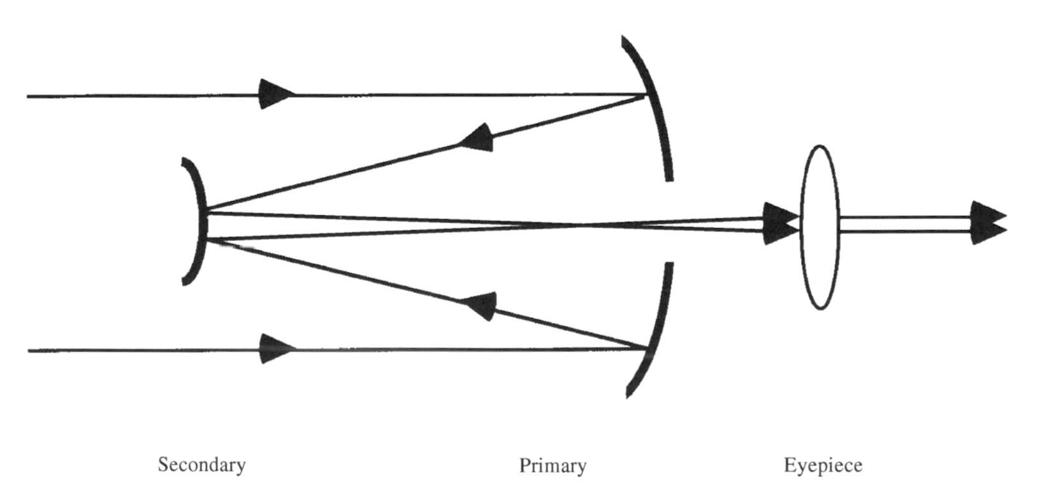

Figure 2.10 Cassegrain (reflecting) telescope. The primary and secondary mirrors act together to mimic the single collecting mirror of the Newtonian telescope and in so doing increase angular magnification while reducing total length. This design is frequently used in large, modern observatories though the 'eyepiece' would rarely be used for looking through. Instead, optics are used to direct the gathered light to an array of electronic instruments

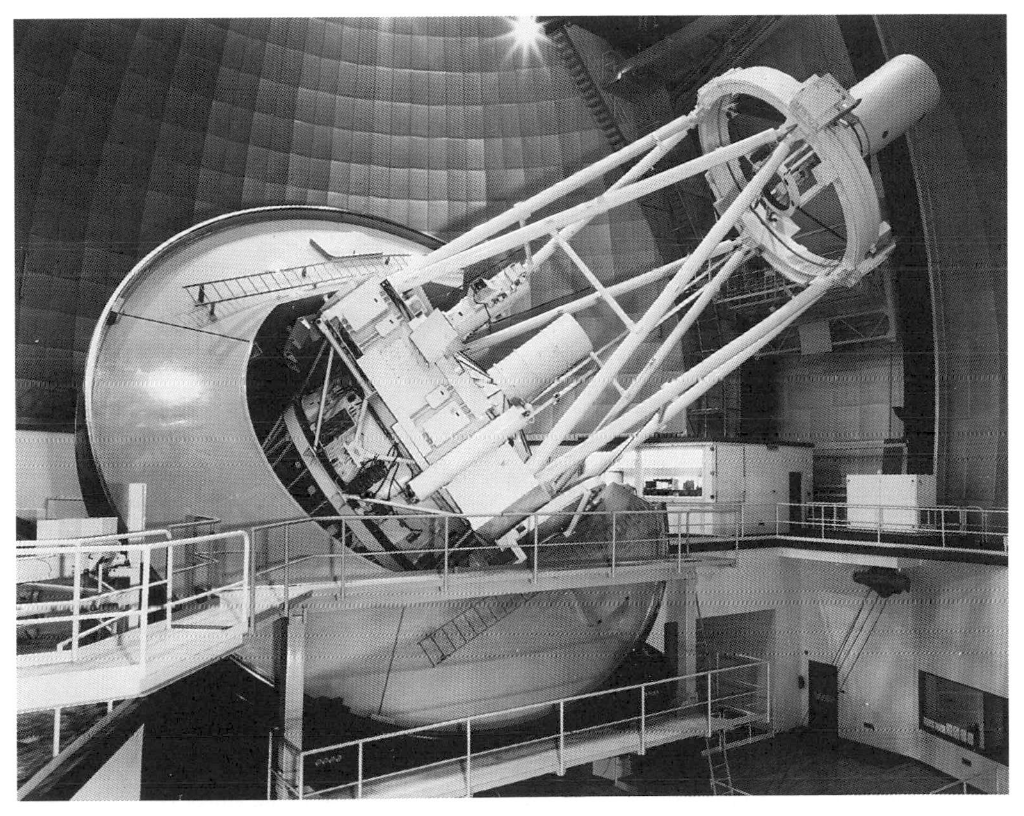

Figure 2.11 A Cassegrain design telescope in a modern major observatory (reproduced by permission of Anglo-Australian Observatory)

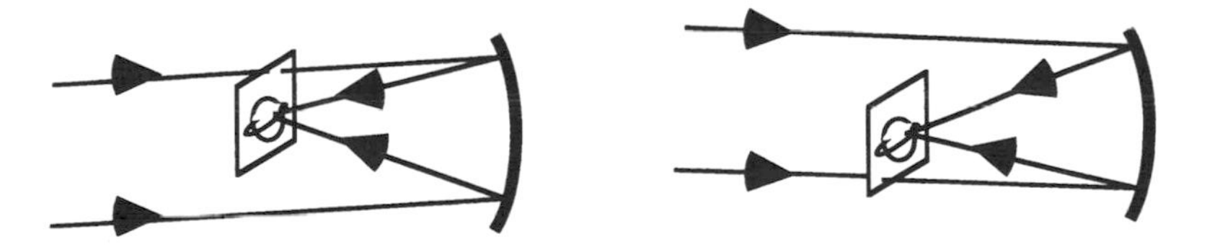

Figure 2.12 Comatic aberration (coma). Non-parallel rays are focused by curved mirrors in different positions. The focus position varies in three dimensions (details not indicated here) according to the entrance angle of the rays

parabolic mirror eliminates spherical aberration but severely limits the field of view. A catadioptric[3] system must be used to increase the field of view without the image suffering from spherical aberration. In a telescope design known as the Schmidt camera (Figure 2.13) a spherically shaped film plate is placed inside and concentric with a spherical mirror. A thin glass correction plate is placed over the front of the telescope. The

[3] A reflecting system is known as catoptric and a refracting system as dioptric. Systems that utilise reflecting and refracting elements are known as catadioptric.

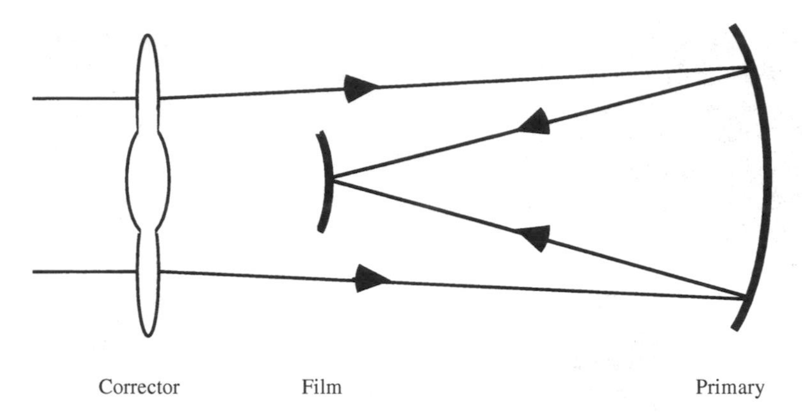

Figure 2.13 The Schmidt camera. A corrector plate (drawn schematically) corrects the incoming light for spherical aberration prior to focusing by the primary, spherical mirror. The final image is produced on a spherical, inner plate

refraction of the rays as they pass through the plate corrects spherical aberration but does not decrease the field of view available. Such instruments are able to take single focused photographs of segments of sky that contain features several degrees apart.

Practical Considerations for Observing

It would appear that any problem can be solved by simply choosing the correct design of telescope. There remain more practical considerations to foil the telescope designer. According to Rayleigh's criterion, in order to obtain the best possible angular resolution the largest possible aperture should be used. Economy limits how large an aperture can be built as the price of optical components goes up rapidly with size. Though aperture is limited by cost it would seem that angular magnification is not limited as it is determined only by the ratio of the effective focal lengths of the two stages of the telescope. Clever design can allow this value to be very large without the length of the telescope having to be so large that it too places unbearable strain on the budget. Difficulties with increasing magnification can be seen by considering various scenarios:

(1) If a single star is being investigated (for instance, it may be desirable to record a spectrum of its light output to determine its temperature) then there is nothing to be gained by attempting to magnify the image size. First, any such attempt would almost always be futile as demonstrated in Worked Example 2.4 and second, all the light will go straight into the spectrometer whether the image is magnified or not. It is important to note that magnification in this sense does not imply amplification of the total light intensity and this is central to the other two scenarios to be considered.
(2) If it is desired to resolve two closely separated stars then angular magnification is clearly a good thing. As the angular magnification increases, however, there will come a point where the diffraction patterns of the stars (see Figure 2.3), as determined by the telescope aperture, will be imaged by the detection apparatus. Any further angular magnification causes both diffraction patterns to be magnified further with

exactly the same overlap between them and no possibility to improve resolution. At the same time the images will become weaker as the light is spread over a larger area.

(3) When the observed object is an extended one, for instance in the case of a planet or nebula, then the effect of angular magnification on image brightness sets in immediately. The larger the image becomes, the dimmer it gets because the light is being spread over a progressively larger area. If a photograph is being taken, for instance, then a faster film or a longer exposure time is required. If lots of time is available to accumulate an image then this appears to be little more than an inconvenience until the next practical problem is considered.

So far, observing conditions have been ignored but any Earth-bound telescope must contend with the effects of the atmosphere on its performance. The degrading influence of the atmosphere on the final image has two main sources. Absorption by the atmosphere prevents some of the light that is incident on the Earth from reaching the surface. The wavelength dependence of this behaviour is caused by the specific absorption bands of atmospheric gas molecules and these are indicated in Figure 2.14. There is strong

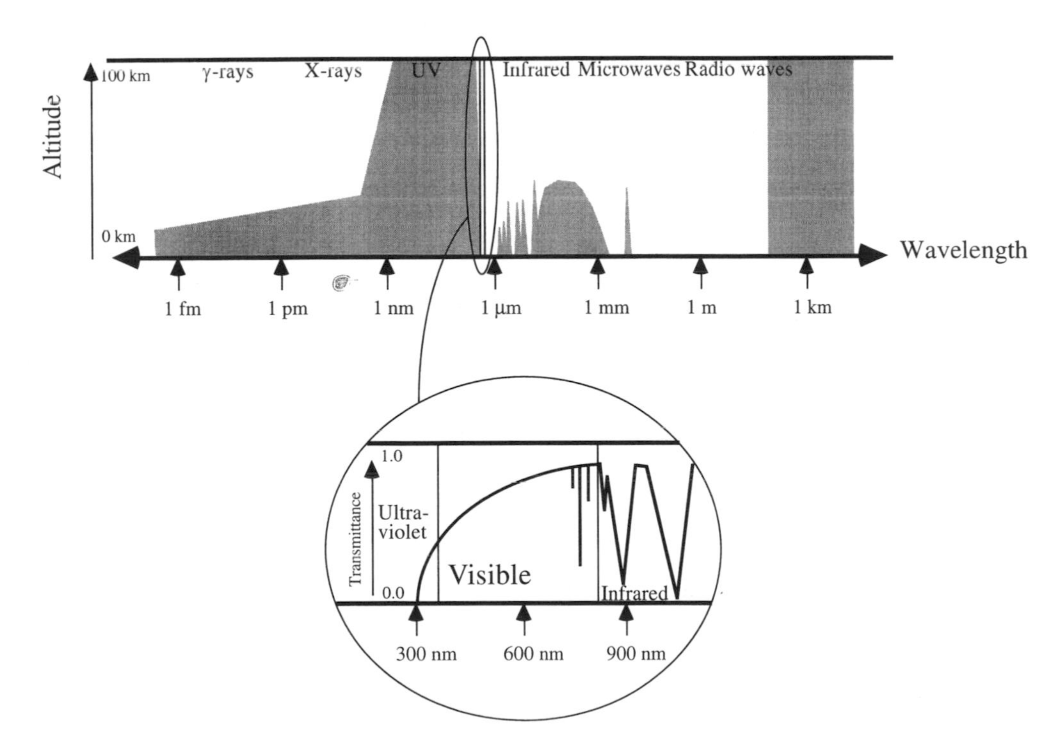

Figure 2.14 Schematic graph of the absorption of the Earth's atmosphere. In the upper trace the altitude at which the atmosphere reduces the intensity of arriving radiation to half its original value is shown. Note the absorption of oxygen and nitrogen in the X-ray and γ- ray regions, the strong absorption of ozone in the ultraviolet and the absorption of water and carbon dioxide in the infrared. Free electrons in the ionosphere (see Chapter 8) reflect very long wavelength radio waves. The visible and near-visible part of the spectrum is shown below (again schematically) plotted as the proportion of light that arrives at sea level. By observing at higher altitudes the effective transmittance is reduced. Note that the best windows of transparency are in the visible and radio parts of the spectrum

absorption of electromagnetic energy in some parts of the spectrum, causing a dimming of the image and preventing accurate spectra of astronomical objects from being recorded. These problems can be overcome. As discussed above, image dimming is not a problem if time is available to accumulate a brighter image. It is possible to correct for the Earth's atmospheric absorption as its spectrum is well known. Actual wavelength characteristics of observed bodies can be obtained by magnifying the image intensity by the factor required to account for atmospheric absorption at all wavelengths. This can be automated using simple computer software.

The most destructive property of the atmosphere for astronomers is its instability. The atmosphere acts as a weak but very thick and constantly changing lens. The most commonplace effect of this is that stars twinkle when viewed by eye. Turbulence in the atmosphere causes the path of light taken by a star to vary and the eye assumes that the star is positioned along the line of sight of the arriving light. As the air shifts position then so does the star. This is known as scintillation and Figure 2.15 illustrates the effect. Even in very still conditions the smallest angle through which a star's position will appear to be oscillating when observed from sea level is about 1 arcsecond[4]. Images of stars produced by telescopes are pulsating and moving. Extended objects shimmer.

WORKED EXAMPLE 2.5

Q. What is the largest telescope aperture that is limited by diffraction rather than by scintillation in good viewing conditions?
A. Take scintillation to limit the resolution of a telescope to 1 arcsecond (4.8×10^{-6} radians). The telescope aperture that provides the same resolution can be found by rearranging equation (2.2). Assuming observation is to take place in the centre of the optical region at around 500 nm:

$$D = \frac{1.22\lambda}{R} = \frac{1.22 \times (500 \times 10^{-9})}{4.8 \times 10^{-6}} = 0.13\,\text{m}$$

The resolution of a telescope with an aperture greater than only 13 cm is thus likely to be limited by the atmosphere rather than by diffraction.

WORKED EXAMPLE 2.6

Q. What would be the advantage of building a very large telescope with an aperture of 6.5 m over a small telescope with an aperture of 0.13 m?
A. Under normal viewing conditions both telescopes would be limited in their resolution by the atmosphere but the large telescope retains a much greater light grasp despite this. From equation (1.1), the ratio of image brightnesses, B, is given by:

$$\frac{B_{\text{large}}}{B_{\text{small}}} = \frac{D^2_{\text{large}}}{D^2_{\text{small}}} = \frac{6.5^2}{0.13^2} = 2500$$

The image produced by the large telescope is thus 2500 times brighter than that of the small one. The observer therefore has the choice of whether to use this advantage to observe much fainter objects or to reduce the light-collection time to help counteract the influence of scintillation.

---

[4] An arc of 360° constitutes a full rotation through a circle. One degree is divided into 60 arcminutes and one arcminute is divided into 60 arcseconds.

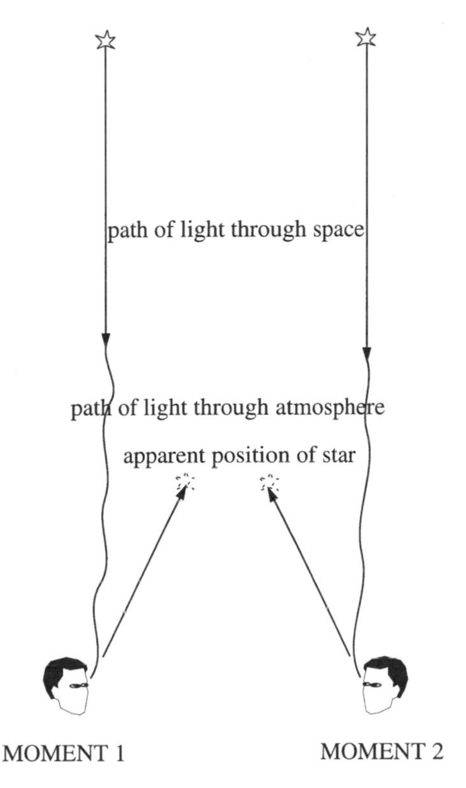

Figure 2.15 Scintillation. Light travels along straight lines through space until it enters the Earth's atmosphere. The air then acts like a thick, weak, but continuously varying lens. The diagram shows two moments during which light enters the observer's eye from different directions (exaggerated) making the star appear to have moved

The atmosphere sets the ultimate limit on the performance of a telescope and so observatory siting is very important. In order to minimise atmospheric disturbance mountain sites are often chosen. Here there is less air between the observer and outer space and the scintillation-limited resolution of the sky may improve a little below 1 arcsecond. Mountain sites also help to minimise a further problem, that of light pollution. This refers to difficulties caused by observing the night sky from a well-lit position such as in or near a city. Though man-made light sources are usually close to the horizon, scattering of light by atmospheric particles causes some of this unwanted light to enter the telescope. A mountain site raises the telescope above and usually a great distance from most sources of light pollution. Timing of observations can also help to ease the problem of scintillation. The best time to observe an object in the sky is when it is directly overhead as this allows the observation to be made through the thinnest layer of air. As the star moves away from overhead, the line between observer and observed extends through more and more of the atmosphere. There are other tricks such as siting the observatory on an island in a lake where cool, still air helps with the problem but the ultimate solution is to mount the telescope on a satellite above the atmosphere (see Figure 2.16). This way the

Figure 2.16 The Hubble Space Telescope during deployment via the Space Shuttle (courtesy of NASA, Jet Propulsion Laboratory and National Space Science Data Center)

theoretical considerations discussed above, only dreamt of from Earth, may once again become relevant. Siting a telescope in space has a large number of other difficulties associated with it but these are problems that can be solved when sufficient money is available.

A recent development for Earth-based telescopes attempting to eliminate the degradation of images due to atmospheric effects utilises flexible mirrors and is known as adaptive optics (see Figure 2.17). The main effect of scintillation is to cause continuously changing and unpredictable movements of the image of an object that is known to be fixed in space. The image of a star would be a single, fixed point if the atmosphere were static. In order to create such an image in real viewing conditions it is necessary to add an extra mirror to the telescope design. When the atmosphere is perfectly still the mirror would be a simple plane mirror and the image would not be altered. When the atmosphere is turbulent it is bent into whatever shape necessary to regain a static point image. As the movements of the image are usually small the mirror does not have to be deformed very much but the process must take place rapidly and accurately. To achieve this, very thin mirrors are used with many actuators beneath the non-reflective surface to push the reflective surface into the desired shape. The desired shape must be calculated very rapidly and this is done using a computer. The image of the star is constantly being electronically fed to the computer so that the necessary mirror shape can be updated continuously. The mirror acts to remove

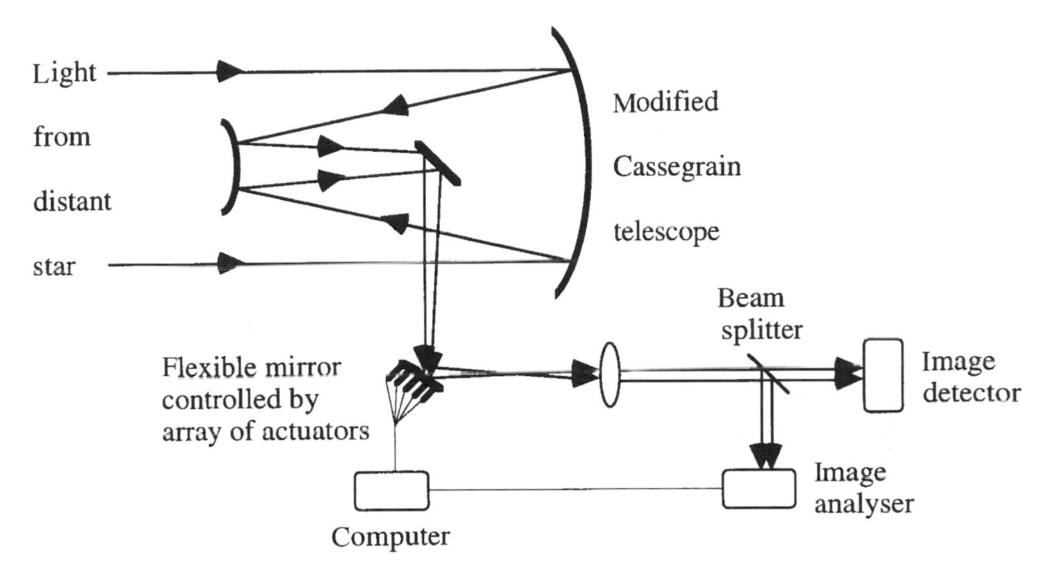

Figure 2.17 Adaptive optics. The image produced by the telescope alone would be blurred by the effects of scintillation. Instead, the light from the telescope is directed into a feedback system in which a computer controls a small flexible mirror in such a way that the image of a bright beacon star in the field of view of the image analyser is always made to appear as a point source of light. In this way the detector retains an image that is corrected for the effects of disturbances in the atmosphere for objects close to the beacon star in the field of view

all image deformations that are being caused by the atmosphere so that the star is focused to a point. Usually, a bright 'beacon star' is chosen as a reference for the feedback process. When the beacon star is focused then other objects in the field of view are similarly steadied. If no bright star is close enough to the object that the observer is interested in, an artificial star can be created using a laser that causes molecules in the upper atmosphere to fluoresce and appear as a point source of light, like a star. In this way, any part of the sky can be studied without the degrading influence of the atmosphere. This technique is currently under development and it will be some time before its achievements match those of orbiting telescopes.

Flexible mirrors controlled by actuators are also being increasingly used as the main collecting element in telescopes. Using a similar feedback system to that described to correct for scintillation, slight deformations in the shape of the main mirror can be corrected as soon as they form.

Astronomy Beyond the Visible

So far, the discussion on seeing into space has been concerned with the visible part of the electromagnetic spectrum but observation at all wavelengths is possible. Indeed, by using the radio part of the spectrum, it is possible to eliminate the problem of atmospheric absorption. Radio astronomy is a comparatively new area of investigation but has become very important. A radio telescope functions in a similar way to an optical telescope but has a much larger size and simpler design. The main component is a concave parabola of a material that reflects radio waves. At the focus of the parabola a radio receiver detects the signal. Radio maps can be produced by both scanning the direction of the dish in a grid across the sky and by varying the reception frequency of the receiver. The main problem with radio astronomy becomes immediately obvious when equation (2.2) is considered. The larger aperture of a radio telescope improves the resolution by as much as two orders of magnitude but the wavelengths of radio waves are around 10^4–10^6 times greater than visible light, leaving the radio telescope with greatly reduced resolving power.

There is a special technique that can be used to improve the resolution of radio telescopes and it is known as interferometry. Above, it was explained how interference arising from diffraction limits resolving power but different effects based on the same principles can be used to improve resolution when several telescopes are used together. The detectors must all detect a definite phase of the signal which means, in the simplest case, that all detectors lock into waves that have their peaks at a given instant. As only a single wavelength is tuned into during an observing period, the signals being collected by the receivers will remain in phase. In order to collect in-phase signals in this way the direction of the source becomes limited, which is another way of saying that the resolution has been improved. For a linear array of telescopes the aperture becomes equivalent to a single telescope having the resolution of the length of the telescope array. The sky can then be scanned by either moving the telescopes together or by altering the relative phase of detection of the telescopes in a definite way. More complicated patterns of radio telescopes are often used and can be spread over several kilometres (see Figure 2.18). Very long baseline interferometry refers to the coupling of very far distant radio telescopes in the same way to improve the resolution still further. This technique requires the use of atomic clocks to synchronise the phase detection of the telescopes but has been successfully demonstrated. Interferometry is theoretically viable at any wavelength but synchronisation of the detectors must be on the scale of the frequency of the signal being detected, which makes optical interferometry considerably more difficult than using the technique with radio signals. Optical interferometry is nevertheless being attempted with some success.

The major difficulty associated with observation in other areas of the spectrum is atmospheric absorption. Various species of oxygen and nitrogen in the upper atmosphere, such as ozone, protect our bodies from the harmful effects of gamma radiation, X-rays and ultraviolet light but this has the obvious consequence that Earth-bound observation is very difficult in these regions. Similar problems occur for infrared telescopes where the radiation intensity at the Earth's surface is drastically reduced by the absorption of water vapour and carbon dioxide in the lower atmosphere. There are a few transparency windows in the absorption of the atmosphere in those regions that allow some observation to take place, especially if the telescopes are sited in warm, dry, elevated places. The best solution is almost always to place the telescope above the Earth's atmosphere on a satellite and this has been done in all these problem regions so that astronomical observations throughout the electromagnetic spectrum have been made.

Figure 2.18 The Very Large Array radio interoferometer in New Mexico, USA. The 25 radio telescopes, each with an aperture diameter of 20 m, can be moved over several kilometres along railway tracks to alter the resolution and field of view of the system (by permission of NRAO/AUI)

WORKED EXAMPLE 2.7

Q. An infrared telescope with an aperture area of 1.0 m^2 is placed in orbit about the Earth where the temperature of the telescope is $-73°$ C. At what wavelength does the telescope emit most radiation, how much radiation does it emit and how much radiation does it collect from a Sun-like star 4 light years away?

A. Consider the telescope to behave in a similar way to a blackbody. The temperature of the telescope is 200 K and so the peak output wavelength can be calculated using Wien's displacement law (equation (1.4)):

$$\lambda_{\text{max}} = \frac{2.9 \times 10^{-3}}{T} = \frac{2.9 \times 10^{-3}}{200} = 1.5 \times 10^{-5}\,\text{m}$$

Using the absolute temperature of the telescope it is possible to calculate the power radiated per unit area using the Stefan–Boltzmann Law (equation (1.3)),

$$\text{For } T = 200\,\text{K}, \qquad P_{\text{A}} = \sigma T^4 = (5.7 \times 10^{-8}) \times 200^4 \approx 100\,\text{W m}^{-2}$$

The telescope emits about 100 W m^{-2} at a peak wavelength of 15 μm, in the middle of the infrared part of the electromagnetic spectrum.

Using the technique of Worked Example 1.4, it is possible to calculate the amount of energy from the observed star collected by the telescope. The star's power (4×10^{26} W) is radiated equally in all directions and so, at the telescope–star separation, the power is spread over a sphere with a diameter of 8 light years ($= 8 \times 10^{16}$ m) and so,

$$\text{Power per unit area} = \frac{\text{Total power}}{\text{Total area}} = \frac{4.0 \times 10^{26}}{\pi(8.0 \times 10^{16})^2} = 2.0 \times 10^{-8}\,\text{W m}^{-2}$$

The telescope therefore collects 2.0×10^{-8} W from the star as it is has an aperture area of 1 m^2.

The difficulty of observing in the infrared region of the electromagnetic spectrum is illustrated by this example. The telescope itself is radiating enormously more power than can be collected from a nearby star. When observing in the infrared this problem is serious as the star's output peaks in the optical region but the telescope peaks in the infrared. When the telescope is sited on the ground the problem is compounded by the output of all nearby objects and by atmospheric absorption.

Perhaps the ultimate solution to the problems caused to the astronomer by living on Earth would be to site a major observatory on the Moon where atmospheric effects could be avoided in all regions of the spectrum and the telescopes could be firmly placed on the Moon's solid surface. The inclusion of a radio telescope coupled to Earth-bound detectors could produce an interferometric apparatus with sufficient resolution to observe details of nearby stars.

Light Detection

Whether the electromagnetic radiation is collected on the Earth or elsewhere it must also be measured quantitatively so that the brightness of different objects can be compared. The eye was historically relied upon to make such comparisons but the invention of photography allowed this task to be performed scientifically for the first time. By the end of the nineteenth century, photographic plates had overtaken the eye in ability to detect faint objects and quantitative estimates of star brightness could be made from these plates. Light catalyses a chemical reaction that causes a change of colour (in the early years only from white to black) in the photographic plates or film. By measuring the image density and size, the brightness of the star can be determined.

In modern observatories the photomultiplier tube is more commonly used. In these devices light is directed through a window onto a photoelectric material. Such materials eject electrons when exposed to photons of sufficiently high energy, the rate of electron emission being proportional to the intensity of light. The emitted electrons are electrically accelerated to a dynode, a metal plate that ejects more electrons when bombarded by the incoming electrons. Each electron causes the ejection of several secondary electrons. The photomultiplier tube has a chain of dynodes so that a swarm of electrons results from the detection of a single photon. The electrons are directed to an external circuit and measured as an electrical current. Photomultiplier tubes are therefore very sensitive but cannot be used as a discrete imaging instrument. To build up an image using a photomultiplier tube, each part of the image must be measured sequentially. Image intensifiers work on the same principle as photomultiplier tubes but are built in a more complex way, so that the ejected electrons remain collimated and can thus retain spatial information. In an image intensifier the output electrons are incident on a fluorescent screen so that the image may be viewed directly by eye.

Observation of very faint distant objects on screen in real time can also be achieved using a more modern advance, the charge coupled device (CCD) which consists of an array of very small detectors on a silicon chip. The electrical charge that accumulates in each position on the array is proportional to the incident light upon it. This charge can be accessed pixel by pixel and may be measured directly by a computer. In this way data can be stored for future reference or displayed directly on a screen. The journey of the light from the stars to the astronomer is complete and its message lies waiting to be interpreted in computers around the world.

Questions

Problems

1 (a) Rewrite equation (2.2) so that entering the telescope aperture in metres and the observing wavelength in nanometres gives the resolving power in arcseconds.
(b) What is the resolving power of a telescope of aperture 0.12 m, observing at 500 nm?

2 (a) What is the resolving power of a telescope of aperture 5 m, observing a heavenly body that emits light at a frequency of 600 THz?
(b) What would the separation of a pair of Sun-like stars in a binary system at a distance of 10^{14} km have to be in order for the telescope in part (a) to be able to resolve them?
(c) What would the separation of a pair of asteroids (small rocky planetoids that are seen through their reflection of sunlight) in a binary system at a distance of 300 million km have to be in order for the telescope in part (a) to be able to resolve them?

3 Estimate the aperture of a telescope if it is required to be able to resolve craters with diameters of 2 km on the Moon. The Moon orbits the Earth at a distance of 385 000 km and the Sun's peak energy output is at 500 nm.

4 A telescope is to be built in the Scottish Highlands (latitude 57° N, longitude 4° W) as cheaply as possible.
(a) If atmospheric conditions limit the resolution available to 0.8 arcseconds and bright stars are to be studied what aperture size should be used given that observations are to be made at 500 nm?
(b) If more money became available and it was possible to use a 1 m aperture, how many times brighter would the image be?

5 The largest telescope in the world is a fixed radio reflector with a 300 m aperture built in a natural bowl in Puerto Rico.
(a) What is the resolving power of the telescope when it is receiving signals at 1 GHz?
(b) Why is it more likely that the telescope can achieve this resolution than an optical telescope can achieve its theoretical resolving power?

Teaser

6 Do stars tend to twinkle most when overhead or when close to the horizon? Why?

Exercises

7 What is the effect that limits the resolution of a large telescope when observing a bright star? What causes this phenomenon and how can it be prevented?

8 Discuss the siting of optical telescopes with regard to optimising their performance.

9 An amateur astronomer decides to build a simple Newtonian reflector to observe the night sky. Explain the difficulties that might be encountered with spherical and chromatic aberration in the design of the primary mirror and eyepiece and how they can be overcome.

10 What is the usual design of optical telescope employed in major, modern observatories and why is this design chosen above others?

3 The View from Earth

Of key importance to the observational astronomer is knowing where to look in the sky to find an object of interest. If it is a bright and familiar star that is to be studied then it is a simple matter to look up at the sky and pick it out but when the object is faint and hidden among a mass of apparently similar bodies, how does the astronomer know where to look? Once the object has been located, how will it move across the sky? It is the purpose of this chapter to answer such questions.

The Observer's Position on Earth

If an observer in the United Kingdom were to observe a new comet directly overhead in the sky one evening and wanted to alert fellow astronomers in Spain and South Africa, for instance, it would not be sufficient to pick up the phone and tell them to look straight overhead. The Earth is a sphere and so 'up' is a different direction depending on where you are standing on it. A quick glance at Figure 3.1 shows that the observer in South Africa would be unable to see the comet at all whereas the Spanish astronomer would have to look in a position closer to the horizon. The position of the observer is clearly important when seeking an object in the sky.

The surface of any object is two dimensional and therefore two coordinates are required to define position upon it. For the Earth's surface, these coordinates are latitude and longitude, both of which are angles measured with respect to great circles. A great circle is a line which divides the surface of a sphere into two equal halves.

In order to define latitude an obvious choice of great circle presents itself as a zero marker. The Earth spins in space and the points at which the spin axis intersect the Earth's surface are called the poles (north and south). The line that bisects the Earth, half-way between the poles, is the equator and this is used in the definition of latitude. All points on the equator have a latitude of zero degrees. The latitude of any other point on the Earth's surface is defined as being the smallest angle subtended at the centre of the Earth between that point and the equator. This may sound complicated but Figure 3.2 shows that it is not. Latitude varies from -90° at the south pole to $+90^\circ$ at the north pole.

The concepts associated with latitude were used as the first method for determining the size of the Earth by Eratosthenes over 2000 years ago. He knew that at noon on midsummer's day in Syene the Sun is directly overhead as it casts no shadows. In his city of Alexandria at noon on the same day he measured the shadow created by a pole sticking vertically out of the ground to be about one eighth of the length of the pole. This

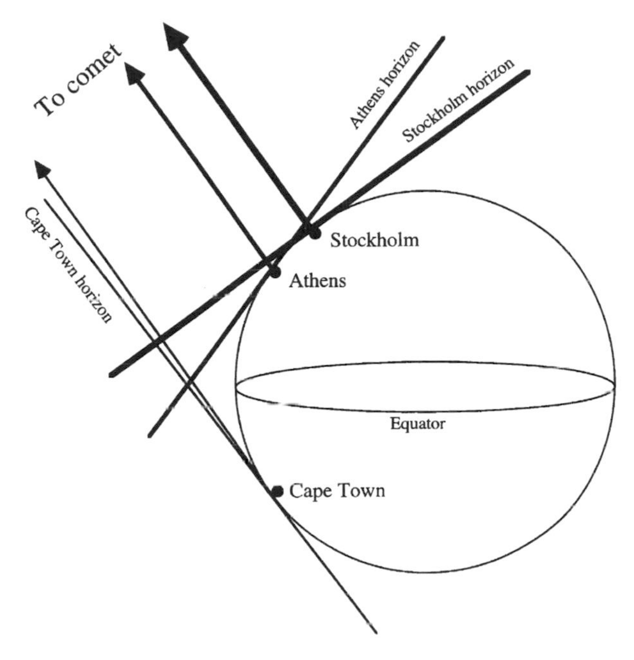

Figure 3.1 The position of a far-distant object in the sky as observed from different positions on Earth. The comet is observed directly overhead in Stockholm (thick lines representing horizon and direction of comet are perpendicular to each other). Viewed from Athens the comet is a few degrees north of overhead (the medium thickness lines are not quite perpendicular to each other). From Cape Town (thin lines) the comet cannot be seen as it is below the horizon

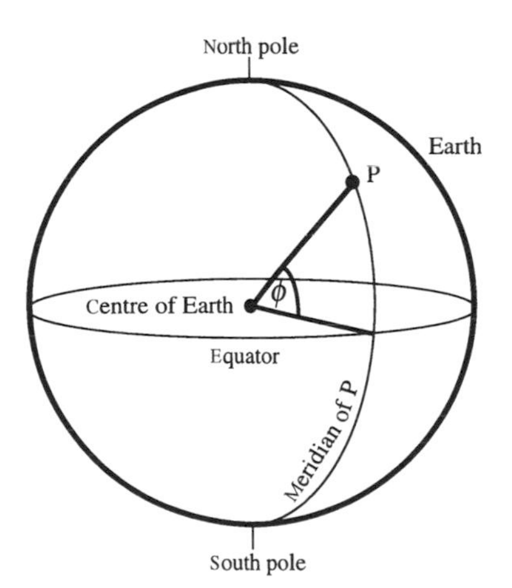

Figure 3.2 The latitude, ϕ, of a position, P, on the surface of the Earth is given by the smallest angle subtended at the centre of the Earth between the equator and P (therefore measured along a meridian)

means that the Sun is about 7.5° from being directly overhead. This implies the north–south angular separation of Syene and Alexandria to also be 7.5° (see Figure 3.3). As it was known that Alexandria is almost due north of Syene then the circumference of the Earth could be calculated as being 48 (= 360/7.5) times greater than the distance between the two cities. The result of Eratosthenes' calculation was remarkably close to the actual value known today.

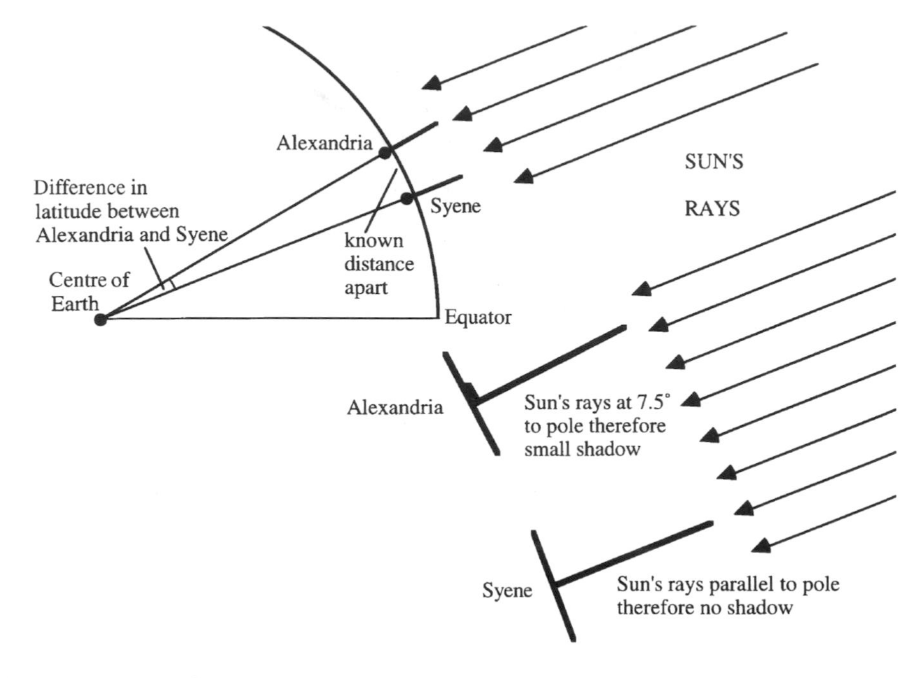

Figure 3.3 Eratosthenes' method for measuring the size of the Earth. A pole perpendicular to the ground in Syene casts no shadow at the same moment a small shadow is cast in Alexandria (lower-right). The angle that the shadow subtends to the pole at Alexandria is therefore equal to the difference in latitudes between Syene and Alexandria (upper-left). If the distance between the two cities is known then the size of the Earth can be inferred

---

WORKED EXAMPLE 3.1

Q. Two 1 m poles are sticking out of the ground perpendicular to the ground, one in Cleethorpes and the other in London. At a given moment, the shadow is 57.7 cm long in Cleethorpes and 53.2 cm long in London. If Cleethorpes is 223 km due north of London, what is the radius of the Earth?

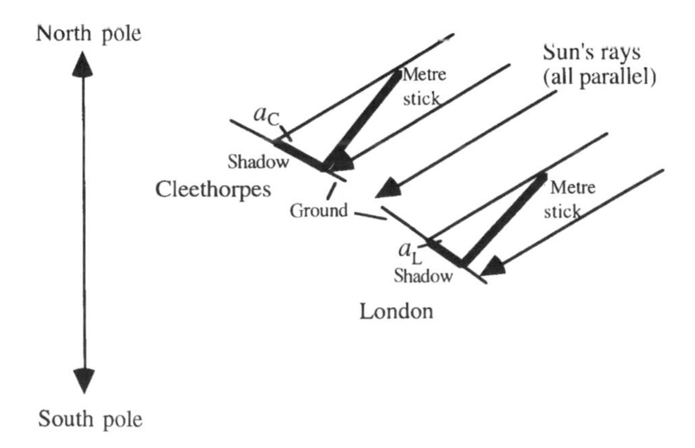

A. The difference in the Sun's angle above the horizon (altitude) as viewed from two positions that have different latitudes but the same longitude must be equal to the difference in the observing latitudes. To see this it is necessary to realise that the two observers will be looking at the same point in space but that their viewing horizons will be tilted from each other by the latitude difference. The Sun's altitude can be found from the length of the shadow. If the sticks are perpendicular to the horizon then the Sun's altitude, a, is given by the inverse tangent of the ratio of the stick length to shadow length.
Thus altitude of Sun as viewed from Cleethorpes, a_C, is $\tan^{-1}(1/0.577) = 60.0°$
and altitude of Sun as viewed from London, a_L, is $\tan^{-1}(1/0.532) = 62.0°$.
So the difference in latitude between London and Cleethorpes is 2.0°. To travel once around the world, from pole to pole, the latitude must change through 360° (i.e. there are 360° in a great circle). If 2.0° corresponds to 223 km then one circumference corresponds to $(223 \times 360/2.0) = 40\,100$ km. The circumference is equal to the radius multiplied by 2π so that the Earth's radius is $(40\,100/2\pi) =$ 6 380 km.

The second coordinate on the Earth's surface requires, as a zero, a great circle that passes through both poles, known as a meridian. Meridians are lines of equal longitude. There is nothing that distinguishes any two meridians on the surface of a spinning sphere and so an arbitrary zero is required. This is the Greenwich meridian which passes through a specific point in London. The Greenwich meridian defines the zero of longitude and this coordinate for any other position on the Earth is given by the angle through which the Greenwich meridian must be rotated to reach the position of interest. Actually only half of the Greenwich meridian is required to define the zero for longitude, from the north pole through London to the south pole. The other half has a longitude of 180°. Other lines of longitude are between 0° and 180° but must be defined as being east or west of the Greenwich meridian. For instance, Tokyo has a longitude of about 139° E and Los Angeles has a longitude of about 118° W.

The measurement of longitude is illustrated in Figure 3.4 and a comparison with Figure 3.2 shows the difference between the ways in which latitude and longitude are defined. Lines of equal longitude are all great circles passing through the poles and divide the Earth into segments, like oranges. Lines of equal latitude are not great circles (except for the equator) and they divide the Earth in the same way that a chef might slice an onion. Any point on Earth can be uniquely defined by stating its latitude and longitude.

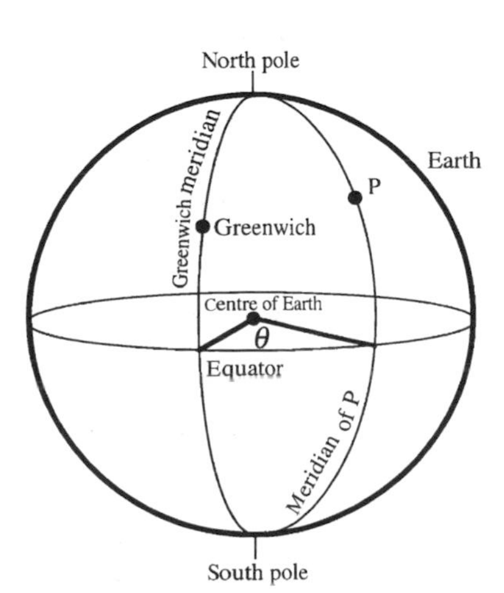

Figure 3.4 The longitude, θ, of a position, P, on the surface of the Earth is given by the angle subtended at the centre of the Earth between the meridian on which P lies and the Greenwich meridian

WORKED EXAMPLE 3.2

Q. Consider the following approximated (latitude, longitude) coordinates of European cities; Paris (48° N, 2° E), Munich (48° N, 12° E), Rome (42° N, 12° E) and Barcelona (42° N, 2° E). Compare the separation of cities of equal latitudes with those of equal longitudes.

A. The separation of places that have the same longitude is determined only by their latitude difference and so the separation of Paris and Barcelona is equal to the separation of Munich and Rome. This is easy to calculate as latitude measurements are made along meridians and these are great circles. An angular separation of 6° thus automatically corresponds to a distance equal to $(6°/360°) = 1/60$th of the Earth's circumference which is about 700 km. Thus Paris is about 700 km from Barcelona and Munich is about 700 km from Rome.

The same argument cannot be applied to cities of equal latitude because lines of equal latitude are not great circles. Thus, because Paris and Munich are further north than Barcelona and Rome, the separation of the former pair is smaller than that of the latter. The difference in the longitudes of both pairs is 10°. At the equator (the only line of equal latitude that is a great circle) the separation of points 10° apart would be $(10°/360°) = 1/36$th of the Earth's circumference which is 1100 km. This separation becomes smaller with increasing (positive or negative) latitude so that at the poles it has reduced to zero. At 42° N it is about $(1100 \times \cos 42° =)$ 800 km whereas at 48° the separation has reduced to about $(1100 \times \cos 48° =)$ 700 km. Thus Paris is about 700 km from Munich but Barcelona is about 800 km from Rome.

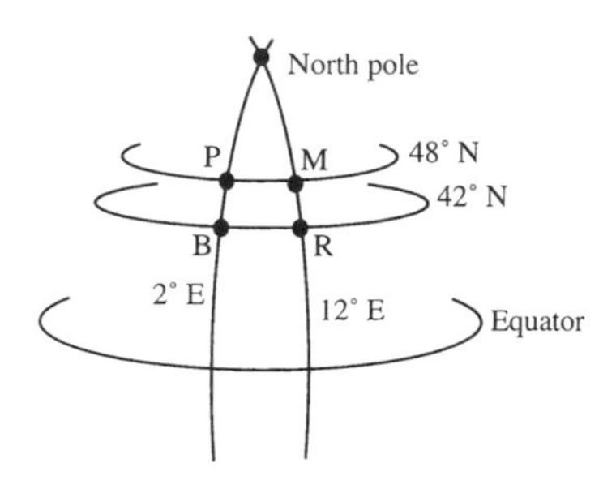

Defining Position in the Sky

It is important to know from where on Earth an astronomer is observing because the sky looks different as viewed from different locations, as has already been illustrated by Figure 3.1. It is therefore also necessary to define position in the sky and this can be done using methods entirely analogous to those used to define position on the Earth's surface. The sky must be imagined to be a hemispherical dome to which the stars are fixed at any given moment. The horizon defines a great circle on the hemisphere and can be used as a zero in much the same way as the equator. As mentioned in Worked Example 3.1, the angle of a given point above the horizon is known as altitude and is explained more fully in Figure 3.5. A star directly overhead has an altitude of 90° and is said to be at the zenith. Altitude is the first coordinate on the sky and is measured in an analogous way to latitude on the Earth's surface.

Azimuth is the second coordinate required to define position in the sky and is directly analogous to longitude. A zero must be defined and this is the observer's meridian, that

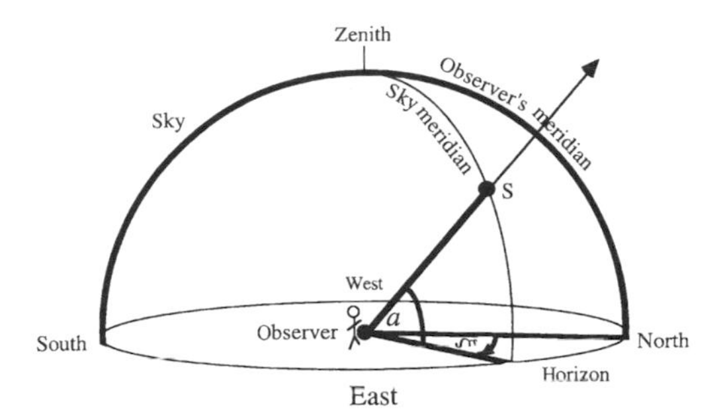

Figure 3.5 The altitude, a, and azimuth, ξ, of a star, S, in the sky. The altitude (analogous to latitude on the Earth's surface) is given by the smallest angle subtended at the observer between the horizon and S (therefore measured along a meridian). The azimuth (analogous to longitude on the Earth's surface) is measured eastwards from the observer's meridian (due north) and is the angle subtended at the observer to the meridian on which S lies. The altitude and azimuth of all stars (except those at the celestial poles) are constantly changing as they move about the sky

part of the great circle that runs from the zenith to the horizon due north. The azimuth of a star at any given moment is the angle between the sky meridian on which it lies and the observer's meridian and runs from 0° to 360°, always measured in an easterly direction.

WORKED EXAMPLE 3.3

Q. An observer is standing on a beach 5 km due north-west of a mountain that has a height of 3000 m. The Moon appears to be directly behind the peak of the mountain. What is the Moon's altitude and azimuth?
A. The observer is due north-west of the mountain and so the Moon must be due south-east of the observer. Measurement of azimuth begins at due north and proceeds in an easterly direction so that due east has an azimuth of 90° and south-east a further 45° at 135°. The mountain's angular height is equal to the Moon's altitude. Simple trigonometry shows that the tangent of this angle is given by the ratio of the mountain's height to its distance at $(3000/5000) = 0.6$. The Moon's altitude is therefore $(\tan^{-1}0.6) = 31°$. At this moment the Moon has an azimuth and altitude of 135° and 31° respectively which uniquely defines its position in the sky. Its position and therefore its azimuth and altitude vary with time, however.

The Celestial Sphere

Because the Earth is spinning on its axis once a day the stars and Sun appear to move across the sky with this period. In other words, the sky is never still. Stars are so far away, however, that their position in the sky relative to each other, even if they are actually moving through space at quite high speeds, are practically invariant. If the Earth stopped spinning, stars in the sky would be static. It is useful to define a third sphere. This sphere defines the direction of bodies in space as viewed from the Earth independently of the Earth's motion. This map of directions in space is known as the celestial sphere.

Again it is necessary to use two coordinates to define position on the celestial sphere. A convenient starting point for this is to project the Earth's polar axis into space (see

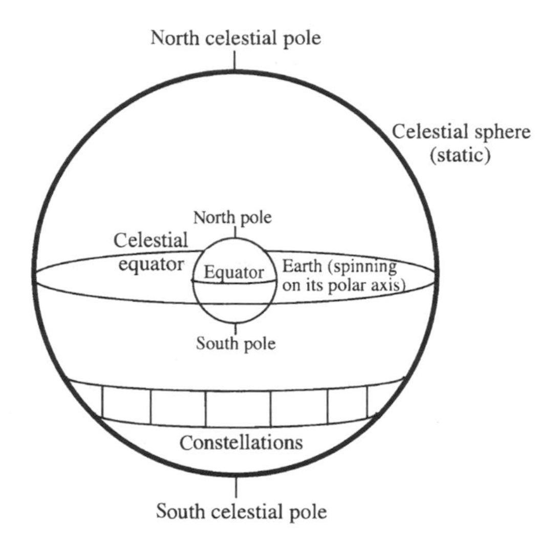

Figure 3.6 The celestial sphere. The north and south celestial poles are projections of the Earth's spin axes into space. The celestial equator is a projection of the Earth's equator into space. All stars are considered to be fixed to the celestial sphere and the Earth spins inside it causing the stars to appear to move about the sky. The stars are at various distances from the Earth but the celestial sphere is a useful device for defining their directions. The celestial sphere is divided into 88 constellations, some of which are schematically indicated in the southern hemisphere

Figure 3.6). Because the Earth is spinning on its axis, the axial projection does not move across the sky as all other points on the celestial sphere do. These two static points are known as the north celestial pole and the south celestial pole. By coincidence, a bright star (Polaris) closely marks the position of the north celestial pole. Polaris can therefore be seen in the same position in the sky at all times from any given latitude. It remains in this position during the day though, of course, it cannot be seen due to the brightness of the sunlight scattered by the Earth's atmosphere.

A projection of the Earth's equator on to the celestial sphere can also be made. The great circle that it draws on the celestial sphere is known as the celestial equator. This is also fixed in the sky but different parts of it are seen at different times of day and night. For instance, if an observer is standing on the Earth's equator then the celestial equator always arcs from the east to the west horizon, passing through the zenith, and as the Earth spins on its axis the sequence of stars that lie on the celestial equator pass overhead. The celestial equator is used as the zero for the analogous coordinate to latitude and altitude, known as declination and denoted by δ (see Figure 3.7). The north celestial pole has a declination of $+90^\circ$ and the south celestial pole is at -90°.

The second coordinate on the celestial sphere is measured in the same way as longitude and azimuth and is known as right ascension, α. Again, an arbitrary zero has to be chosen and this is a meridian that passes through a point on the celestial sphere known as the vernal equinox, the position on the celestial equator that the Sun passes across in March, to be discussed in more detail in the following chapter.

It is therefore possible to identify the position of any star on the celestial sphere using the coordinates declination and right ascension. This is equivalent to pointing out the

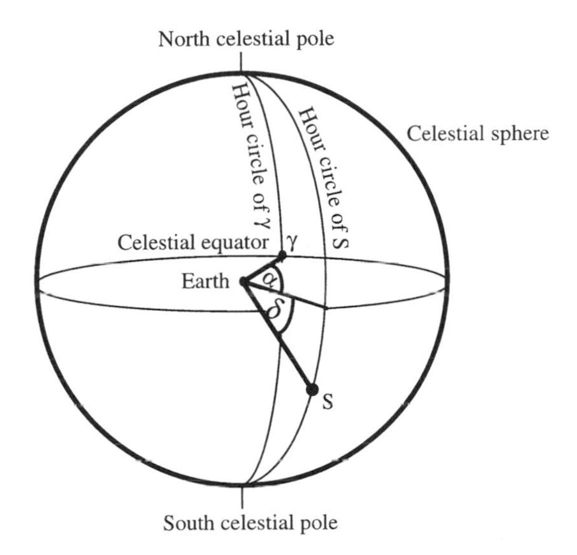

Figure 3.7 The declination, δ, and right ascension, α, of a star, S, on the celestial sphere. The declination (analogous to latitude on the Earth's surface and altitude in the sky) is given by the smallest angle subtended at the Earth between the celestial equator and S (therefore measured along a meridian known as the hour circle). The right ascension (analogous to longitude on the Earth's surface and azimuth in the sky) is measured eastwards from the vernal equinox, γ, and is the angle subtended at the Earth to the meridian on which S lies (its hour circle). The declination and right ascension of all stars are fixed

direction of the star in space in an absolute way rather than where it appears to be to an arbitrarily oriented observer. The celestial sphere is divided into 88 areas. These areas are called constellations and are based on the shapes produced by the brightest stars in the sky that were imagined to be animals and gods by the early stargazers. The modern constellations are centred on these shapes but are used principally to indicate different areas of the sky. They cover the whole of the celestial sphere so that they are useful for naming and numbering fainter stars, galaxies, radio sources, etc. that fall within any given constellation.

Coordinates and special positions have now been defined on three spheres; the Earth's surface, the sky and the celestial sphere. Table 3.1 brings together the terminology for analogous measurements;

Table 3.1

	'Up/down' coordinate	'Up/down' zero	'Up/down' top	'Up/down' bottom	'Around' coordinate	'Around' zero
Earth's surface	Latitude	Equator	North pole	South pole	Longitude	Greenwich meridian
Sky	Altitude	Horizon	Zenith	Nadir	Azimuth	Observer's meridian
Celestial sphere	Declination	Celestial equator	North celestial pole	South celestial pole	Right ascension	Vernal equinox

WORKED EXAMPLE 3.4

Q. From where on the Earth's surface are the declinations of all stars equal to their altitudes?
A. Declination is measured relative to the celestial equator whereas altitude is measured relative to the observer's horizon. Only when the horizon and the celestial equator are equivalent are declination and altitude equivalent. Remembering that the celestial equator is a projection of the Earth's equator, then standing on the Earth's equator will mean that the celestial equator runs in an arc overhead. As the observer moves towards either pole the celestial equator will form an arc that has a progressively lower and lower maximum altitude. At the poles the horizon and celestial equator will follow the same line. Hence all stars have equal altitudes and declinations when observed from the Earth's poles.

Note:
At the south pole, stars with negative declinations are observed (the stars of the south celestial hemisphere) though their altitudes will be positive. Thus the stars have equal altitudes and declinations but with opposite sign.

The azimuth and right ascension of stars are not equal (except for a moment once per day) when observed from the poles (or anywhere else) due to the spin of the Earth on its axis.

The Rotating Sky

It is useful to think of the celestial sphere as being static with the Earth spinning inside it. The appearance of the sky depends on the time and where the observer is situated on Earth. The three spheres of the Earth's surface, the sky and the celestial sphere are thus dependent on each other. Star maps are drawn using the celestial sphere and the observer must use these with a knowledge of his position on Earth to find objects of interest in the sky. The time is also important as the stars make their daily rotation around the sky. It is instructive to consider the view from a representative position on Earth, say York in the United Kingdom, at a latitude of 54° N and a longitude of 1° W. The place in the sky that is invariant when observing from the northern hemisphere is the north celestial pole. When standing at the north pole the north celestial pole is obviously directly overhead. Standing in York implies that the observer is tilted at $(90° - 54°) = 36°$ to an observer at the north pole and so Polaris will be viewed 36° from overhead, that is, at an altitude of 54°. The altitude of the celestial pole is thus equal to the observer's latitude and the same argument can be applied to any position on the Earth. Of course, it is the south celestial pole that is observed from the Earth's southern hemisphere.

In order to understand the path that stars take across the sky it is only necessary to remember that the Earth spins once a day upon its axis. This means that an observer's view of the sky is swung through 360° every day. As the observer is not aware of their own motion, the stars appear to circle the sky. Consider a star that is situated a few degrees from the celestial pole. At a particular time the star will be directly above the celestial pole, that is, it will have a larger altitude but the same azimuth. Half a day later the observer's vantage point will have been rotated through 180°. The position of the celestial pole will not have changed but the star of interest will now appear directly under the pole. In between, the star will trace out a circle in the sky with the celestial pole at the centre. All stars move in this way, describing circles in the sky around their nearest celestial pole (see Figure 3.8).

Figure 3.8 An eleven-hour time exposure of the sky. Stars move just less than one half of their journey around the (south) celestial sphere in this time. Note that some stars in this photograph rise during the eleven-hour exposure, some set and some are clearly circumpolar. That the camera is steady relative to the Earth during the exposure is indicated by the fact that the image of the dome of the observatory in the foreground is not blurred (photograph by David Malin; © The Anglo-Australian Observatory)

The size and position of the circles in the sky described by the stars are again dependent on the position of the star on the celestial sphere and the observer's position on Earth. It is easy to see that stars close to the celestial pole will travel in small circles around the pole. Such stars will reach their highest position in the sky, known as upper culmination, when they cross the observer's meridian. At this moment, their altitude is given by the altitude of the celestial pole (equal to the observer's latitude, ϕ) plus the difference in declination between the star and the celestial pole. The north celestial pole has a declination of 90° and so the altitude of a high declination star at upper culmination is given by

$$a_{\max} = \phi + (90^\circ - \delta) = 90^\circ + \phi - \delta \tag{3.1a}$$

Stars that are further from the north celestial pole have smaller declinations and describe larger circles in the sky. As the maximum altitude, as given by equation (3.1a), increases it will eventually become greater than 90° (when $\delta < \phi$). This actually marks the declination at which the star's upper culmination occurs south of the zenith. The angle as given by equation (3.1a) is correct as measured from the northern horizon (the northern cardinal point) but altitude must be measured from the closest point on the horizon (see Figure 3.9). This is now the southern cardinal point and is thus given by

$$a_{\max} = 180^\circ - (90^\circ + \phi - \delta) = 90^\circ - \phi + \delta \tag{3.1b}$$

In both cases the star's minimum altitude also occurs on the observer's meridian but directly below the north celestial pole. As the motion is circular $a_{\min}$ is separated from the pole by the same angle as $a_{\max}$ thus:

$$a_{\min} = \phi - (90^\circ - \delta) = \phi + \delta - 90^\circ \tag{3.2}$$

The question now arises as to what happens when the observer's latitude plus the star's declination ($\phi + \delta$) is less than 90° so that the star's minimum altitude is less than zero. This is the case of a star 'setting' which means that it dips below the observer's horizon. The path of such a star across the sky is still circular but part of the circle is hidden by the Earth.

Those stars that never leave the sky (even though they are not visible during the day due to the scatter by the atmosphere of the Sun's rays) are said to be circumpolar. The condition for a star to be circumpolar is:

$$\delta \geqslant 90^\circ - \phi \tag{3.3}$$

A quick inspection of equation (3.3) shows that when the observer is positioned at the north pole ($\phi = 90^\circ$) the same stars remain in the sky at all times. The stars that are visible are those of the north celestial hemisphere and they all circle parallel to the horizon about the north celestial pole which is at the zenith. As the observer travels south more stars become visible but some stars are no longer circumpolar and disappear from view for periods of time (see Figure 3.10). For instance, observing from York allows all of the northern hemisphere plus the southern hemisphere down to -36° ($= 54^\circ - 90^\circ$) to be observed. Only stars with declinations greater than $+36^\circ$ are circumpolar and the path of

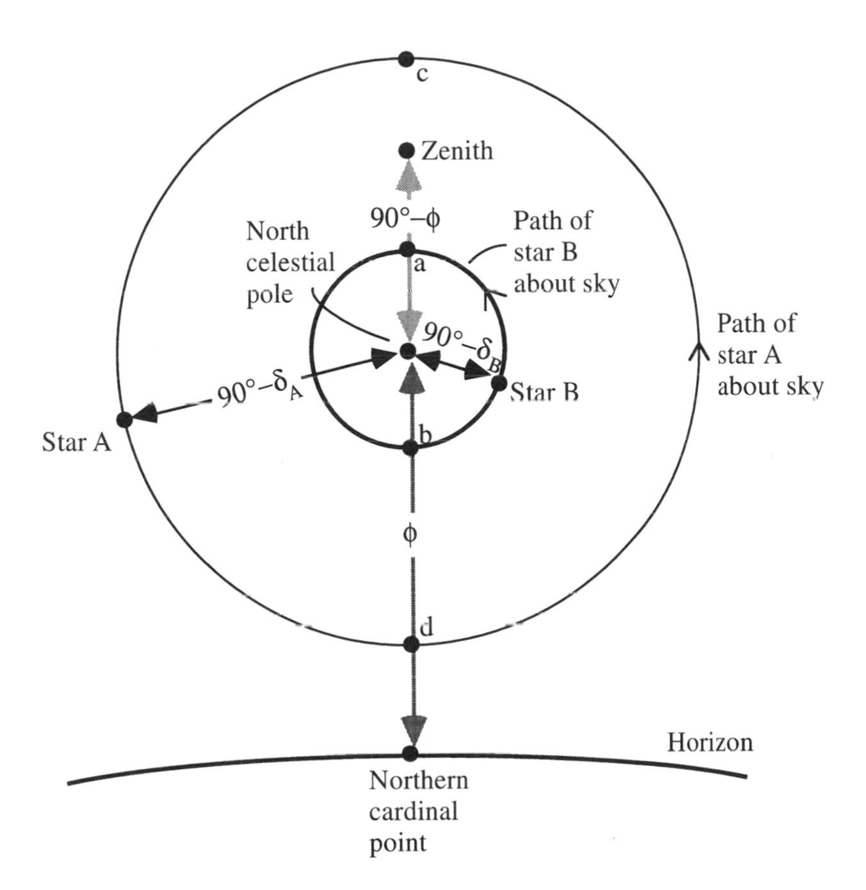

Figure 3.9 The path of two stars about the north celestial pole (at an altitude of ϕ). Star B has a (large) declination of δ_B and is therefore $90° - \delta_B$ from the celestial pole at all times. Its upper and lower culminations occur at positions a and b which are, by inspection, at altitudes of $\phi + (90° - \delta_B)$ and $\phi - (90° - \delta_B)$. Star A has a (smaller) declination of δ_A and is therefore $90° - \delta_A$ from the celestial pole at all times. Its upper and lower culminations occur at positions c and d which are, by inspection, at altitudes of $\phi + (90° - \delta_A)$ and $\phi - (90° - \delta_A)$. However, position c is beyond the zenith ($a = 90°$) so that star A is actually closer to the southern cardinal point at upper culmination. Its altitude is then $180° - (\phi + (90° - \delta_A)) = 90° - \phi + \delta_A$.

all stars are circular about the north celestial pole which is at an altitude of 54°. For an observer on the equator the north and south celestial poles lie at the north and south cardinal points on the horizon. The celestial equator loops overhead and all stars describe perfect semi-circular paths across the sky parallel to the celestial equator. There are no circumpolar stars but the whole of the celestial sphere appears in the sky throughout the course of each day. This means that during the course of a year, as the Earth circles the Sun so that different parts of the celestial sphere appear in the night sky, the whole of the celestial sphere can be seen from the equator. As the observer travels south, less and less of the northern celestial sphere is visible and stars circle the southern celestial pole. At the south pole, the south celestial pole is at the zenith and all stars are circumpolar, orbiting the sky parallel to the horizon, but only the southern celestial hemisphere can be seen.

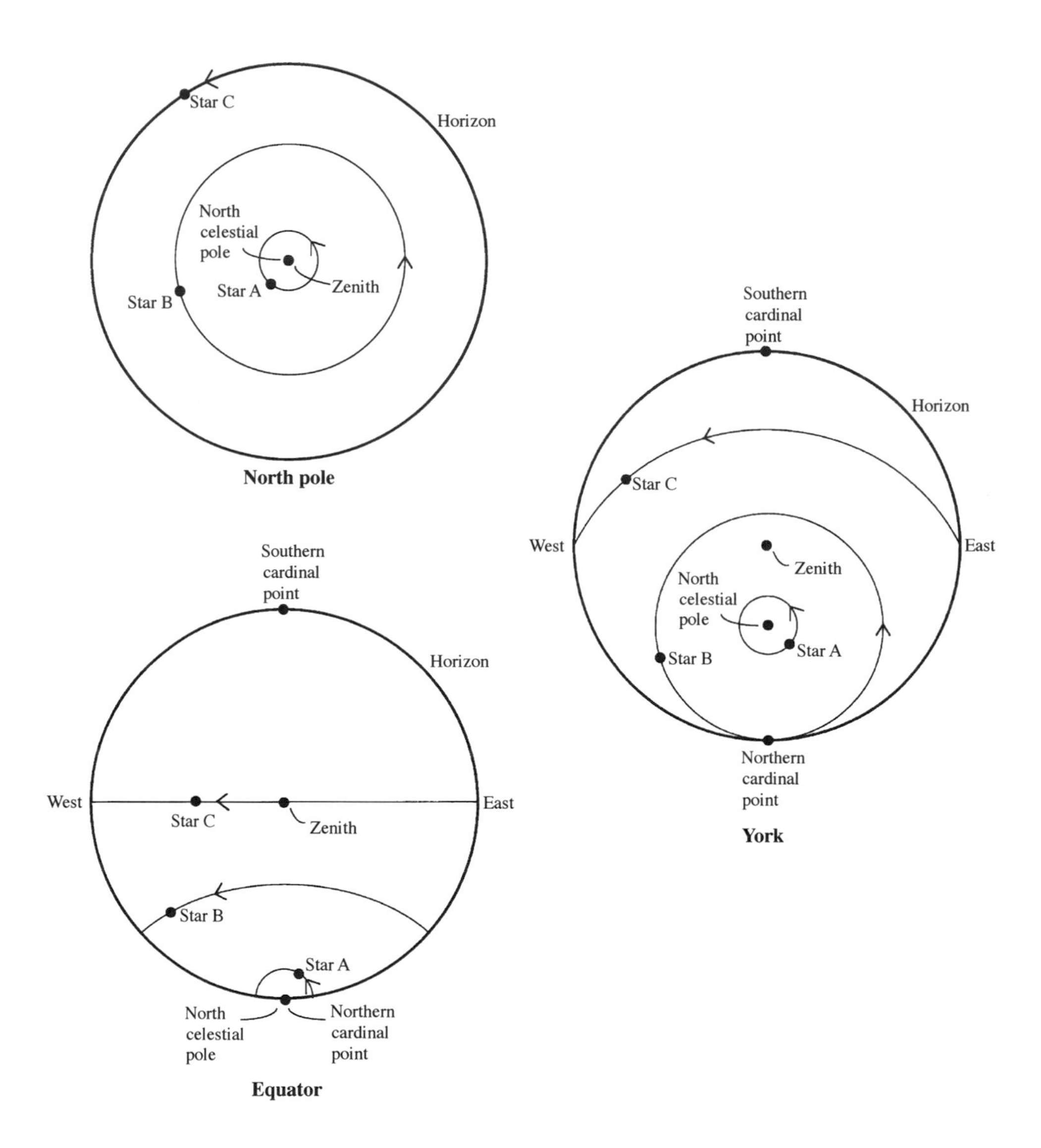

Figure 3.10 The motion of three particular stars about the sky as observed from the north pole, a position on the northern hemisphere (York) and the equator. The observer is lying on their back with the zenith at the centre of their field of view and the horizon forming a circle all around. The north celestial pole occupies a different position in the sky depending on where it observed from but the three stars must remain separated from it by the same angle at all times and observed from all places. Note that at the north pole all observable stars (those of the northern celestial hemisphere) are circumpolar, as observed from York some stars rise and set and some are circumpolar whereas, observed from the equator, all stars rise and set

WORKED EXAMPLE 3.5

Q. As observed from York (latitude 54° N, longitude 1° W), what will be the altitude of Kochab (right ascension 6°, declination +77°) when it crosses the observer's meridian? Will Kochab ever disappear from the sky and, if not, what will its minimum altitude be?
A. A star crosses the observer's meridian at upper culmination. To answer questions relating to maximum and minimum altitudes, longitude and right ascension are not required. Because $\delta > \phi$, equation (3.1a) should be used:

$$a_{\max} = 90^\circ + \phi - \delta = 90^\circ + 54^\circ - 77^\circ = 67^\circ$$

A quick check with equation (3.3) shows that δ (77°) is greater than $90^\circ - \phi$ ($90^\circ - 54^\circ = 36^\circ$) so that Kochab is circumpolar as viewed from York. To determine its lowest position in the sky equation (3.2) is used:

$$a_{\min} = \phi + \delta - 90^\circ = 54^\circ + 77^\circ - 90^\circ = 41^\circ$$

So Kochab remains in the York sky at all times with an altitude that varies between 41° and 67° (at all times north of the zenith).

WORKED EXAMPLE 3.6

Q. Repeat Worked Example 3.5 for Vega (declination +39°).
A. In this case, upper culmination takes place south of the zenith ($\delta < \phi$) so that equation (3.1b) is used:

$$a_{\max} = 90^\circ - \phi + \delta = 90^\circ - 54^\circ + 39^\circ = 75^\circ$$

Checking with equation (3.3) shows that δ (39°) is larger than $90^\circ - \phi$ ($90^\circ - 54^\circ = 36^\circ$) so that Vega is circumpolar as viewed from York. To determine its lowest position in the sky equation (3.2) is used:

$$a_{\min} = \phi + \delta - 90^\circ = 54^\circ + 39^\circ - 90^\circ = 3^\circ$$

So Vega remains in the York sky at all times describing a large circle about the northern celestial pole that takes it from 3° above the horizon due north at its lower culmination to an altitude of 75° at upper culmination at which time it is due south of the zenith.

WORKED EXAMPLE 3.7

Q. Repeat Worked Example 3.5 for Spica (declination 11°).
A. In this case, upper culmination again takes place south of the zenith ($\delta < \phi$) so that equation (3.1b) is used:

$$a_{\max} = 90^\circ - \phi + \delta = 90^\circ - 54^\circ - 11^\circ = 25^\circ$$

Checking with equation (3.3) shows that δ (-11°) is smaller than $90^\circ - \phi$ ($90^\circ - 54^\circ = 36^\circ$) so that Spica is not circumpolar as viewed from York. So Spica, as viewed from York, describes a large part-circle about the north celestial pole that takes it to 25° above the horizon due south at upper culmination. It will rise in the eastern sky and set towards the west. It should be noted that the star might more appropriately be thought of as circling the south celestial pole even though this point is not visible from York.

WORKED EXAMPLE 3.8

Q. The upper culmination of a star as measured from Glasgow (latitude 55° 50′ N) is at an altitude of 65° at which time the star is north of the zenith. What altitude does the star have at lower culmination?

A. It is possible to extract the answer to this problem by using equations (3.1a) and (3.2) sequentially. It is simpler, however, to consider the problem logically. The star is always the same angular distance from the north celestial pole (which has an altitude equal to the observer's latitude) and this angle is easy to calculate as being ($65° - 55° 50' =$) $9°10'$ at upper culmination. At lower culmination the star must be the same angle below the north celestial pole, still north of the zenith, at ($55° 50' - 9° 10' =$) $46° 40'$.

WORKED EXAMPLE 3.9

Q. What is the range in declination of stars that can be observed from New Orleans (latitude 30° N)?
A. As the north celestial pole is always in the New Orleans sky, the maximum declination is easy to state as being +90°. The minimum declination for a star to just reach the horizon in New Orleans at the moment of upper culmination can be calculated by rearranging equation (3.1b) (clearly $\delta < \phi$) setting a_{max} to 0°:

$$\delta = a_{max} + \phi - 90° = 0° + 30° - 90° = -60°$$

So the range of stars observable from New Orleans is between −60° and +90°. Generally, this result can be stated that an observer can see all stars in their own celestial hemisphere plus those with declination magnitudes less than their colatitude ($90° - \phi$) in the other celestial hemisphere.

Proper motion

A final complication is that there are a few stars whose lateral motion is significant compared to their distance from Earth. These stars are therefore not fixed to the celestial sphere but drift across it. This movement across the celestial sphere is known as proper motion and is extremely slow compared to the rate at which stars move across the sky due to the Earth's spinning motion. The star with the largest proper motion is Barnard's star with a value of just one 350th of a degree per year. Almost all stars have proper motions significantly less than this value and so the celestial sphere can truly be considered to be fixed. Its apparent motion is due only to the daily spin of the Earth on its axis and the form of the motion is due to the latitude from which the sky is observed.

Questions

Problems

1 (a) What is Vega's maximum altitude in the sky as observed from the Scottish Highlands (latitude 57° N, longitude 4° W)? Vega's right ascension is 18.5° and its declination +39°.
(b) Will Vega ever disappear from the sky and, if not, what will its minimum altitude be?

2 A couple of lottery winners decide to spend their fortune and the next ten years of their lives producing their own map of the celestial sphere. In order to complete as much of the task as possible would the best site for the observatory be at the north pole, on top of a French alp (latitude 45° N, longitude 6° E) or in Kenya (on the equator)?

3 What is the range in declination of stars that can be observed from the following locations?
(a) Saigon (11° N, 107° E)
(b) Izmir (38° N, 27° E)
(c) Reykjavik (64° N, 22° W)
(d) Rio de Janeiro (23° S, 43° W)
(e) Port Stanley (52° S, 58° W)

4 For the locations listed in Problem 3 determine whether the star Caph (declination +59°, right ascension 00h 06m) is circumpolar or never visible. Where appropriate, state the altitude of the star's upper and lower culmination.

5 An astronomer in Madrid (latitude +40° N, longitude 4° W) observes a supernova directly overhead and immediately contacts colleagues in Glasgow (latitude +56° N, longitude 4° W) and Abidjan (latitude +5° N, longitude 4° W). At what azimuth and altitude should the two colleagues look for the supernova?

6 The upper culmination of a star as measured from Superior, Wisconsin (latitude 46° 40′ N) is at an altitude of 63°. What altitude does the star have at lower culmination?

7 A star which lies on the celestial equator is observed to subtend an angle in the sky of 135° with the vernal equinox when viewed from the Earth's equator. The vernal equinox is to the east of the star. What is the star's right ascension?

8 What is the maximum declination for a star to be circumpolar when observed from Bulawayo, Zimbabwe (latitude −20°)?

Teasers

9 The largest telescope in the world is a fixed radio reflector with a 300 m aperture built in a natural bowl in Puerto Rico. In what way does the fixed nature of this telescope restrict observation? Use a sketch of the celestial sphere to illustrate your answer.

10 An explorer lost in the Australian desert has a watch and a tape measure. At noon on 21 June the explorer's shadow is 2.03 m long. The explorer's height is 1.80 m.
(a) What is the explorer's latitude?
(b) If another explorer were lost in the northern hemisphere and cast the same length shadow at exactly the same moment, what would their latitude be?

4 The Sun, The Stars and Time

Systems of time depend intimately on astronomical details. The time it takes the Earth to spin once on its axis is called a day and is simply detectable as the Sun rises and sets once every 24 hours. The hour and the units that subdivide it, seconds and minutes, are arbitrarily chosen. The year is the other physically defined time period and is based on the time the Earth takes to make one revolution about the Sun. The period of a year is perceived by the passing of the seasons.

The Sun's Position in the Sky

The main reason for the variation in temperature experienced between seasons is the difference in orientation between the Earth's spin axis and its revolution axis. So far only the Earth's axial spin has been considered and, as far as observing stars is concerned, this is effectively the only motion that matters. This is not true of the Sun's position in the sky as the Sun is very much closer and occupies a position inside the Earth's orbit. Figure 4.1 shows why this is of fundamental importance. Consider an inhabitant of the northern hemisphere. On the right of Figure 4.1 the Earth's spin axis tilts the northern hemisphere away from the Sun so that the Sun is in the sky for shorter periods during each day and at lower altitudes so that the temperature becomes colder. In terms of the celestial sphere it can be seen that the Sun is below the celestial equator at this time. That is, the Sun is south of a projection of the Earth's equator into space. The Sun is furthest south on the celestial sphere on the shortest day when its declination is equal to the angle between the spin and revolution angles at $-23.5°$. On this day the Sun moves across the sky like any other star with a declination of $-23.5°$. From equation (3.1b), the highest altitude in the sky that the Sun reaches as observed from York, at a latitude of 54° N, is $(90° - 54° - 23.5° =)12.5°$. So the Sun is below the horizon for most of the day and remains low in the sky after it rises. This moment of minimum declination is known as the winter solstice in the northern hemisphere. For latitudes greater than $(90° - 23.5° =)67.5°$ equation (3.1b) shows that the Sun remains below the horizon for the whole day at this time. This line of latitude is known as the Arctic Circle.

Three months later the Earth will have completed one quarter of its revolution about the Sun. This position is also indicated in Figure 4.1. The Sun now occupies a position in

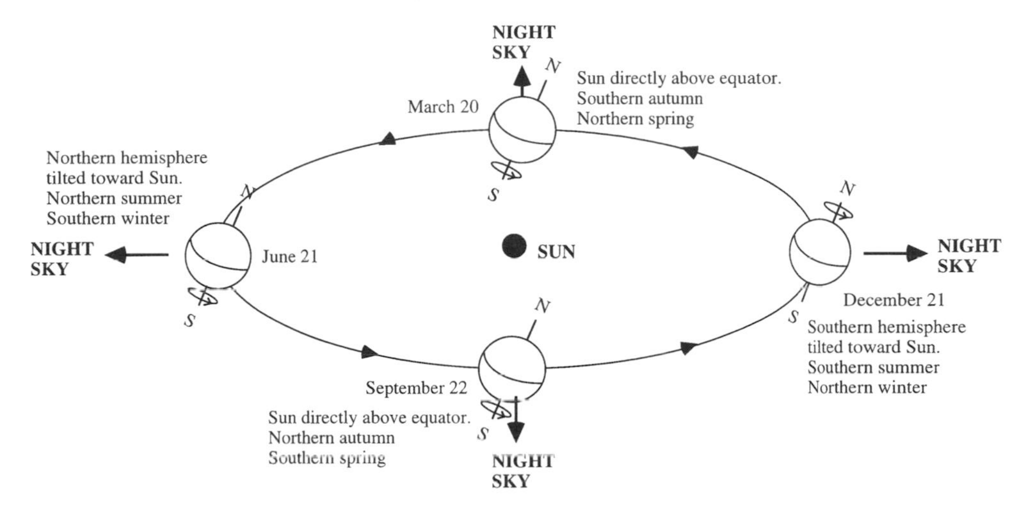

Figure 4.1 The orbit of the Earth about the Sun showing how the tilt of the Earth's axis affects the seasons and why the night sky varies throughout the year

space that is an extension of the Earth's equator. In other words, the Sun is on the celestial equator and has a declination of 0°. The precise position that the Sun occupies on the celestial sphere at this moment is known as the vernal equinox and is the zero point for the measurement of right ascension as discussed in the previous chapter. The Sun will be directly overhead at noon at the equator on this special day and will reach an altitude of (90° – 54° – 0° =)36° in York.

After a further three months the spin axis of the Earth tilts the northern hemisphere towards the Sun. The Sun is now 23.5° above the celestial equator and remains in the sky, as viewed from the northern hemisphere, for much longer. The temperature of the northern hemisphere thus increases and summer arrives. From York, the Sun now reaches an altitude of (90° – 54° + 23.5° =)59.5°. On the timescale of a day the Sun now behaves exactly like a star with a declination of +23.5°.

The Sun follows an annual path across the celestial sphere known as the ecliptic, illustrated in Figure 4.2.

WORKED EXAMPLE 4.1

Q. How does the Sun's highest position in the sky vary throughout the year in Quito which is on the equator in Ecuador, South America, and in Libreville, which is on the equator in Gabon, Africa?
A. On 21 December (or close to it, depending on the calendar for that year) the Sun's position on the celestial sphere reaches its furthest position south, at a declination of –23.5°, when the Earth's southern hemisphere is tilted towards the Sun. The observer's latitude (for both Quito and Libreville) is 0° and so upper culmination takes place south of the zenith ($\delta < \phi$) at this time and equation (3.1b) can be used:

$$a_{\max} = 90° - \phi + \delta = 90° - 0° - 23.5° = 66.5°$$

As the year progresses, the Sun's declination steadily increases towards zero (decreases magnitude) which it reaches on (or close to) 20 March. At this time equations (3.1a) or (3.1b) can simply be used to show that the maximum altitude is 90°. The use of equations is not strictly necessary for this

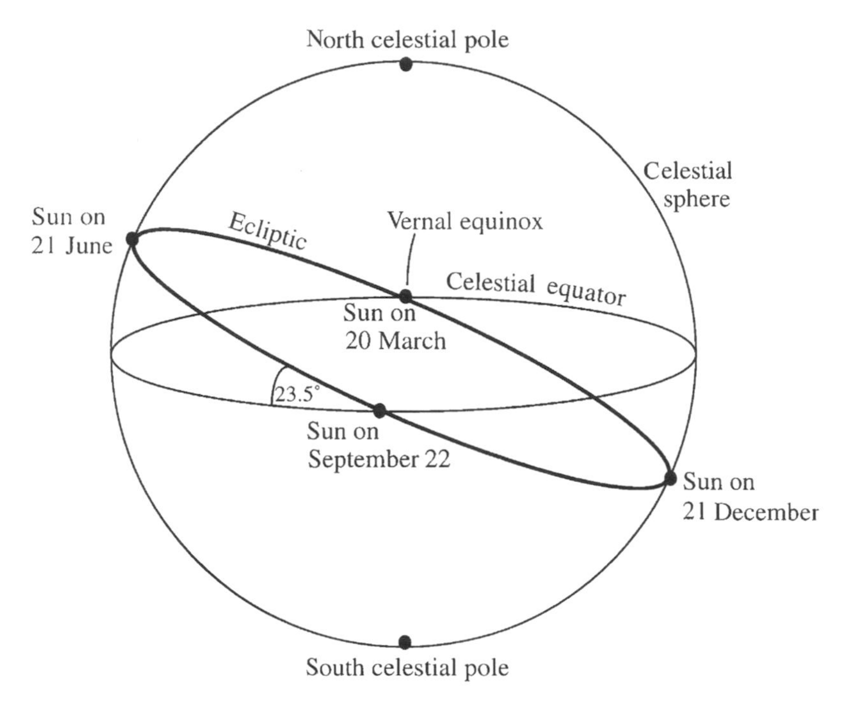

Figure 4.2 The motion of the Sun on the celestial sphere along the ecliptic (shown in bold) during the year

scenario as it should be easy to see that when the Sun lies on the celestial equator, which is simply a projection of the Earth's equator, the Sun must pass directly overhead all points on the equator as the Earth spins on its axis. During April and May the Sun continues to increase its declination until it peaks on the summer solstice on (or close to) 21 June. At this time the Sun's declination is 23.5° so equation (3.1a) can be used ($\delta > \phi$):

$$a_{\max} = 90^\circ + \phi - \delta = 90^\circ + 0^\circ - 23.5^\circ = 66.5^\circ$$

The Sun's peak altitude is as for the winter solstice but the Sun is now on the observer's meridian north of the zenith rather than south. The Sun's declination continues to decrease, reaching 0° on (or close to) 22 September, when the Sun once again crosses the sky from due east to due west passing directly overhead.

Throughout the year then, the Sun's peak altitude describes an oscillation between 66.5° above the northern horizon to 66.5° above the southern horizon, passing through the zenith in March and September. Note that the Sun passes directly overhead at the equator during the times at which non-tropical regions of the Earth are experiencing spring and autumn. As the Sun is always high in the sky and peaks twice per year, the seasons in tropical regions are more difficult to define and tend to be referred to in terms of more apparent weather phenomena such as rainy periods.

There is essentially no difference in the way in which the Sun travels across the sky in Quito and in Libreville except that there is a time offset. When the Sun is directly overhead in Libreville (longitude 10° E) it will be just rising in Quito (longitude 79° W). Six hours later the Sun will be directly overhead in Quito and setting in Libreville. This leads to the need to have different time zones for regions of the Earth at different longitudes.

WORKED EXAMPLE 4.2

Q. How often does the Sun rise and set at the north pole?
A. From Worked Example 3.4, the declination of an object on the celestial sphere is equal to its altitude in the sky at the north pole. This means that the Sun's altitude in the sky will vary as its declination varies during the year. The Sun makes one lap of the celestial sphere per year, rising above the celestial equator once, in March, and dipping below once, in September. The Sun will thus rise once per year at the north pole, stay in the sky for six months and then set again. While the Sun is in the sky its altitude will only vary by an average of $(23.5^\circ \times 2/365) = 0.13^\circ$ per 24-hour period so that the Sun will appear to orbit the sky in circles parallel to the horizon, slowly edging up to a peak altitude of 23.5° and then back down again. The same phenomenon will be observed at the south pole where the Sun will be in the sky between September and March.

The Sun and the Night Sky

In the previous chapter the effect of the Sun on star visibility was ignored but Figure 4.1 also shows why the night sky varies throughout the year. During the day the Sun is so bright that its light, scattered by the atmosphere, is sufficient to swamp the light arriving from other stars and they are therefore invisible. At night the Sun has set and the stars become visible. As the Earth travels around the Sun, the dark side of the Earth points towards different directions in the universe. In other words, different parts of the celestial sphere are obscured from view as the year progresses so that the night sky changes slightly from night to night.

WORKED EXAMPLE 4.3

Q. How does the night sky in Madrid (latitude 40° N, longitude 4° W) differ from that in New York (latitude 40° N, longitude 73° W) on any particular night?
A. It doesn't! As the two locations have the same latitude then the portion of the sky that is visible from each is the same. The only difference is that the two places are spun away from the Sun's dazzling influence to reveal (the same) half of the celestial sphere with a time lag of a few hours. In this time the Earth's position with respect to the Sun and stars has changed by a negligible amount.

If the night sky slowly varies from night to night that implies the stars and the Sun are moving about the sky at different rates. There are two ways to see why this is the case. The first is by considering the motion of a spinning, revolving body from outside its orbit and from in it. If the directions of rotation and revolution are the same then there appears to be one less rotation of the body when observed from the centre than from the outside. Consider the simplest example of a body that revolves in such a way that the same point on the body always points towards the centre of the orbit. An observer at this facing position on the revolving body will always have the orbital centre at the zenith and anything in that position will appear to be fixed. This is the case for the Moon's orbit about the Earth. From any position on the Moon, the Earth's position in the sky is constant. When viewed from outside the orbit the body will obviously make one rotation for every revolution. In the case of the Moon, the stars will rotate about the sky once per revolution. In the case of the Earth – Sun system, during one revolution the Earth rotates 366.26 times relative to the stars and 365.26 times relative to the Sun. A second way of thinking of this situation is in terms of the celestial sphere. As explained above, the Sun moves around the celestial sphere, along the ecliptic, once per year. If a given point on the celestial sphere moves across the sky 366.26 times per year then the Sun must complete one less rotation than this, that is, 365.26 rotations.

Solar Time and Sidereal Time

Clearly we must live our lives according to the rising and setting of the Sun and a time system known as solar time is used, based on 365.24 days per year[1]. An astronomer is usually more interested in the movement of the stars and so a second system known as sidereal time is defined and is based on having 366.26 days per year. Each type of day is divided into seconds, minutes and hours but the solar version of each is (366.26/365.24) = 1.0028 times or 0.28% longer than the sidereal equivalent. The Sun moves across the sky once every solar day and the stars once every sidereal day.

WORKED EXAMPLE 4.4

Q. If sidereal and solar clocks show the same time at midnight on 31 December then what time will the sidereal clock show at midnight on 18 August?
A. The time period of interest is 230 solar days. If there are 366.26 sidereal days in 365.24 solar days then the number of sidereal days in 230 solar days must be in the same proportion:

$$\frac{366.26}{365.24} \times 230 = 230.642 \text{ sidereal days} = 230 \text{ sidereal days, 15 sidereal hours, 25 sidereal minutes}$$

The sidereal clock will therefore read 15:25 at midnight on 18 August.

There is an obvious problem with the fact that there are not a whole number of days in a year, particularly in terms of solar time. Everyday considerations demand that solar time has an integer number of days in a year. This is chosen to be 365, the closest integer to 365.24, but that means that every new year the Earth is 0.24 of a day's orbit from where it was the previous year. Slowly, the Earth's calendar will drift from the alignment of the Earth and Sun in space. To correct for this, leap years are added on a regular basis. After four years, the Earth is 0.96 of a day out of position and so an extra one is added (29 February) to knock it forward almost into position. There are more complicated rules to compensate for the odd 0.04 of a day (see Question 7) and occasionally even leap seconds are added to or subtracted from the clock.

WORKED EXAMPLE 4.5

Q. An imaginary planet in an imaginary system rotates on its axis in the opposite direction to the direction of its revolution about its parent star. If 279.33 rotations of the planet take place during one revolution, how many 'solar' days are there per year? How often is a leap day required?

---

[1] If the direction of the Earth's spin axis were completely static in space then there would be precisely one less solar day per year, at 365.26, than the number of sidereal days. This is not quite true. The direction of the Earth's spin axis slowly varies (an effect known as precession, discussed at the end of this chapter). As a result, the number of solar days per year must be reduced slightly so that the solstices and seasonal variations take place at the same time every year.

A. The parent star moves about its ecliptic on the celestial sphere once during the period of revolution but, unlike the Sun, the motion has the same sense as the celestial sphere's rotation about the planet's sky. In other words, the parent star moves faster about the sky than the distant stars of the night sky so that there is one more 'solar' day than sidereal day. There are 280.33 'solar' days per year. The obvious choice would be to have 280 days per regular year. The parent star's position would therefore drift by 0.33 of a day per year and so a leap year would be required once every (1/0.33 $\approx$) 3 years. As for the Earth, other corrections would be required to account for the small fractions that are uncorrected by the general rule that a leap year is required once every three years.

It may seem that solar time is the more straightforward system but in fact sidereal time suffers fewer complications. Any star can be identified in a particular position and 24 sidereal hours later it will be back in that same position. As has been explained above, the Sun varies its altitude on a day-to-day basis, due to the tilt of the Earth's spin axis, but there is a second effect that causes a variation in the length of the solar day. The Earth does not revolve about the Sun in a perfectly circular orbit. Consequently, the Sun's path across the celestial sphere is not at a constant rate. To prevent solar time from having to speed up and slow down throughout the year a mean Sun is defined. The mean Sun tracks across the celestial sphere at a constant rate and is used for our time systems. In fact a clock based on the position of the actual Sun would never be more than 17 minutes different from that based on mean time.

Different positions on Earth require different time zeroes. It is necessary for it to be light at noon in both Los Angeles and Beijing even though they are on opposite sides of the Earth. It is convenient to define local solar time on the basis of noon being the moment at which the mean Sun crosses the observer's meridian and reaches its highest altitude. When observing from 0° longitude, on the Greenwich meridian, local solar time is known as Greenwich mean time (GMT) or Universal time. This is often used as a standard by which other local timescales are defined. It is possible to define a local mean time for any given observer's longitude but that would be unwieldy and so the Earth is roughly segmented into 24 time zones of about $(360°/24) = 15°$ each. All of the 15° segment centred on Greenwich uses GMT. The next segment east is always one hour ahead of GMT and the next segment west is always one hour behind. The segment edges are distorted to allow for various political borders but the system results in workable time zones for everyone in the world. For very few people does the Sun reach its highest position precisely at noon but it is rarely more than half an hour or so away. Such imperfections do not affect people's lives and few would ever notice.

WORKED EXAMPLE 4.6

Q. What is the time difference between London (latitude 51° N, longitude 0°) and Sydney (latitude 39° S, longitude 151° E)?

A. Latitude is irrelevant to this calculation even when the two places are in opposite hemispheres. The difference in longitude is $(151° - 0°) = 151°$. Time zones are generally about 15° across and so Sydney should be $(151/15) = 10.07$ hours ahead. Zones are always rounded to the nearest hour so that Sydney is ten hours ahead of London. Actually, the difference in hemispheres does make a difference in this case because the two nations that contain London and Sydney each employ a daylight-saving system during their respective summers that allows an extra hour of light during the evening at the expense of one in the early morning. Though the average time difference across the year is ten hours the actual time difference is nine hours in the northern summer and eleven hours during the southern summer.

Sidereal time must be measured precisely. This system of time is used for tracking stars across the sky to great precision and so it is not sufficient to package areas for which the local sidereal time is similar into zones. A knowledge of the local sidereal time for any particular location is required.

Sidereal time works in exactly the same way as solar time except that it is measured by the passage of the vernal equinox across the sky rather than the mean Sun. The local sidereal time is defined as being 0h 0m 0s when the vernal equinox passes the observer's meridian. It will cross the same point in the sky 24 sidereal hours later which is equivalent to 23h 56m 4s in solar time. In the interim period, that point in the sky is occupied successively by every other star that lies on the celestial equator. The rate at which the celestial sphere rotates across the sky is $(360°/24 \text{ hours}) = 15°$ per hour. The coordinate that describes how far a point on the celestial equator is from the vernal equinox is right ascension and so it is convenient to use time units instead of angular units for this coordinate. Right ascension can thus be defined as being the time between the vernal equinox crossing the observer's meridian and the point of interest doing the same. The conversion is 15° per hour but time is the unit that is normally used.

WORKED EXAMPLE 4.7

Q. As observed from Glasgow (latitude 55 50′ N, longitude 5 20′ W) Fomalhaut has its upper culmination one hour before the vernal equinox crosses the observer's meridian. What is the right ascension of the star?

A. In this case, the information given about the observer's location is irrelevant. Right ascension can be defined as the time period between the vernal equinox and the position of interest crossing the observer's meridian. For Fomalhaut the vernal equinox crosses the observer's meridian $(24 - 1 =)$ 23 hours before the star. Fomalhaut's right ascension must therefore be 23h.

Sidereal Time and the Night Sky

It is now a simple matter to find a star in the sky if the observer's position, the local sidereal time and the star's coordinates are known. From the observer's latitude the position of the celestial pole can be found and the angular separation of the star from the celestial pole determined from the star's declination. How far around its lap of the celestial pole the star has progressed can be determined from the local sidereal time and the star's right ascension.

Figure 4.3 illustrates the local hour angle of a star. As the star orbits its closest celestial pole the local hour angle is said to be the angle (also expressible in time units) through which the star has moved since its upper culmination. The angle is therefore measured from the observer's meridian and is subtended at the celestial pole. A determination of the local sidereal time can be made by measuring the position of the vernal equinox in the sky relative to the observer's meridian or the local hour angle of a star of known right ascension thus:

$$\text{local sidereal time} = \text{local hour angle} + \text{right ascension} \quad (4.1)$$

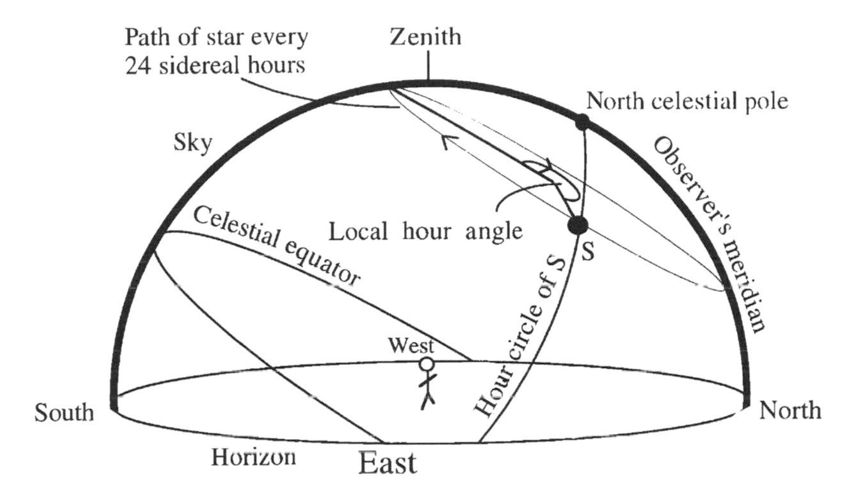

Figure 4.3 The local hour angle. All stars revolve in the sky about their local celestial pole once every 24 sidereal hours. The local hour angle indicates how far the star has revolved since its upper culmination and may be given as an angle (from 0° to 360°) or a time (from 0 to 24 hours). The local hour angle of the vernal equinox defines the local sidereal time

---

WORKED EXAMPLE 4.8

Q. At 0h GMT the Greenwich hour angle of the vernal equinox on a particular day is 12h 46m.
(a) What is the Greenwich sidereal time?
(b) What is the local sidereal time at Flagstaff, Arizona (latitude 35° 11′ N, longitude 118° W)?
(c) What is the local hour angle of Betelgeuse (right ascension 5h 53m, declination 7° 24′) measured from Flagstaff?
(d) What is the local hour angle of Betelgeuse measured from Flagstaff at 0h GMT the next day?
A. (a) By definition, the local hour angle of the vernal equinox gives the local sidereal time so the Greenwich hour angle of the vernal equinox gives the Greenwich sidereal time. This is 12h 46m.
(b) Flagstaff is 118° W of Greenwich which translates to time at 15° per hour as being (118/15 =) 7 hours 52 minutes behind. The solar time is therefore likely to be 8 hours behind but sidereal time must be calculated exactly as being (12h 46m − 07h 52m =) 04h 54m.
(c) Using equation (4.1), the local hour angle of Betelgeuse is given by;

$$\text{local hour angle} = \text{local sidereal time} - \text{right ascension} = 04\text{h}\ 54\text{m} - 05\text{h}\ 53\text{m} = -0\text{h}\ 59\text{m}$$

Times are always recorded as being positive and so the local hour angle of Betelgeuse in Flagstaff at this time is (24 h − 0h 59m =) 23h 01m.
(d) One solar day later is (366.26/365.24 =) 1.0028 sidereal days or 24h 04m later so the local sidereal time has increased by 4 minutes since the same solar time the day before and the local hour angle of Betelgeuse must have increased by four minutes also (as α is a constant) to 23h 05m. More physically, the star has rotated a little over one revolution of the sky in 24 solar hours and so it has shifted its position in the sky slightly. This 4-minute shift means that the same sky pattern can be seen four solar minutes earlier every night, slowly shifting the night sky into daytime and vice versa.

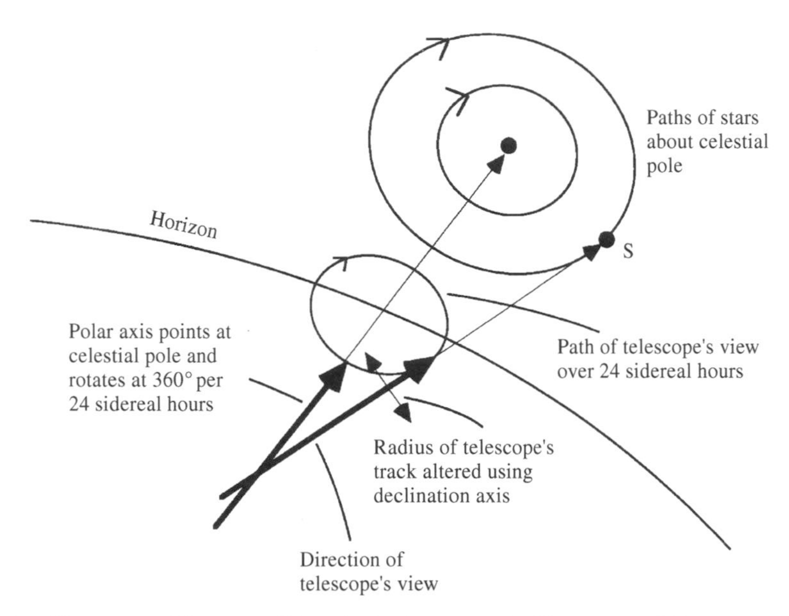

Figure 4.4 The equatorial mount. The principal mount axis (the polar axis) is permanently aligned towards the celestial pole. The telescope is mounted so that it can be moved perpendicularly to this axis to reach the correct declination. The polar axis can then be rotated so that the telescope points at the star of interest. To remain fixed on the star the polar axis is rotated at a rate of once per 24 sidereal hours

Telescope tracking

An important application of spherical astronomy is in telescope control. When observing faint objects in the sky it is often essential to gather data for substantial periods of time. It is therefore important for the telescope to track across the sky with the star. The simplest way to do this is to mount the telescope on a polar axis (Figure 4.4). Such an axis is pointed at the local celestial pole and rotated with a period of 24 sidereal hours. This is known as an equatorial mount. Once the telescope has been guided onto the position of interest the field of view will remain constant. Technological improvements have allowed an alternative design to be implemented in major, modern observatories. The altazimuthal mount allows the telescope to be held more firmly in place mechanically but offers no simple way of smoothly tracking the telescope as, in this case, the planes of movement are parallel and perpendicular to the horizon. As explained above, stars execute circular motions through the sky that have no simple relationship with the altitude and azimuth coordinates and this results in the field of view of the telescope rotating with time (see Question 9). Computer software and precise drive mechanisms now render these difficulties of little importance.

Precession

There is one final complication to consider. The Earth's spin axis is not static in space; its direction is slowly rotating or 'precessing'. The spin axis remains at an angle of 23.5° to

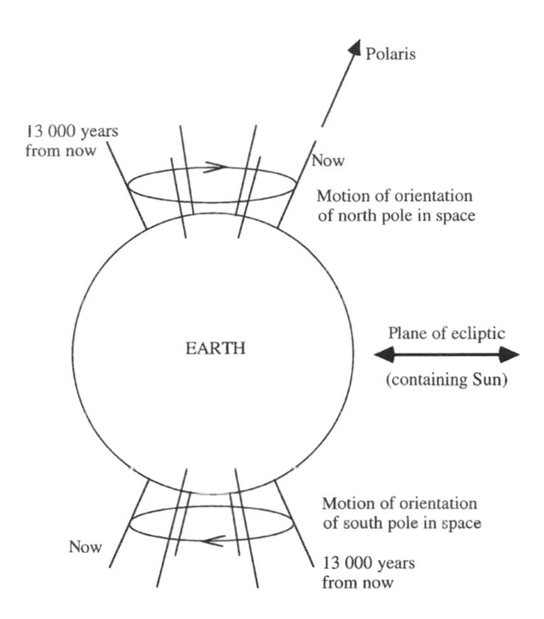

Figure 4.5 Precession. How the Earth's spin axis varies in space with time

the axis of revolution[2] but the direction of the angle is constantly rotating (see Figure 4.5) with a period of 26 000 years. As referred to above, this means that it is necessary to change the definition of the length of a solar year (or 'tropical year') by about 0.02 days[3] so that the Sun crosses the celestial equator at the same times every year and solar time does not lose synchronicity with physical observables. The slow pace of change of the orientation of the Earth's spin axis means that the night sky is not noticeably affected, even within an entire human lifetime, but precession will have the effect of gradually shifting the coordinates around the celestial sphere. The position of the celestial poles will follow a circular path across the celestial sphere and take the coordinates with them. About half-way through the Earth's precessional period, in around AD 15 000, Vega will become the north pole star and it will be Polaris that appears to orbit about Vega in the night sky.

Precession also causes some embarrassment for astrologers. Star signs are customarily based on the constellation that the Sun occupies on the day of a person's birth. As the Sun makes its annual lap of the celestial sphere it traditionally passes through 12 constellations, each for about one month (though the dates used are already incorrect). As the direction of the spin axis changes then the constellations through which the Sun passes will also vary and it will become necessary to invent a whole new superstition.

---

[2] Actually, the spin axis angle also oscillates over a few degrees with a long period.

[3] To be more precise, the difference is 0.0142 days, corresponding to about 20 minutes. After 26 000 years, the sum deficit is $(0.0142 \times 26\,000) =$ one year, in other words, the alignment of the Earth's spin axis in space has returned to its position of 26 000 years previously.

Questions

Problems

1 A radio source located on the celestial sphere at right ascension 5h 53m, declination $+18^\circ$ is observed from Puerto Rico (latitude 18° N, longitude 67° W) at 0h Greenwich Mean Time. At this time the Greenwich Hour Angle of the vernal equinox is 12h 46m.
(a) What is the Greenwich sidereal time?
(b) What is the local sidereal time in Puerto Rico?
(c) What is the local hour angle of the radio source measured from Puerto Rico?

2 What is the altitude of a star of declination $+11^\circ\ 50'$ as it crosses the observer's meridian in Glasgow (latitude $55^\circ\ 50'$ N)?

3 As observed from Glasgow (latitude $55^\circ\ 50'$ N, longitude $5^\circ\ 20'$ W) a star has its upper culmination one hour before the vernal equinox crosses the observer's meridian. What is the right ascension of the star?

4 From what latitude ranges would the Sun be visible at midnight on the following dates?
(a) 20 March
(b) 21 June
(c) 22 September
(d) 21 December

5 At 5h 57m Greenwich Mean Time the Greenwich hour angle of the vernal equinox on a particular day is 3h 12m.
(a) What is the Greenwich sidereal time?
(b) What is the local sidereal time in Tokyo (latitude $35^\circ\ 40'$ N, longitude $139^\circ\ 45'$ W)?
(c) What is the local hour angle of Merak (right ascension 10h 59m, declination $56^\circ\ 39'$) measured from Tokyo? Is Merak in the sky? Is it visible?
(d) What is the local hour angle of Merak measured from Tokyo at 5h 57m GMT the next day? Is Merak in the sky? Is it visible?

Teasers

6 Why is the altitude of the Sun in the sky important to local temperature rather than just the period that the Sun is in the sky?

7 Leap years take place when the year is divisible by four except when the year is divisible by 100 but not by 400. For instance, the following years were or will be leap years:

...1892, 1896, 1904, 1908...1992, 1996, 2000, 2004, 2008...

If this were to correct precisely for the Earth's orientation relative to the Sun, how many days would there be in a solar year?

8 Does twilight last longer on a sunny evening in England or Brazil? Why?

Exercises

9 The brightest stars in the constellation Cassiopeia form a 'W' shape. Compare the images that would be produced by two telescopes being driven, respectively, on equatorial and altazimuthal mounts over a long observation period.

10 Explain the difference between a solar day and a sidereal day and why the two periods are different. Ignore the effects of precession.

5 Observation of the Solar System

Each sidereal day that goes by sees the stars make the same journey across the sky. With the exception of very slow changes caused by effects such as proper motion and precession and very rare events such as supernovae the stars remain unchanged, keeping strictly to their 24 sidereal hour schedules. In terms of solar time there are small shifts in position of the stars from day to day but the pattern still repeats once every year. It is the close proximity of the Sun that causes its motion across the sky to be different from that of the very far distant stars. The other components of the Solar System are also relatively nearby and so their motion across the sky does not follow that of the stars. The simultaneous revolution of the Earth and the planets about the Sun means that planetary movement across the celestial sphere can be quite complicated. The diversity of the bodies of the Solar System also means that their appearance varies greatly. In this chapter the motion and appearance of the planets, their satellites and other Sun-orbiting objects are discussed and explained and in so doing the basic nature of the bodies elucidated.

The Sun dominates its orbital system, containing 99.9% of the mass of the whole and providing more than that proportion of the energy output. Despite this, it is important to note that the Sun is an unremarkable star in an unremarkable galaxy. It is only special because we are so close to it, because we are gravitationally tied to it and because it provides the Earth with the energy that allowed life to evolve.

The Earth is just one of the nine known planets that revolve around the Sun. The planets vary extensively but there are common characteristics. Each planet is composed mainly of condensed matter in solid and/or liquid form at surface temperatures much lower than that of the Sun. Blackbody radiation given off by the planets is consequently very weak in the visible part of the spectrum and the planets can only be seen by the sunlight that they reflect. Due to their close proximity to the Earth this is often sufficient to make them appear to outshine the brightest stars.

Planetary Orbits

The planets all revolve about the Sun in the same direction. Strictly speaking, the orbits are elliptical but most can be approximated quite well by circles. The orbits mostly share the same plane as the Earth, known as the ecliptic. Consequently, the planets follow paths across the sky similar to that of the Sun but at different rates. Mercury and Pluto have the orbits that are both least well approximated by a circle[1] and furthest from the plane of the Earth's orbit. They are the smallest planets and, respectively, the closest to and furthest from the Sun. They also revolve about the Sun at the fastest and slowest rates respectively because period of revolution increases with distance. The Sun itself rotates on its axis in the same direction as the planets revolve. The planets also rotate on spin axes, usually in this same direction. Here Venus, Uranus and Pluto are the exceptions. Venus rotates in the opposite (retrograde) direction while Uranus and Pluto rotate on axes close to the ecliptic plane, that is, sideways to the other planets. The data that illuminates the similarities between the planets is gathered in Table 5.1 (also see Appendix 5).

Table 5.1

	Mercury	Venus	Earth	Mars	Jupiter	Saturn	Uranus	Neptune	Pluto
Direction of revolution[a]	Same	Same	Same	Same	Same	Same	Same	Same	Same
Direction of rotation on axis[a]	Same	Opposite	Same	Same	Same	Same	Sideways	Same	Sideways
Orbital eccentricity	0.21	0.01	0.02	0.09	0.05	0.06	0.05	0.01	0.25
Angle of plane of revolution to ecliptic	7.0°	3.4°	0[b]	1.8°	1.3°	2.5°	0.8°	1.8°	17.1°

[a] Relative to the direction of the Sun's rotation.
[b] Definition.

Planetary Diversity

Similarities between the planets are heavily outweighed by differences. For instance, the masses of Jupiter and Pluto are respectively more than 300 times greater and almost 500 times less than the mass of the Earth. The period of revolution about the Sun (sidereal period) for Pluto is more than a thousand times greater than Mercury's. The surface temperature on Mercury can vary by 600 K whereas on Saturn it hardly varies at all. Pluto's surface temperature is only 40 K whereas on Venus it is 700 K. Saturn has 18 moons (discovered so far) and a full ring system whereas Venus has neither. Venus is surrounded by a dense atmosphere mainly composed of carbon dioxide, Uranus a dense atmosphere of mainly hydrogen and helium whereas Mercury has almost none. The variations between the planets are almost endless and these are what make study of the Solar System so fascinating. Data that quantifies this diversity is gathered together in Table 5.2 (also see Appendix 5).

The Empty Solar System

Before examining the planets' appearance more carefully it is instructive to consider a scale model of the Solar System based on everyday life. If the Sun were to be shrunk to a

[1] The measure of the deviation of elliptical orbits from circles is known as eccentricity. A mathematical explanation of eccentricity is given in Appendix 3.

Table 5.2

	Mercury	Venus	Earth	Mars	Jupiter	Saturn	Uranus	Neptune	Pluto
Diameter (mean)[a]	0.383	0.950	1.00	0.532	11.0	9.14	3.99	3.86	0.179
Rotation period[a]	58.7	243	1.00	1.03	0.41	0.43	0.72	0.67	6.4
Distance from Sun[a]	0.387	0.723	1.00	1.52	5.20	9.52	19.2	30.1	39.5
Sidereal period[a]	0.241	0.615	1.00	1.88	11.9	29.5	84.0	165	248
Surface temperature (K)	100-700	730	260–310	190–240	110-150	97	58	58	50
Mass[a]	0.0552	0.814	1.00	0.107	318	95.2	14.5	17	0.013
Mean density[a]	0.984	0.942	1.00	0.712	0.241	0.125	0.239	0.297	0.371
Moons	0	0	1	2	16	18	17	8	1
Albedo[b]	0.06	0.72	0.39	0.16	0.70	0.75	0.90	0.82	0.15
Synodic period[b] (days)	116	584	n/a	780	399	378	370	368	367
Main colour	Grey	White	Blue	Red	Yellow	Yellow	Blue	Blue	Red
Ring system	No	No	No	No	Yes	Yes	Yes	Yes	No

[a] Relative to the Earth (diameter = 12 740 km, distance from Sun = 149.6 million km, mass = 5.98×10^{24} kg, density = 5520 kg m^{-3}).
[b] Defined below.

sphere with a diameter of 3 m, about the size of a car, then the nearest planet, Mercury, would be more than 200 m away, at the opposite end of a large sports stadium. It would have a diameter of just 10 mm, about the size of a marble. Venus would follow at a distance of almost 500 m with a size of less than 30 mm, about the size of a squash ball. Earth would have a similar size but would lie at a distance of 650 m, at the opposite end of a village. On this scale even the largest planets remain small. Jupiter would have about the size of a football but would be removed to a distance of over 3 km, at the opposite end of a small town. The smallest and remotest planet, Pluto, is just 4 mm in diameter and, on average, at a distance of more than 25 km, about the size of a large city like Manchester. It is clear, therefore, that if the Sun is represented by a car, the nine planets never get bigger than footballs, and the whole is spread across a large city, then most of our local space is empty. Having said that, it is still tightly packed compared to the region beyond as, in this scale model the next nearest star would be more than 100 000 km away, equivalent to two and a half laps around the Earth's surface.

Table 5.2 shows that the period of revolution of the planets increases with distance from the Sun. This causes the motion of each planet across the celestial sphere to be unique. The apparent motion of the Earth's near neighbours is particularly complicated due to the similarity in these planets' sidereal periods. The planets will now be dealt with in turn to explain their basic appearance.

Mercury

Mercury is a very difficult planet to observe from the Earth. Though it is our third closest neighbour it is also the closest planet to the Sun. This means that Mercury must always be close in the sky to the Sun (Figure 5.1). Of course, the Sun obliterates almost everything from the sky when it is visible and so Mercury can only ever be seen low in the sky just before dawn when it is in a position to rise shortly before the Sun or, at other times, just after sunset when its position in space is such that it sets shortly after the Sun. The time between these two circumstances is about 58 days and is not equal to half of Mercury's

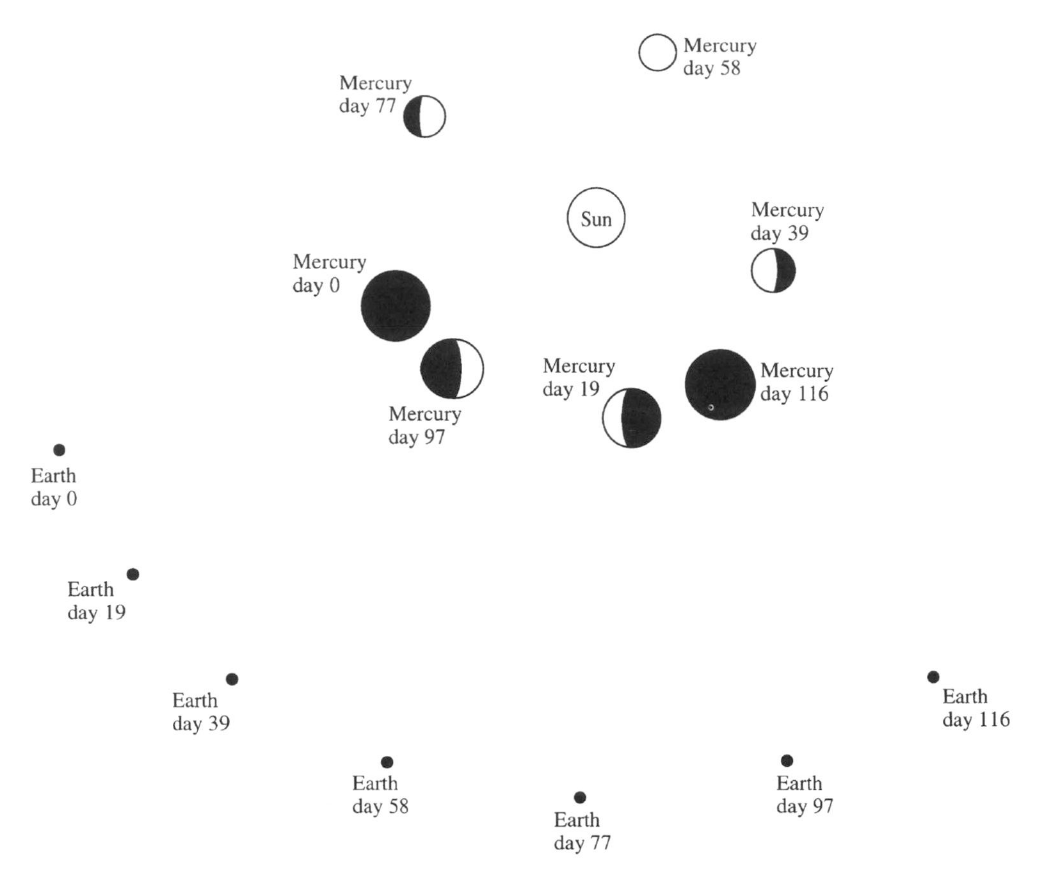

Figure 5.1 One synodic period for the Earth/Mercury/Sun system. Between day 0 and day 116 the Earth makes about one third of a revolution about the Sun while Mercury makes about one and one quarter revolutions. This is the period of time required to bring the system back into alignment. In between, Mercury is viewed through a full sequence of phases (from inferior conjunction to inferior conjunction in the example illustrated). The telescopic image of Mercury as viewed from Earth at the various times indicated is shown in Mercury's position. Note the change in Mercury's apparent (angular) size throughout the synodic period

sidereal period which is 88 days. The period of 58 days is half of Mercury's synodic period. A synodic period is the amount of time between similar arrangements of the Earth, the Sun and a third body. As the Earth and the third body are revolving about the Sun then both motions influence the synodic period. The concept is illustrated for Mercury in Figure 5.1.

WORKED EXAMPLE 5.1

Q. What is the maximum angle in the sky by which Mercury can be separated from the Sun, as observed from Earth? Assume Mercury's orbit to be circular.

A. The moment at which the maximum angle occurs can be best seen by considering a simple diagram of the motion of Mercury while considering the positions of the Earth and Sun to be fixed. This is allowed because only the relative position of Mercury in the sky is of interest. On such a diagram Mercury would move with a period equal to the synodic period of the system.

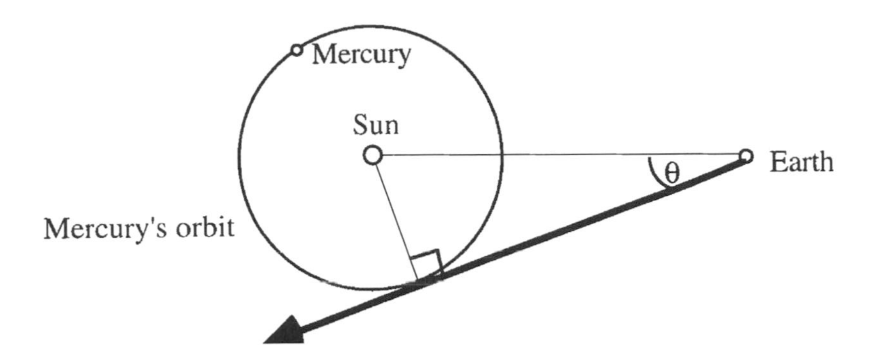

The arrowed line in the diagram indicates the position of maximum angle, θ, when the line of observation from the Earth just touches Mercury's orbit. Such a line is a tangent and must therefore be at right angles to the line joining the circle to its centre. The angle is therefore easy to calculate as $\sin\theta = \text{opposite/hypotenuse} = 0.387/1.00$ (from Table 5.2) $= 0.387$, so that $\theta = 22.8°$. If the orbits of the Earth and Mercury were genuinely circular then Mercury would never be further than 23° from the Sun in the sky. Actually, Mercury has an unusually large eccentricity (somewhat elliptical orbit, see Appendix 3) as a result of which it can occasionally be seen a few degrees further away in the sky. Nevertheless, considering the fact that the Earth's atmosphere scatters light sufficiently to keep the sky bright even when the Sun is several degrees below the horizon, Mercury is rarely observable in a dark sky.

Mercury also has a very low albedo. This term refers to the proportion of light that the planet reflects[2]. At 0.06 (6%), Mercury has the lowest albedo of all the planets. This is due to its surface of dark rock and its lack of atmosphere or clouds. Nevertheless, Mercury can still be observed when conditions are right as the solar light flux (insolation) on Mercury is very large due to its close proximity to the Sun. Observation of Mercury reveals a heavily cratered surface. The craters are generally flat with secondary craters (formed by the impact of material dislodged during primary cratering events) near particularly large impact sites. The largest impact site on Mercury is the Caloris Basin which is 1300 km across (see Figure 5.2). There are also series of very long ridges called scarps and some mountain ranges.

Venus

Venus is much easier to observe than Mercury and is usually the brightest of the planets. As the second closest planet to the Sun it receives strong illumination and, as it is usually our closest neighbour, enough of that light is reflected in our direction to give Venus the impression of being a very bright, white star with the naked eye. The albedo of Venus is more than ten times that of Mercury at 0.76 due to a complete covering of highly reflective clouds composed mainly of sulphuric acid. Venus orbits between the Earth and the Sun and so, like Mercury, must be in the same part of the sky as the Sun. However, Venus is about twice the distance from the Sun as Mercury and is therefore usually separated by a

[2] There are various ways of defining albedo. This text uses the 'bond albedo' which considers all wavelengths of light, whether visible or not.

Figure 5.2 A photomosaic of Mercury taken by the Mariner 10 space probe. Note Mercury's dark, crater-covered surface appears to be quite similar to the Moon's (NASA, Jet Propulsion Laboratory and National Space Science Data Center)

much greater angle, that can be as large as 42°. Venus is still observed close to dusk or dawn but for much greater periods than Mercury, allowing the sky to become much darker.

WORKED EXAMPLE 5.2

Q. Calculate the ratio of reflected light from Venus to that reflected by Mercury.
A. Referring to Worked Example 1.4, the amount of light incident per unit area upon an object is inversely proportional to its distance from the light source. The ratio of separations from the Sun for Mercury and Venus is (from Table 5.2) (0.72/0.39=) 1.85 so that Mercury receives (1.85^2=) 3.42 times more light per unit area. Half of the surface area of each planet is turned towards the Sun at any given moment and this constitutes the reflecting surface. The collecting area of this hemisphere is given by $\pi d^2/4$, where d is the planetary diameter. The reflecting surfaces are therefore proportional to d^2. The ratio of reflecting surface areas for Mercury and Venus is therefore (from Table 5.2) ($0.38^2/0.95^2$=) 0.16. Finally, the proportion of light actually reflected from the surfaces is given by the ratio of albedos which for Mercury and Venus is (from Table 5.2) (0.06/0.76 =) 0.079. The actual amount of light reflected is therefore given by the product of the three ratios at ($1.85 \times 0.16 \times 0.079$ =) 0.023. Venus therefore reflects (1/0.023 =) 43 times more light than Mercury. As Venus is also more often observed in a dark sky, it appears as a much brighter object.

The surface of Venus can never be seen through its clouds though it has been mapped by radar and visited by landing probes. The latter technique may appear to be the best way of obtaining information on planetary surfaces but the extreme conditions on Venus make successful missions very difficult despite Venus' proximity to the Earth. Even if a

Figure 5.3 An image returned from the Venera 14 lander (parts of which are visible in the foreground). Note the flat basaltic rock. This picture is strongly curved so that the horizon is visible at the top-left and top-right (NASA, Jet Propulsion Laboratory and National Space Science Data Center)

landing probe can survive a passage through clouds of sulphuric acid it must still be able to withstand surface temperatures of over 700 K and atmospheric pressures 90 times those on Earth's surface. A few photographs have been obtained and these have shown small, flat basaltic rocks strewn about the surface (Figure 5.3). Orbiting spacecraft have now completed radar surveys of the Venusian surface with sub-kilometre resolution. The principle of radar (RAdio Detection And Ranging) is based on bouncing radio waves from a surface and measuring how long they take to return. Such radio waves pass through the Venusian atmosphere and allow the surface to be 'seen' (compare Figures 5.4

Figure 5.4 Venus laid bare. A global view of the Venusian surface made by mapping radar images produced by the Magellan space probe onto a computer-simulated globe. The bright region, just below the centre of the image, are the Maxwell Mountains that include the highest point on Venus (NASA, Jet Propulsion Laboratory and National Space Science Data Center)

Figure 5.5 The swirling clouds that completely cover Venus. This picture (taken by the Pioneer Venus Orbiter) focused the ultraviolet light reflected by Venus's clouds to emphasise their structure (NASA, Jet Propulsion Laboratory and National Space Science Data Center)

and 5.5). Complex techniques involving wave interference have been developed to interpret signals returning from different parts of the surface and the topography of Venus is now well known. It is a planet with significant cratering but only small variation in its surface height distribution, having only a few areas of highlands. Information on the planet's spin was also obtained using radar techniques. In this case the reflected signal must be wavelength resolved. The Doppler effect (see Chapter 1) causes the reflected radio waves to be shifted in frequency depending on whether the reflecting part of the planet is rotating away from or toward the space probe. The size of the shift is proportional to the rate of rotation of the planet according to equation (1.5). From such experiments it was determined that Venus rotates in a retrograde direction very slowly, taking 243 days per complete spin.

The sidereal period of Venus is about two thirds of the Earth's and so during one Venusian revolution of the Sun, the Earth has moved through two thirds of a revolution itself. By the time Venus catches up, the Earth has made almost two revolutions. It is interesting to observe Venus throughout its synodic period as Venus exhibits phases similar to those of the Moon (Figure 5.6). At any given moment half of Venus is illuminated by the Sun. The proportion of the illuminated half that faces the Earth determines how the telescopic image of Venus appears. When it is on the far side of the Sun (at superior conjunction) the whole illuminated face is inclined towards the Earth and so Venus appears as a full circle. The circle is relatively small as the planets have maximum separation at this time and Venus may be difficult to see due to its proximity in

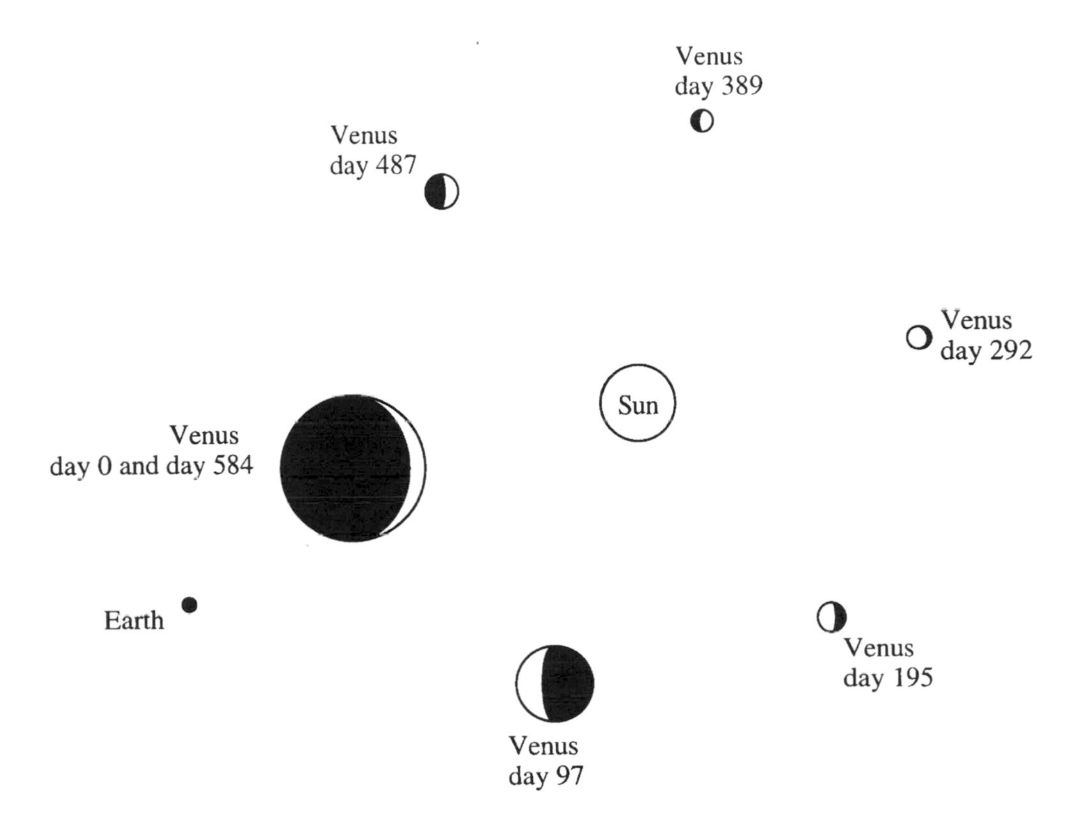

Figure 5.6 One synodic period for the Earth/Venus/Sun system. Between day 0 and day 584 the Earth makes almost two revolutions about the Sun but is shown as being static here, that is, the diagram is drawn relative to the Earth–Sun alignment. In 584 days Venus varies through a full sequence of phases as its position relative to the Earth and the Sun changes. The telescopic image of Venus, as viewed from Earth at the various times indicated, is shown in Venus' position. Note the change in Venus' apparent (angular) size throughout the synodic period

the sky to the Sun. As the synodic period progresses, less and less of the illuminated face of Venus faces the Earth, fading to a half and then a crescent. The period close to Venus appearing as a semi-circle represents its furthest angular separation from the Sun and the crescent represents its largest angular size. In the latter case it is once again hard to observe due to its proximity in the sky to the Sun.

The Earth and the Moon

The Earth is often regarded as the sister planet of Venus, being of a similar size and, on the scale of the Solar System, a similar distance from the Sun. There are few other similarities (see Figure 5.7). The Earth has a much lower surface temperature, much lower atmospheric pressure and much sparser covering of cloud. Together with a richness in water and oxygen, the Earth becomes a hugely more suitable habitat for life than Venus, or indeed any other planet in our star's system. While the Earth has a similar composition of

Figure 5.7 The Earth seen from space (by the Galileo space probe). The features that set the Earth apart from other planets can be seen in this picture; continental land masses (see Africa and Arabia), large water oceans (see the Atlantic and Indian Oceans) and a moderate atmosphere (evidenced by light water cloud cover) (NASA, Jet Propulsion Laboratory and National Space Science Data Center)

metals and rock to its near neighbours in space, it is its propensity for life that really sets it apart.

The Earth is the closest planet to the Sun to be orbited by a natural satellite, the Moon. The Moon revolves about the Earth once every 27.3 days and, like Venus (and Mercury), completes a full sequence of phases from having its illuminated half facing square towards the Earth (full Moon) to having its dark side facing towards the Earth (new Moon). The dark side of the Moon is often referred to but not always in the correct context. The Moon's rate of rotation on its axis is equal to its rate of revolution about the

Earth. That means that the same half of the Moon faces the Earth at all times. The other half, that can never be seen from the Earth, is referred to as being the dark side though it receives light from the Sun for precisely the same proportion of the time as the side that faces us. Figure 5.8 demonstrates how this takes place as well as demonstrating the similarities and differences to Venus's pattern of phases. There are two main differences. First, the Moon has a much shorter synodic period, at 29.5 days. This is because its sidereal period is so much shorter than the Earth's that the Earth has moved only a short distance around the Sun when the Moon has completed its 27.3-day orbit. The second difference is caused by the fact that the Moon remains almost equidistant from the Earth and so retains the same angular size in the sky.

In fact the Moon's path about the Sun is not perfectly circular and this means that it is not strictly true to say that precisely the same face always points towards the Earth. The small eccentricity of the Moon's orbit allows observers on Earth to see a little 'around the edges' of the Moon. This is known as libration in longitude and causes our perspective of the Moon to vary by about 7°. Two similar phenomena also allow different parts of the Moon to be viewed from a single position on the Earth. Libration in latitude is due to the

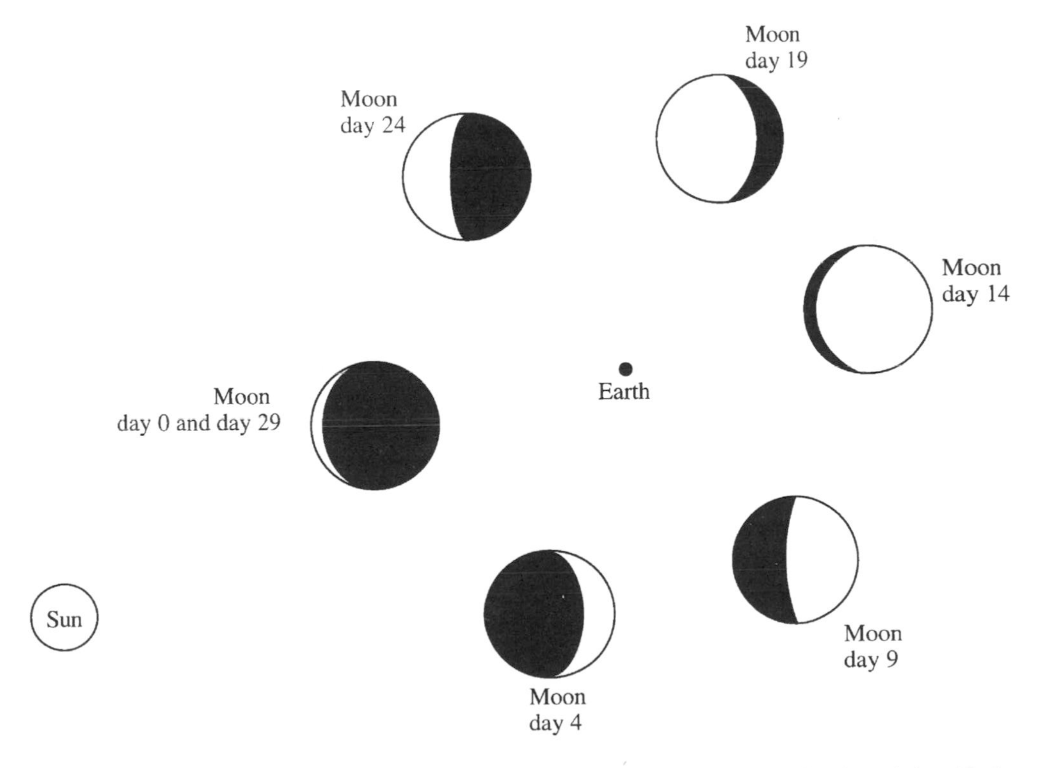

Figure 5.8 One synodic period for the Earth/Moon/Sun system. Between day 0 and day 29 the Earth moves a small distance relative to the Sun but is shown as being static here, that is the diagram is drawn relative to the Earth–Sun alignment. In 29 days the Moon varies through a full sequence of phases as its position relative to the Earth and the Sun changes. The telescopic image of the Moon as viewed from Earth at the various times indicated is shown in the Moon's position. The Sun's position in this diagram indicates only its direction relative to the Earth and the Moon

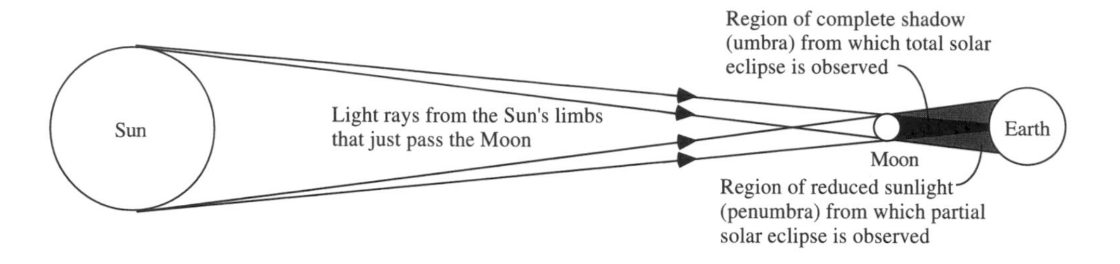

Figure 5.9 The alignment of the Sun, Moon and Earth during a solar eclipse

Moon's orbit being in a plane at 5° to the ecliptic and its spin axis being at 1.5° to this. The Moon can therefore appear to tilt by up to 6.5° in a north–south direction. Diurnal libration is a more subtle effect that also causes a variation in the east–west perspective but on a more rapid timescale than libration in longitude. Diurnal libration is due to the Earth's rotation on its axis which causes the Moon to be viewed from positions 13 000 km apart (if on the equator) every 12 h.

By coincidence, the Moon and the Sun have almost the same angular sizes as viewed from the Earth. Occasionally the Moon moves exactly between the Earth and the Sun and the Sun is blocked for a brief time, giving a few minutes of darkness during the day. This is known as a solar eclipse. As the angular sizes of the two bodies are so similar the cone of complete darkness, known as the umbra, has a small size at the Earth and this region of darkness sweeps out only a small strip across the Earth as the Moon passes in front of the Sun. A much broader strip experiences partial darkness where only a part of the Sun is blocked by the Moon. This region is known as the penumbra. The whole scenario is illustrated in Figure 5.9.

WORKED EXAMPLE 5.3

Q. If a solar eclipse takes place when the Sun (diameter, 1.392×10^6 km) and Moon (diameter, 3480 km) are separated from the Earth (diameter, 12 800 km) by 1.530×10^8 km and 3.79×10^5 km respectively, what is the width of the strip on the Earth's surface from which a total eclipse is observed?

A. The solution to this problem is best seen by considering the figure which illustrates the size of the shadow cast on the surface of the Earth by the Moon. For the purposes of calculation it is easier to work with the half-shadow which is cast by half of the Moon obscuring half of the Sun and produces a shadow of length $A''C''$ on the surface of the Earth (as the distance $A''C''$ is small the curvature of the surface can be ignored). The triangles ABC, $A'BC'$ and $A''BC''$ (points marked by spots) are all right-angled and enlargements of each other so that they all subtend the same angle, θ, at **B**. The position of **B** is initially unknown but can be calculated from the information available from triangles ABC and $A'BC'$:

$$AC = BC \tan\theta \qquad \text{and} \qquad A'C' = BC' \tan\theta$$

But AC is the Sun's radius at $(1.392 \times 10^6/2) = 6.96 \times 10^5$ km and $A'C'$ is the Moon's radius at $(3480/2) = 1740$ km. The distance CC'' is the Earth–Sun separation at 1.530×10^8 km and the Earth–Moon distance is 3.79×10^5 km, so that CC' is $(1.530 \times 10^8 - (3.79 \times 10^5 =)\ 1.526 \times 10^8$ km. The distance CC' is equal to $BC - BC'$. Three equations are now available:

$$6.96 \times 10^5 = BC \tan\theta, \qquad 1740 = BC' \tan\theta, \qquad BC - BC' = 1.526 \times 10^8$$

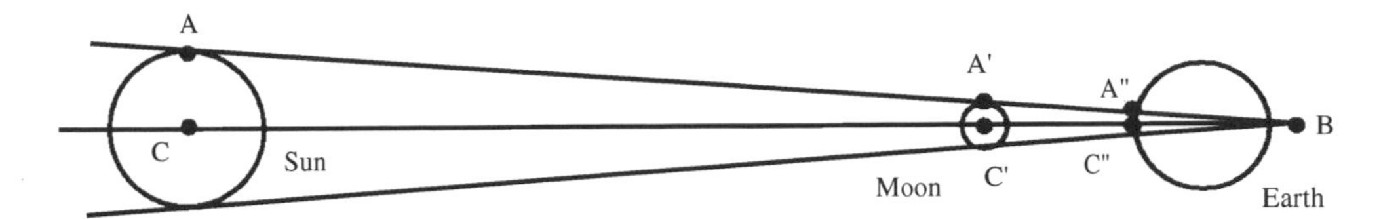

As there are only three unknowns, the equations can be solved to give $BC' = 3.83 \times 10^5$ km and $\tan\theta = 4.55 \times 10^{-3}$ (BC is not required). This means that B is within the Earth's volume, slightly further away from the Moon than the Earth's centre. Its separation from the Earth's centre is ($3.83 \times 10^5 - 3.79 \times 10^5 =$) 4000 km. The distance BC'' is therefore 4000 km plus the Earth's radius (6400 km) which is 10 400 km. The size of the half-shadow on the Earth's surface is given by:

$$A''C'' = BC'' \tan\theta = 10\,400 \times (4.55 \times 10^{-3}) = 47 \text{ km}$$

The path of the total eclipse is thus only about 100 km across. It should be noted that this calculation is subject to quite large rounding errors as the scales involved vary from thousands of kilometres to hundreds of millions of kilometres. When such numbers are subtracted from each other, second and third most significant figures of the smaller numbers can be lost even though they may have some influence on the final shadow size. Small changes in the separation of the three bodies, caused by the elliptical nature of the orbits, result in large changes in the final shadow size due to the similarity in angular size of the Moon and Sun as observed from the Earth. Under many circumstances, when the Moon is further than average from the Earth in its orbit and the Sun closer than average, the Moon's angular size is actually smaller than the Sun's. A Solar eclipse that takes place at such a time would not cause complete darkness as the Sun's outer limb would never disappear from the sky. The Sun would appear as a thin, bright ring in the sky.

Solar eclipses are very important opportunities for observation. The Sun's intensity is blocked for a moment and so its outer atmosphere can be studied. More exotically, the apparent position of stars on the celestial sphere when they are close to the Sun in the sky vary slightly from their normal position. Of course, this is measurable only during a solar eclipse when the Sun's dominating rays are blocked. The magnitude of the shifts, caused by the Sun's enormous mass bending local spacetime, have been used to experimentally verify Einstein's theory of general relativity.

A Lunar eclipse occurs when the Earth blocks the Sun's light from reaching the Moon and the Moon disappears from the night sky. As the Earth is much larger than the Moon, its angular size as viewed from the Moon is much larger than the Moon's angular size as viewed from the Earth (see Figure 5.10). Consequently, the Earth's umbra can envelop the Moon for periods much longer than the totality of a solar eclipse and total lunar

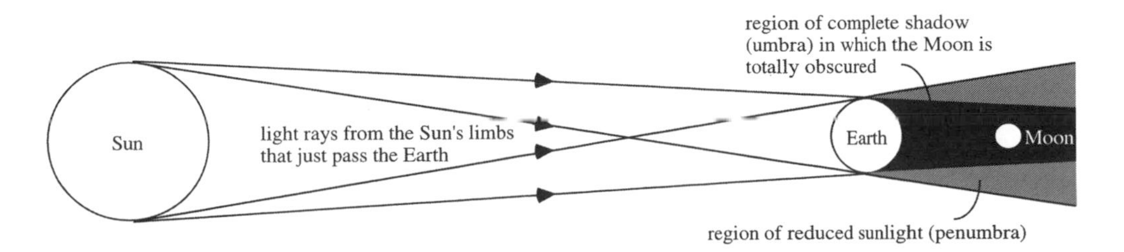

Figure 5.10 The alignment of the Sun, Earth and Moon during a lunar eclipse

eclipses often last for longer than an hour. There is also a period of partial eclipse as the Moon moves from the Earth's penumbra to its umbra and back again.

The Moon is the only extraterrestrial body on which surface features can be identified with the naked eye. The most obvious of these are the large, smooth blotches that are known as seas due to their apparent likeness to the way one of the Earth's oceans might appear as viewed from space. Actually the seas are simply solidified lava flows, formed around the same time as the Moon's extensive cratering events. To see the Moon's craters it is better to use a small telescope or binoculars so that a range of sizes can be seen, the larger of which are often accompanied by radiating surface lines known as rays. These rays correspond to the material that was ejected during the crater-forming impact of an extralunar object (see Figure 5.11). Secondary craters, formed in a similar way, can also be seen close to very large craters. Craters are more numerous in areas away from the seas. Space probes that have been able to examine the far side of the Moon have revealed it to be devoid of seas but heavily cratered.

The Moon and Mercury resemble each other in many ways: they are of a similar size, heavily cratered, have ranges of mountains, very low albedos and almost no atmosphere. Even though Mercury is a small planet it is significant that the Earth has a natural satellite of planetary size. With the exception of Pluto, which is effectively a binary planet, the Moon is the largest satellite in the Solar System in comparison to its parent planet. The Moon's mass is 1.2% of the Earth's whereas for Saturn's moon Titan, which comes next on the list at 0.024%, the proportion is 50 times smaller.

Even though astronauts did not land on the Moon until 1969 it has been known for many years that its atmosphere is very sparse. The lack of surface weathering provided

Figure 5.11 An astronaut collecting rock samples from a boulder on the Moon. Many of the boulders that are strewn across the Moon's surface are impact crater ejecta. Crewed trips to the Moon were limited to the period 1969–72 (NASA, Jet Propulsion Laboratory and National Space Science Data Center)

good evidence but a technique known as occultation gave confirmation. As the Moon moves across the celestial sphere it blocks the light from a succession of different stars. Just before occulting a star the light that arrives in a telescope on Earth must travel very close to the Moon's surface and therefore through any atmosphere that it might have. When this happens it is observed that there is no additional absorption or scintillation so that the Moon's atmosphere must be much less dense than the Earth's.

Mars and its Moons

The last of the so-called terrestrial planets is Mars. Like Venus, Mars appears in the sky like a bright star as observed from Earth. It has only a weak atmosphere (first detected using occultation) and cloud system and so its coloration is due to its surface being covered in a sand that has a large iron oxide content. Iron oxide is better known as rust and this makes Mars red. When viewed through a telescope, enormous dust storms can occasionally be observed raging on the surface of the planet. The effect of these storms is to weather many surface features thus making the planet appear quite smooth, though it still possesses significant quantities of craters and a number of large, extinct volcanoes. Early observations of the planet indicated there to be large canal-like features running across the surface, suggesting there to be intelligent life present. These observations turned out to be optical illusions and, while there is evidence that fluids may once have flowed on the planet's surface, there remains no evidence of higher lifeforms on the planet. The most striking features on the surface of Mars are its polar caps (Figure 5.12), apparently similar to the Earth's, except that they vary greatly in size with the Martian season.

Figure 5.12 A view of Mars taken by the Hubble Space Telescope showing the polar regions of dry ice (NASA, Jet Propulsion Laboratory and National Space Science Data Center)

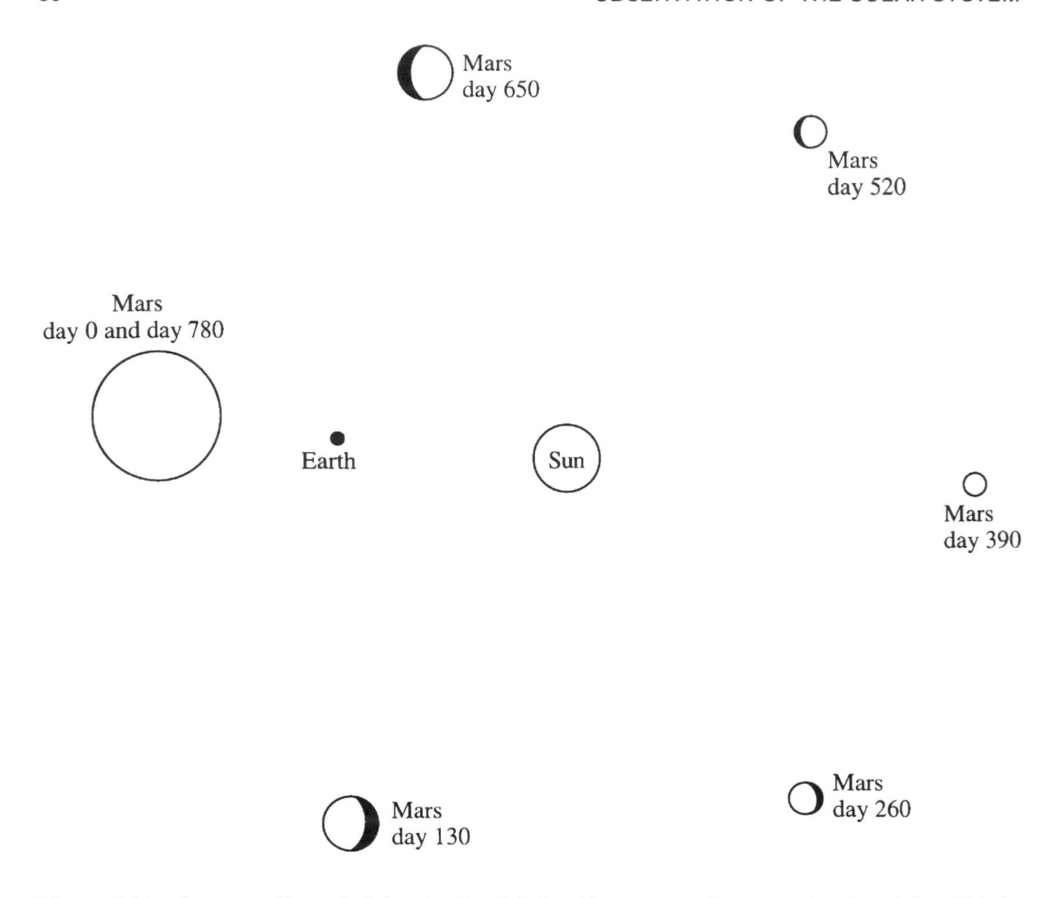

Figure 5.13 One synodic period for the Earth/Mars/Sun system. Between day 0 and day 780 the Earth makes more than two revolutions about the Sun but is shown as being static here, that is, the diagram is drawn relative to the Earth–Sun alignment. In 780 days the phase of Mars varies through a full cycle as its position relative to the Earth and the Sun changes. As it orbits outside the Earth more than half of its illuminated hemisphere points towards Earth. The telescopic image of Mars as viewed from Earth at the various times indicated is shown in Mars' position. Note the change in Mars' apparent (angular) size throughout the synodic period

The Martian year lasts 687 days and Mars has the longest synodic period of all planets at 780 days (see Figure 5.13). As Mars is beyond the Earth's orbit then part of Mars' illuminated hemisphere must always point towards us. Mars is brightest when the Sun, the Earth and Mars are aligned in this order but Mars' illuminated face also faces us when Mars is on the opposite side of the Sun to the Earth. Mars exhibits smaller phases at intermediate positions but always appears as being more than half-illuminated.

WORKED EXAMPLE 5.4

Q. What is the minimum proportion of Mars' illuminated hemisphere that is oriented towards the Earth?

A. Consider Mars to be fixed in space as the Earth revolves about the Sun. The smallest proportion of the Sun-illuminated hemisphere of Mars observable from Earth occurs when the Sun, Mars and

Earth are positioned in the shape of a right-angled triangle as can be seen from the diagram. If the Earth moves in either direction from its indicated position, the angle, θ, will decrease. This angle is given by:

$$\sin\theta = \frac{\text{Earth} - \text{Sun}}{\text{Mars} - \text{Sun}} = \frac{1.00}{1.52} = 0.66$$

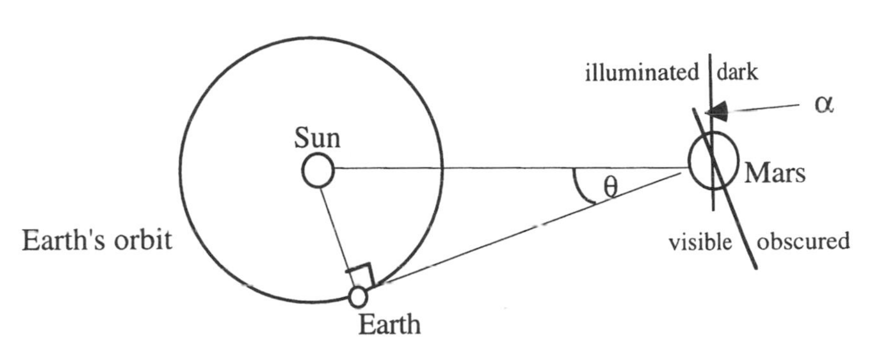

The Earth–Sun and Mars–Sun distances are found from Table 5.2 and are entered in astronomical units so that θ (= $\sin^{-1}$0.66) is 41°. The illuminated and dark hemispheres of Mars are indicated in the diagram and the dividing line between the two is perpendicular to the line between Mars and the Sun. Similarly, the line that delineates the hemisphere of Mars that can or cannot be seen from Earth is perpendicular to the line between Earth and Mars. The angle α marked on the diagram represents the angular proportion of the Martian surface that cannot be observed from Earth. As the lines that form it are both perpendicular to lines that meet at the angle θ, then $\alpha = \theta$. Thus, 180° − θ of the 180° of the illuminated surface of Mars can be observed from Earth. Ignoring small effects due to the curvature of the Martian surface, this means that the minimum percentage of the planet's illuminated surface that is turned towards the Earth is (100 × (180° −41°)/ 180° =) 77%.

Mars has two tiny natural satellites that orbit very close to the planet. Both Phobos and Deimos are irregularly shaped ellipsoids, each having three different-sized axes. Phobos orbits only 6000 km above Mars' surface and has dimensions between 19 and 27 km while Deimos is a little over half of that size and orbits 14 000 km further from Mars. Due to their small size and very low albedo of just 0.02 they are very difficult to observe from the Earth. Seen from Mars both moons present their same faces at all times in a similar way to our Moon. Observation of Phobos from Mars would be particularly curious. As discussed for the planets, closer orbiting bodies have shorter periods of revolution than those that lie farther out. Phobos is so close to Mars that its rate of revolution is faster than the speed at which Mars rotates on its axis whereas Deimos' angular speed is less than that of Mars. Phobos and Deimos both revolve around Mars in the same direction as Mars spins on its axis but would appear to revolve in different directions as observed from the planetary surface as Phobos 'overtakes' the observer on the planet below (see Figure 5.14).

Internal Structure of the Terrestrial Planets

The planets discussed so far, Mars, Earth, Venus and Mercury, are together known as the terrestrial planets as their bodies have similar anatomies. There is some variation in size

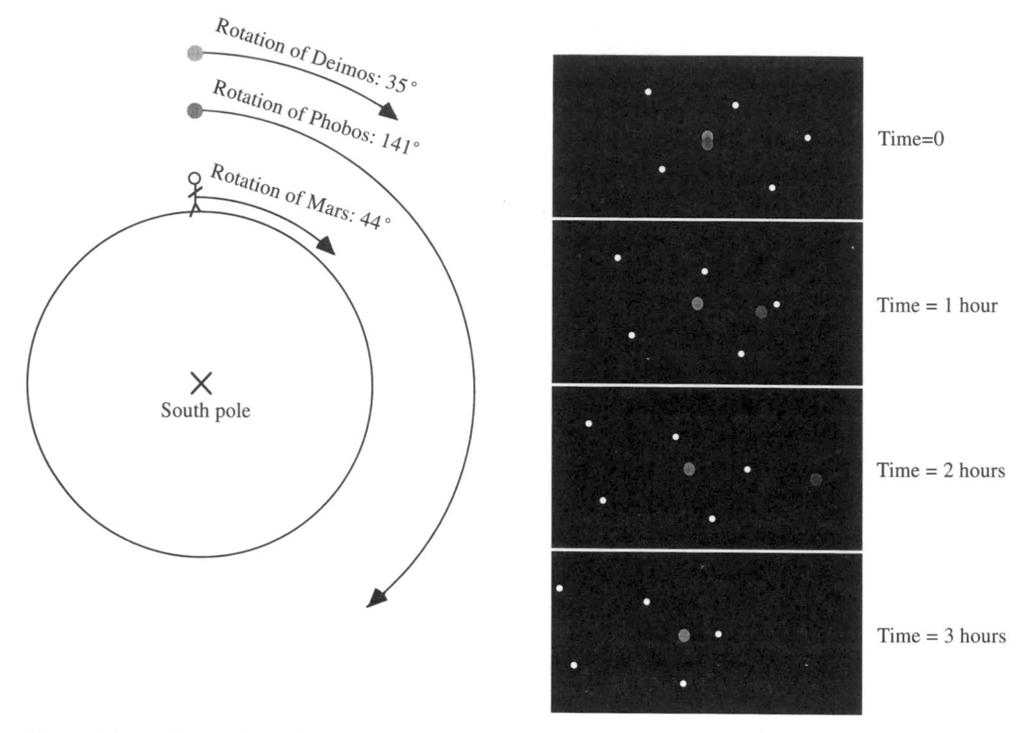

Figure 5.14 The motion of the Martian moons across the sky as observed from the surface of Mars during a 3-hour period. Both moons revolve in the same direction as Mars' rotation (left) but Phobos appears to move in the opposite direction to the background of stars (right) as it has a greater angular speed than Mars itself

and chemical composition but all have similar basic constitutions. At the centre of each planet is a metallic core that may be in either a solid or a liquid state (or zones of each in the case of the Earth) depending on the prevailing temperature and pressure. The relative size of this core is thought to decrease going away from the Sun though measurements for Venus and Mars have still not been interpreted entirely unambiguously. Around the core is a rocky layer known as the mantle. This is largely composed of silicates, in particular olivine, an iron and magnesium silicate. Such are the conditions in the mantle, hot and at high pressure, that it is somewhat plastic and able to flow slowly. The outermost thin layer is the solid, cool crust. Of course, in the case of the Earth two thirds of the crust is covered in liquid water but the total volume of the oceans is very small in comparison to the bulk of the Earth.

The Asteroids

The next planet from the Sun is Jupiter but in between lie a large number, at least 5000, of lumps of rock also orbiting the Sun. These bodies are known as asteroids and together they form a region known as the asteroid belt. Many asteroids resemble Phobos and Deimos in terms of their irregular shapes, low albedos and sizes (see Figure 5.15) but there

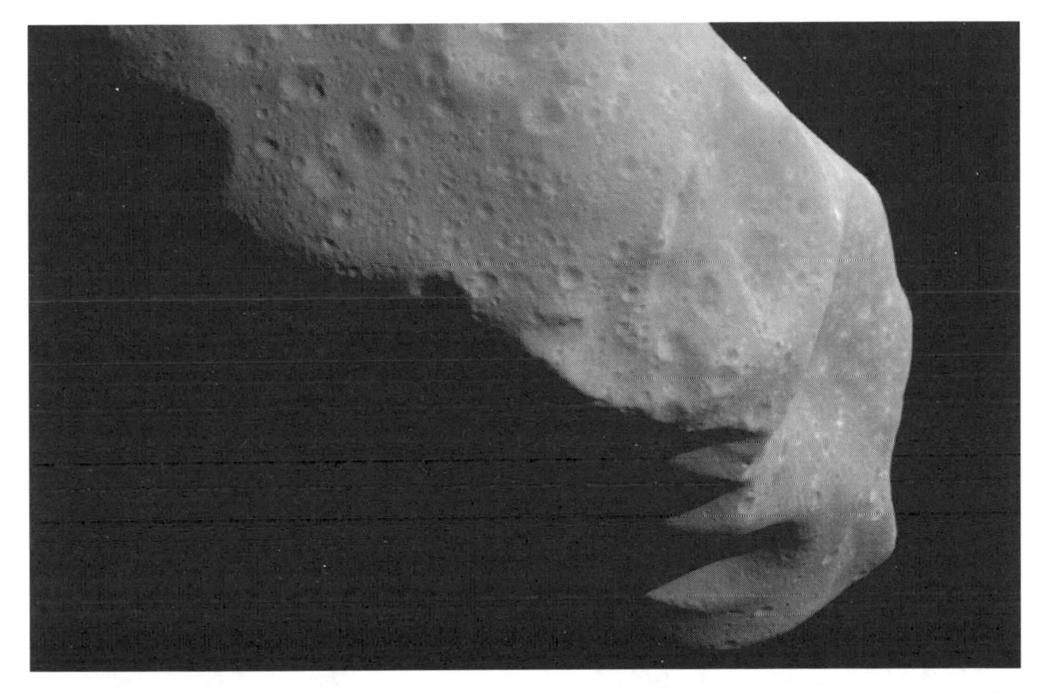

Figure 5.15 A typical asteroid. Ida is an irregular, cratered body with a length of 52 km (NASA, Jet Propulsion Laboratory and National Space Science Data Center)

is a great deal of variation. For instance (much) the largest, Ceres, has a diameter of about 900 km. The size of the smallest is not known as such small dark objects are very difficult to observe. Most orbit the Sun at distances of two to four times further from the Sun than the Earth. There are a few asteroids that have somewhat higher albedos, up to 0.1, which makes viewing a little easier but the only occasion on which good observations can be made are from space probes or when one passes by close to the Earth. This happens from time to time as some of the asteroids have quite eccentric orbits reaching inside Mercury's orbit or beyond Jupiter's.

Jupiter and its Moons

It is appropriate that the asteroids provide a break between Mars and Jupiter because it is at this distance that planets change character entirely. The diameter of Jupiter is more than ten times that of the Earth, the largest body discussed previously. Its size and its 50% albedo mean that it is often observable with the naked eye from Earth despite the fact that it orbits the Sun at a distance of 5.2 astronomical units[3]. Its large distance from the Sun implies two important facts for the observation of Jupiter. First, it must always have the vast majority of its illuminated face inclined towards the Earth so that its phases are only slight variations from being full. Second, Jupiter's much greater distance from the Sun

[3] An astronomical unit (AU) is equal to the mean Earth–Sun separation at 149.6 million km.

Figure 5.16 Jupiter and Io (bottom-left) as photographed by Voyager 1. The Great Red Spot is visible in Jupiter's southern hemisphere on the right-hand edge of the planet's surface. (NASA, Jet Propulsion Laboratory and National Space Science Data Center)

means that it has a large period of revolution compared to that of the Earth so that after the Earth has completed an orbit of the Sun, Jupiter has not moved very far. The synodic period is therefore close to the Earth's revolution period at 399 days. For more outlying planets the phases become less and less variant from full and the synodic period gradually approaches 365 days.

The telescopic appearance of Jupiter is also radically different from that of the terrestrial planets. Jupiter is seen as a slightly equatorially oblate sphere with coloured bands running parallel to the equator (Figure 5.16). Between the bands are occasional large spots, the most famous of which is the Great Red Spot, half-way between the equator and south pole. It is Jupiter's atmosphere that is observed. Light and dark bands are known, respectively, as zones and belts and are simply regions of the upper atmosphere that have different characteristics such as temperature and chemical composition. The spots are also essentially atmospheric disturbances though they may be very long-lasting. The Great Red Spot has been in place since observations began several hundreds of years ago. Such features can be seen to rotate at great speeds and allow Jupiter's rotation period to be determined as being about 9 hours, 50 minutes. This figure cannot be given precisely as higher latitudes have slower rotation periods, increasing by about 5 minutes from the equator to the poles. Such differential rotation and the planet's oblateness are strong indicators that Jupiter is not a solid body. In fact Jupiter is essentially a huge ball of hydrogen and helium that becomes denser and more strongly compressed going from the outer atmosphere towards the centre. Eventually the gases liquify to define a nominal planetary surface, though this can never be seen due to the high opacity of the 1000 km thick shell of gas that surrounds it.

Jupiter is the centre of its own orbital system, having at least 16 moons and a series of rings surrounding it. The moons are divided into two very distinct sets according to size. The four so-called Galilean moons are of a similar size to the Earth's Moon while the rest resemble the tiny moons of Mars. The latter range in radial size from less than 10 km to a little more than 100 km, have irregular shapes and are strewn at distances between 120 000 and 24 million km from Jupiter, resulting in orbital periods between less than a day to more than two years. The two innermost moons, Metis and Andrastea, are less than one Jovian radius from Jupiter's surface, are surrounded by Jupiter's ring system and, like Phobos, revolve more rapidly than their parent planet rotates on its axis, particularly remarkable in Jupiter's case due to its rapid rotation rate. Metis and Andrastea orbit in almost the same path as each other and average a speed of more than 110 000 km h^{-1}, about the same speed at which the Earth travels around the Sun. The four outermost moons, Ananke, Carme, Pasiphae and Sinope, actually revolve in the opposite direction to the other moons and therefore in the opposite sense to all the bodies of the Solar System that have been discussed so far. This is known as retrograde motion.

Due to their large size, the Galilean moons are much the easiest to observe and are so named because of their discovery by Galileo using only a primitive telescope. They lie in similar orbits between 400 000 and 2 million km from Jupiter's centre, have similar radii between 1800 and 2700 km and have synchronous rotational and revolutionary periods but are quite different from each other in many ways.

Io is arguably the Solar System's most spectacular body. It has a mottled appearance in orange, black and white and closer inspection shows that these colours are due to chemical deposits laid down by constantly erupting volcanoes (Figure 5.17). These volcanoes are violent. They are able to eject matter at 1 km s^{-1}, spewing dust hundreds of kilometres into the air and producing lakes of lava covering hundreds of square kilometres.

The next most distant of Jupiter's moons is Europa and the contrast is immediate. Europa is the Solar System's smoothest body. It has a pale mottled appearance from afar but using higher resolution its surface can be seen to be covered by dark cracks that extend around the surface for thousands of kilometres. Very few craters can be seen on Europa's surface.

Ganymede is the Solar System's largest moon and the next furthest from Jupiter. It is grey and blue in colour and much more heavily cratered than Io and Europa. Like Europa it has regions that are dominated by grooves and ridges but these areas are contained in certain areas. Surface features are generally flat but not as smooth as Europa.

Callisto is the outermost of the Galilean moons and the closest in appearance to our Moon. It is dark and covered in crater impacts, many of which have rays emanating from them. The largest, Valhalla, is 600 km in diameter and is surrounded by concentric bands with diameters up to 3 000 km.

Jupiter's rings cannot be seen from the Earth. They are virtually transparent due to their extreme thinness (~30 km) and the fact that they are composed of particles only

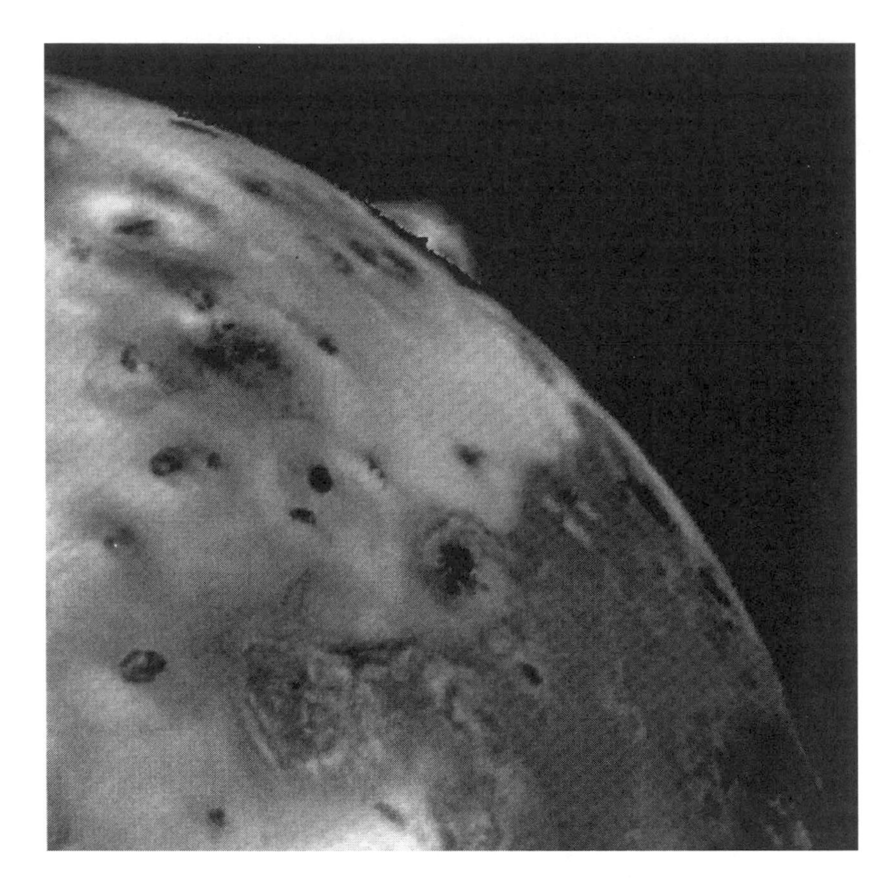

Figure 5.17 A volcano erupting on Io. The plume reaches an altitude of about 50 km and spreads material over distances of up to 300 km. The foreground shows a multitude of calderas (volcanic craters) and an otherwise smooth surface dominated by lava flows, coloured by various compounds of sulphur (NASA, Jet Propulsion Laboratory and National Space Science Data Center)

about 10 μm in size. The extent of the rings is large, however, starting from close to Jupiter's upper atmosphere and extending outwards for some 200 000 km. There is a variation in density with distance from the planetary surface and this gives the impression that there is a series of rings. In fact Jupiter's rings are just a very thin disk of dust.

Saturn and its Moons

Saturn's rings are much more visible and are the feature that most readily distinguishes Saturn from the other planets. Otherwise, Saturn could be regarded as being a little sister of Jupiter. Saturn is about 15% smaller in diameter than Jupiter, has less pronounced banding (see Figure 5.18) and shorter-lasting surface features but such differences are essentially of scale rather than essence. Though it is much further from the Earth than Jupiter, Saturn is still visible by eye and its surface features and rings can be distinguished using a moderately sized telescope.

Figure 5.18 Image of Saturn taken by Voyager 1 showing the planet's smooth banding (NASA, Jet Propulsion Laboratory and National Space Science Data Center)

From the Earth the three rings appear to be solid, known, from out to in, as A, B and C, with a gap between rings A and B known as Cassini's division. As Saturn's spin axis is inclined at 27° to the ecliptic and its rings revolve about Saturn's equator then the rings are presented to the Earth at a constantly changing angle between ±27° with a period equal to Saturn's sidereal period of nearly 30 years. When the rings are viewed sideways-on it becomes clear how thin they are. The A, B and C rings extend from 11 000 km to 80 000 km from Saturn's upper atmosphere but have a thickness of only a few kilometres. Closer inspection by space probe has revealed the rings to be more extensive than is visible from Earth and that their structure is much more complex than originally thought, being composed of many hundreds of ringlets. These ringlets are now known to be strewn from 5000 km to 340 000 km from Saturn's surface and are composed of thousands of individual pieces of rock and ice sized between centimetres and tens of metres. The components of each ringlet orbit in the same path. Space probes also revealed that Cassini's division is not entirely devoid of matter but that it also contains a few ringlets.

Saturn also resembles Jupiter in its number of moons but, at the present count, is slightly ahead with 18. Titan is much the largest of these, being only slightly smaller than Jupiter's Ganymede. Titan is unique among moons in that it has a thick atmosphere, dense enough to prevent direct viewing of the surface. Probes will visit the surface in the near future and it is expected that they will reveal oceans of liquid hydrocarbons.

Saturn's other moons, like almost all in the Solar System, have no known atmosphere. It has four medium-sized moons with diameters about one third the size of our Moon. Rhea, Dione and Tethys are all bright objects due to their icy surfaces while Iapetus has a mottled surface with an albedo that varies from 0.05 to 0.5. The rest of the moons are smaller and become more and more irregular with decreasing size. With the exception of Enceladus (and Titan which is as yet unknown) the surfaces of all of Saturn's moons are very heavily cratered. Enceladus has some cratering but its most prominent features are grooves and ridges that run along its white surface.

The Outer Planets and Moons

The planets discussed so far have been known to be special, that is, different from the stars, for millennia. They are bright objects in the sky and, though telescopes and space probes have been required to make useful observations, they are easily visible with the naked eye. The final three planets of the Solar System are considerably dimmer as they are both further away and smaller than Jupiter and Saturn.

The first of the dim planets is Uranus, which is just at the limit of detection by the naked eye in perfect viewing conditions, not the sort of object one would notice unless straining to look for and not discovered until the era of the telescope. Even viewed through a telescope there is little to see on Uranus (Figure 5.19), again due to a thick atmosphere that gives the planet a greenish-blue colour. Space probes have revealed some weak structure and clouds in the upper atmosphere. Like Jupiter and Saturn, the atmosphere is dominated by hydrogen and helium but beneath the atmosphere it is thought that the planet is dominated by icy materials, such as water, some of which could be in a fluid state.

The inclination of the spin axis of Uranus to its plane of revolution is close to 90° so that its poles lie in its plane of revolution. This means that during one 84-year revolution there is one very long solar day and one very long solar night ($\sim$20 years each) with a series of much shorter days in between (see Figure 5.20). As the Sun is very far from Uranus and the planet has a thick atmosphere the temperature gradient between hemispheres is not too severe.

The moons of Uranus orbit in the planet's equatorial plane, that is, perpendicular to the orbits of almost everything else in the Solar System. There are 17 such moons, ten of which have radii of less than 100 km. The outer five moons have radii between 200 and 800 km, about the size of Saturn's medium-sized natural satellites. The moons have a

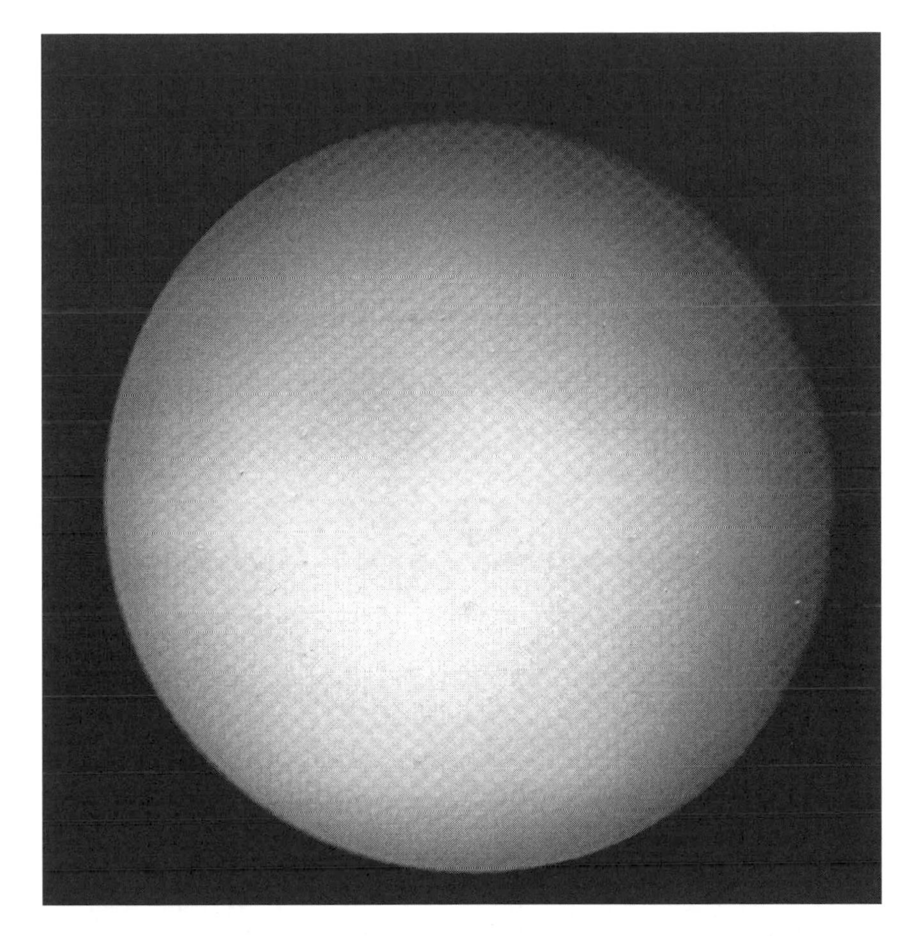

Figure 5.19 Uranus. The only feature visible is a slight difference in shading of the face that points toward the Sun (NASA, Jet Propulsion Laboratory and National Space Science Data Center)

range of surface features, apparently covered by discoloured ice. Most show cratering though some, in particular Miranda, show large grooves, ridges and cliffs.

Uranus also has a ring system, accidentally discovered through star occultation observations while attempting to measure something else. The rings of Uranus are very difficult to see due to their very low albedo of around 0.05 and to the fact that the extent of each of the 11 rings varies between only a few kilometres and about 100 km while their thickness is also very small. Uranus's rings are about one planetary radius above the planetary surface.

Though Uranus and Neptune, the next planet out from the Sun, are very faint as observed from Earth it should be noted that they are still sizeable bodies, both having diameters about four times greater than the Earth's. Neptune resembles Uranus in other ways. It is thought to have the same type of interior and has a similar atmosphere, resulting in Neptune also having a blue colour. Unlike Uranus, Neptune has some weak banding with occasional storm features such as the Great Dark Spot (observed by the Voyager space probe in 1989 but found to have disappeared by the Hubble Space Telescope in 1997) and a few wispy white clouds (see Figure 5.21).

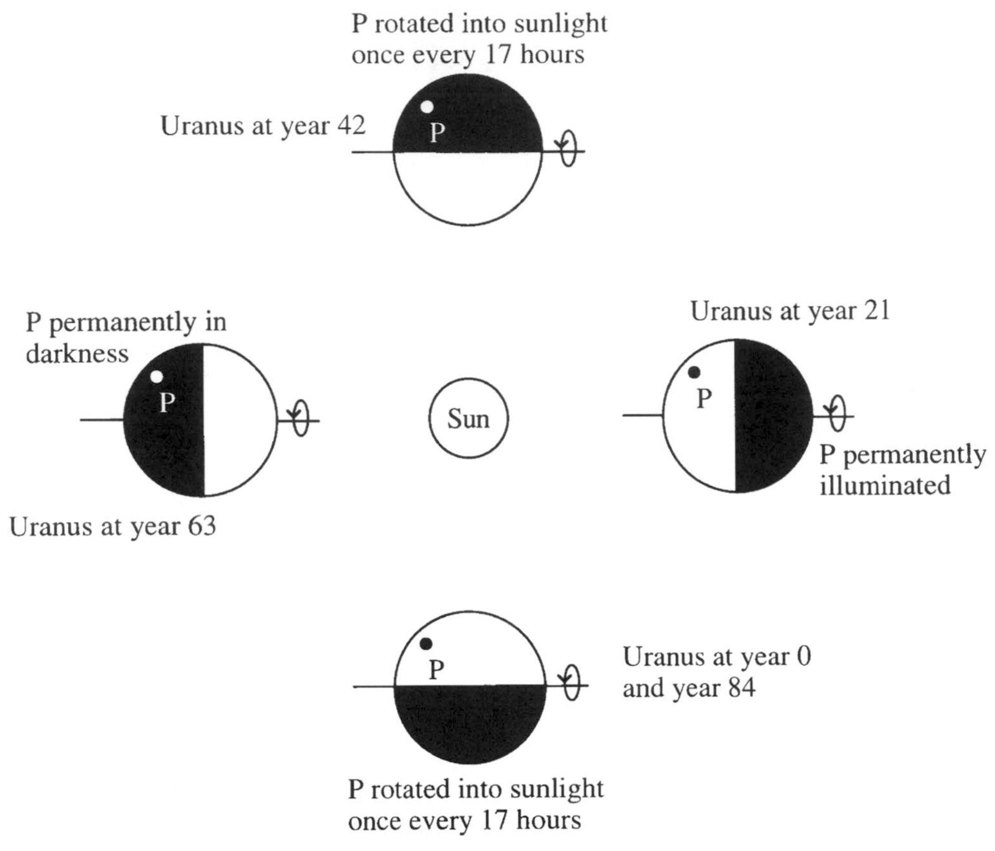

Figure 5.20 How the length of a Uranian day varies. As Uranus rotates on its side the length of a day at an arbitrary position P will vary depending on where the Sun is relative to Uranus. The variation will be from several Earth years to just a few hours

Neptune has eight known moons, all of which are small except for Triton. Triton is another moon that has a synchronous orbit about its planet but it is unique among large moons in that its motion is retrograde. One of Neptune's smaller moons, Nereid, has a highly eccentric orbit, its distance from Neptune varying from 1.4 to 9.6 million km.

Neptune's ring system orbits at a similar distance to that of Uranus and is composed of four distinct sections each of which is somewhat thicker than those of Uranus, making them a little easier to observe from Earth.

Under normal circumstances Pluto lies at the edge of the Solar System. However, the high eccentricity of its orbit means that there are times when it is closer to the Sun than Neptune, for instance between 1979 and 1999. This represents about one twelfth of a sidereal period for Pluto. Its extreme distance and very small size make observations from Earth very difficult and space probes have not yet gone close to it. However, a little is known. Of most interest is its moon, though twin planet might be a more appropriate description due to Charon's large relative size (see Figure 5.22). Charon's radius is about half that of its parent planet. The rotation period of each body is equal to the period of revolution of the pair about each other. This means that the same faces of the two bodies always point at each other. This period is very short at just over six days, implying that

Figure 5.21 Neptune as photographed by Voyager 2. Note the subtle banding, large storm and small numbers of white clouds. The storm feature was called the Great Dark Spot but proved not to be persistent as is the Great Red Spot of Jupiter (NASA, Jet Propulsion Laboratory and National Space Science Data Center)

they are very close together. Only Phobos, a very small moon, orbits closer to its planet. Like Uranus, Pluto spins on its side, meaning that Charon executes an orbital motion perpendicular to the ecliptic. Twice per orbit, as viewed from Earth, Pluto and Charon, completely or partially eclipse each other. This provides a good opportunity to observe the differences between them. As Pluto takes nearly 250 years to orbit the Sun such eclipses occur only once every 125 years. Nevertheless, such an occasion has allowed the surfaces of the planets to be distinguished as being red in Pluto's case and grey in Charon's. Both are thought to be covered by various ices though Pluto is thought to contain a higher proportion of rock.

Figure 5.22 Hubble Space Telescope image of Pluto and Charon (top-right). Pluto, at 1160 km, is less than twice the radial size of Charon at 640 km (NASA, Jet Propulsion Laboratory and National Space Science Data Center)

Comets and Meteoroids

There remain two other sets of objects in the Solar System, both of which are very small on a planetary scale. The first of these are the comets. These are small bodies with sizes similar to small asteroids that orbit the Sun with highly eccentric orbits. It is thought that there are two reservoirs of such bodies, both beyond the orbit of Pluto. The least populous of these is known as the Kuiper disk and extends outwards from the orbit of Neptune. This contains the short-period comets, such as Halley's that returns to the Earth's sky every 76 years as it travels to and from its perihelion[4] at 0.6 astronomical units. The much more populous Oort Cloud is considerably further from the Sun, at around 5000 to 100 000 astronomical units, and is thought to contain many billions of comets distributed almost spherically. Occasionally a comet is shifted into a new orbit, bringing it closer to the Sun, though many of these make only one journey, breaking up en route, burning up in the Sun or being spun off into outer space never to return.

---

[4] Perihelion is the position of closest approach of a body to the Sun. The position of greatest distance is known as aphelion.

Figure 5.23 The nucleus and inner tail of Comet Kohoutek photographed during its visit to the inner Solar System in 1974 (NASA, Jet Propulsion Laboratory and National Space Science Data Center)

In the night sky comets appear as a bright spot, often accompanied by a spectacular tail (see Figure 5.23). It should be noted that comets do not move across the celestial sphere on the timescale of an evening's observation. They appear static in space, not like watching a rocket speeding across the upper atmosphere! Space probes to bodies such as Halley's Comet reveal a solid nucleus made of ices and rocks and a tail, that increases in size close to the Sun, composed of gas and dust. The tail always points away from the Sun regardless of the direction of the comet's motion.

The final dregs of the Solar System are meteoroids, small objects with a size of up to only a few metres. Such bodies can be found throughout the volume of the Solar System. With such small sizes they can only be seen when they enter the Earth's atmosphere (thus changing name to meteors) and then only through the emission of light as they are vaporised by air friction. Such events last only a second or less and can be seen as sudden streaks of light across ten or so degrees of the sky. Consequently meteors are often known as shooting stars. At certain times of year increased incidences of meteors are observed and these are known as meteor showers. A small proportion reach the Earth's surface and these are known as meteorites, sometimes forming craters. Under these circumstances observation is turned on its head. The heavens are sending material to us so that we can pick it up, experiment on it, put it under the microscope and stare at it until it reveals its secrets. What such observations and all the others discussed in this chapter tell us about the Solar System will be explored in the following four chapters.

Questions

Problems

1 Calculate the maximum angular separation of Venus from the Sun as observed from Earth assuming that Earth and Venus follow circular orbits of radius 150 and 108 million km respectively.

2 Relatively, how much more solar power per unit area, on average, does Mercury receive compared to Pluto? How does this compare to the value for Earth?

Teasers

3 What is the fundamental difference between the phase sequence of inferior and superior planets?

4 How many Venusian solar days are there in a Venusian year?

5 (a) Where in the Solar System is the speed of a planetary surface highest due to the rotation of a planet on its axis and what is this speed? Assume all planets are spherical.
(b) Which planet has the highest speed through space due to its orbital motion about the Sun and what is this speed? Assume all orbits are circular.

6 The Moon is held in a captured orbit, in that its periods of rotation on its axis and revolution about the Earth are the same.
(a) How does the Earth move across the sky with time as observed from the Moon? Take libration into account.
(b) What is the synodic period of the Earth as viewed from the Moon?
(c) How does the Earth appear during a solar eclipse when viewed from the Moon?
(d) Explain how the Sun appears when viewed from the Moon as a Lunar eclipse passes from partial to total and back to partial again.

Exercises

7 Why is Venus often the brightest planet observable in the night sky? Why is it sometimes not?

8 Why is Mercury only observed close to sunset and sunrise but Mars can sometimes be seen in the middle of the night?

9 True or false?
(a) Mercury is mainly composed of hydrogen and helium
(b) Mars is red mainly due to the presence of iron oxide (rust) on its surface
(c) Jupiter is heavily cratered
(d) Pluto has no moon

10 Which of the planets have solid surfaces?

6 Gravity and the Solar System

The role of gravity in the universe is paramount. In this chapter, the action of this force will be derived and discussed in the context of our local environment in the Solar System. The ideas developed here will be used throughout the remainder of the book.

Once the heliocentric (Sun at the centre) model of the Solar System had been established by Copernicus in the sixteenth century it was necessary to perform accurate observations of the motions of the planets in order to produce evidence to explain why such objects move as they do. Kepler provided these observations and Newton, with a few observations such as the apocryphal apple landing on his head, interpreted them in terms of a physical rule of nature. In what follows, Newton's law of gravitation is derived from Kepler's laws of planetary motion though not necessarily using the precise logic of Newton.

Kepler's Laws of Planetary Motion

The result of Kepler's observations was the establishment of three laws that describe how the planets orbit the Sun. As described in the previous chapter, the planets follow elliptical orbits, but these orbits do not differ greatly from circles (see Appendix 3). The derivation follows in the same way for both circles and ellipses (as a circle is just a special case of an ellipse with eccentricity = 0) except that the mathematics is simpler for a circle. Kepler's laws are therefore stated for the case of an ellipse and a circle but the mathematics that follows is only presented for the case of a circular orbit. The laws are:

(1) The planets orbit the Sun in ellipses with the Sun at one focus. For a circular orbit, the Sun is at the centre.
(2) The line joining the Sun and a planet sweeps out equal areas in equal times. This implies for a circular orbit that the planet's speed is constant.
(3) The square of the sidereal period of a planet is proportional to the cube of the semimajor axis of the ellipse. Mathematically expressed for a circle:

$$r^3 \propto T^2 \tag{6.1}$$

where T is the sidereal period and r is the distance between the Sun and the planet.

WORKED EXAMPLE 6.1

Q. The Earth's mean distance from the Sun is 150 million km. The period of revolution of Venus about the Sun is 0.62 Earth years. Use Kepler's third law to calculate the mean distance between Venus and the Sun.

A. Kepler's third law relates the orbital radius of a body to its orbital period. For the Earth, the orbital radius, r_E, is 150 million km and the orbital period, T_E, is 1 year. For Venus, the orbital period, T_V, is 0.62 years and r_V is unknown. If an unknown number, k, is used as the constant of proportionality in equation (6.1) then:

$$r_E^3 = kT_E^2 \Rightarrow 150^3 = k1^2 \therefore k = 150^3$$

$$r_V^3 = kT_V^2 \Rightarrow r_V^3 = k0.62^2 \therefore r_V^3 = 150^3 \times 0.62^2 = 1.30 \times 10^6$$

Taking the cube root of 1.30×10^6 gives a value for r_V of 110. Note that units have been almost ignored in the calculation. This is acceptable as k is able to take whatever units are convenient as this constant is used only as an intermediary between the two equations and effectively cancels out. The units of the answer are therefore equal to the units of the quantities that were put into the equation so long as all time units and all distance units are kept the same throughout. Hence Venus is 110 million km from the Sun (retaining just two significant figures as in the data supplied).

Newton's Law of Gravitation

Kepler's third law provides the starting point for a derivation of Newton's law of gravitation. It is possible to substitute the sidereal period of the planet in terms of its speed (which is constant for circular motion according to law 2). The distance travelled during one circular orbit is simply given by $2\pi r$ and so the period can be written in terms of the speed, v:

$$T = \text{distance/speed} = \frac{2\pi r}{v} \tag{6.2}$$

Thus substituting into (6.1);

$$r^3 \propto \frac{4\pi^2 r^2}{v^2}$$

$$\Rightarrow v^2 \propto r^{-1} \tag{6.3}$$

Equation (6.3) shows that planetary speed decreases with separation from the Sun. Thus the outer planets both travel more slowly and have longer journeys to complete an orbit, resulting in the strong dependence on sidereal period with distance.

Newton also developed three laws that relate the overall force experienced by an object to its motion. The first of these states that an object will continue in a straight line at constant speed unless an overall force acts upon it. A planet's direction of motion is constantly changing and so it must be experiencing a force to keep its motion circular (or elliptical) as can be seen by examining Figure 6.1. A useful analogy here is of a ball being swung around in a circle on the end of a piece of string. The tension in the string provides

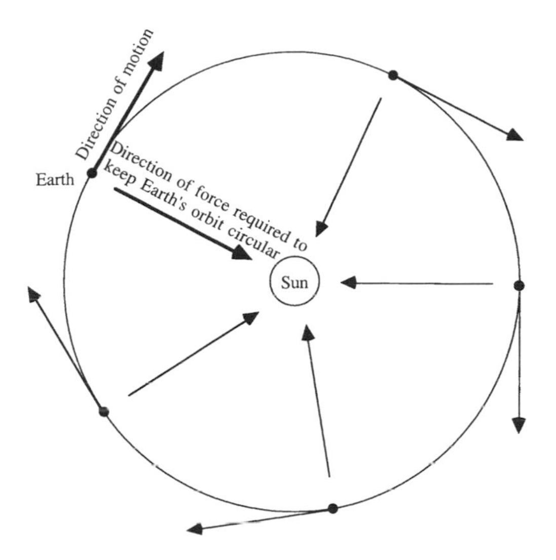

Figure 6.1 Centripetal force. The Sun's gravitational pull provides the force required to keep Earth in its orbit

the force on the ball to keep its direction constantly changing, always pulling inwards. If the string were to suddenly break the ball would fly off in a straight line as prescribed by Newton's first law of motion. The force required to maintain circular motion is known as centripetal force, F_{cp}, and can be expressed in terms of the quantities already discussed and the moving object's mass, m:

$$F_{cp} = \frac{mv^2}{r} \tag{6.4}$$

The square of the velocity of a planet has already been shown to be proportional to r^{-1} and so the centripetal force required to maintain the planet's circular motion can be determined by combining equations (6.3) and (6.4);

$$F_{cp} \propto \frac{m}{r^2} \tag{6.5}$$

The final link to determine the full expression is simply a logical step. When Newton was struck on the head by an apple he is supposed to have realised that the apple was being attracted to the Earth by the same force that holds the Earth in orbit about the Sun, the force of gravity. If the Sun attracts the Earth and the Earth attracts the apple then the Earth must also attract the Sun. If this is the case then every body must have a gravitational field proportional to its mass and the force between two objects is proportional to both (m and M). Hence;

$$F = \frac{GMm}{r^2} \tag{6.6}$$

G is the constant of proportionality, known as the universal gravitational constant, and is equal to 6.67×10^{-11} N m^2 kg^{-2}. The distance, r, is always measured between the centres of the two bodies (for bodies with symmetrical mass distributions which applies quite well to almost all astronomical bodies). In summary, Newton's law of gravitation states that the mutual force between two bodies is proportional to the mass of each and inversely proportional to the square of the distance that separates them.

WORKED EXAMPLE 6.2

Q. Use the data from Worked Example 6.1 and the masses of the Earth, the Sun and Venus at 6×10^{24} kg, 2×10^{30} kg and 5×10^{24} kg respectively, to calculate the ratio between the maximum gravitational force exerted on the Earth by Venus to that exerted by the Sun on the Earth.

A. There are three variables in equation (6.6), the masses of the two bodies that are gravitationally interacting and their separation. In this problem the masses are fixed and the Earth–Sun separation is constant (using the circular approximation). Only the Earth–Venus separation varies. For maximum force the bodies should be at their minimum separation. This occurs when they are aligned on the same side of the Sun. At this time their separation is the difference between their separation from the Sun at $[(1.5 \times 10^8) - (1.1 \times 10^8)] = 4 \times 10^7$ km $= 4 \times 10^{10}$ m (SI units must be maintained throughout this calculation so that the final value is in the SI unit for force, Newtons). Equation (6.6) is applied twice to determine the force on the Earth due to each body:

$$F_{\mathrm{E-S}} = \frac{GM_{\mathrm{S}}m_{\mathrm{E}}}{r^2_{\mathrm{E-S}}} = \frac{(6.7 \times 10^{-11}) \times (2 \times 10^{30}) \times (6 \times 10^{24})}{(1.5 \times 10^{11})^2} = 3.6 \times 10^{22}\ \mathrm{N}$$

$$F_{\mathrm{E-V_{max}}} = \frac{GM_{\mathrm{E}}m_{\mathrm{V}}}{r^2_{\mathrm{E-V}}} = \frac{(6.7 \times 10^{-11}) \times (6 \times 10^{24}) \times (5 \times 10^{24})}{(4 \times 10^{10})^2} = 1.2 \times 10^{18}\ \mathrm{N}$$

The force of the Sun on the Earth is thus 30 thousand times greater than the maximum force due to Venus. The Earth's orbit is consequently about the Sun but small perturbations are caused by the effect of Venus. The Earth's orbit is thus never precisely the same each time around.

Working back from equation (6.6), applying the same arguments in reverse, the constant in equation (6.1) can be found so that the final equation is

$$T^2 = \frac{4\pi^2}{GM} \cdot r^3 \tag{6.7}$$

For a single orbital system only T and r vary but in comparing two different systems the dominant, central mass, M, also varies. Therefore the constant of proportionality varies between systems and, indeed, the central mass can be calculated from observations of T and r. For instance, the mass of Jupiter can be calculated from simple observations of the Galilean moons.

WORKED EXAMPLE 6.3

Q. Given that Io orbits Jupiter at a mean distance of 422 000 km and takes 42 hours 28 minutes to complete one revolution, what is the mass of Jupiter?

A. Unlike Worked Example 6.1, this problem requires a knowledge of the form of the constant in Kepler's third law that is given by equation (6.7). For Jupiter's orbital system the constant of proportionality is inversely proportional to its mass so that if T and r are known for an orbiting body then Jupiter's mass can be determined. Converting km to metres and hours and minutes to seconds (SI units) and then rearranging equation (6.7) gives:

$$M_J = \frac{4\pi^2 r_{Io}^3}{GT_{Io}^2} = \frac{4\pi^2(4.22 \times 10^8)^3}{(6.67 \times 10^{-11} \times (1.53 \times 10^5)^2} = 1.90 \times 10^{27}\,\text{kg}$$

So the mass of Jupiter is just under 2×10^{27} kg which is 0.1% of the Sun's mass. Note that nothing can be said about the mass of Io from the information given. Any body would move in the same way if orbiting at this separation from Jupiter.

Centre of Mass

It is important to realise that the force of gravity is bi-directional. Bodies attract each other with equal force. In the case of an apple being attracted to the Earth and vice versa, one force can effectively be neglected. While both forces are equal, at about 1 N for a 100 g apple (using equation (6.6)), the Earth's mass is so enormous compared with the apple's that the Earth's motion towards the falling apple is minute. The same applies to the orbit of the Earth about the Sun. The Sun applies a force of 3.6×10^{22} N on the Earth and this is enough to keep the Earth in its (almost) circular orbit. The Earth applies this same force on the Sun but, as the mass of the Sun is more than 300 thousand times greater, it remains almost static. There are systems where this is not the case, however. The best example that has been mentioned so far is the Pluto–Charon system. As the masses of the two bodies are quite similar to each other, neither remains fixed relative to the other. They each orbit about their centre of mass. For two objects of masses M and m, the centre of mass (or barycentre) is a distance d_M from the larger mass and d_m from the smaller mass, given by

$$d_M = \left(\frac{m}{m+M}\right) r \tag{6.8a}$$

$$d_m = \left(\frac{M}{m+M}\right) r \tag{6.8b}$$

where r represents, as usual, the separation of the bodies. The centre of mass can be thought of as being the 'balance point' of the two bodies (see Figure 6.2).

WORKED EXAMPLE 6.4

Q. What force does an apple apply to the planets Mercury (mass, 3.3×10^{23} kg, radius 2400 km) and Jupiter (mass, 1.9×10^{27} kg, radius 71 000 km) when placed on their surfaces?
A. The planets pull on the apple with the same force that the apple pulls on the planet. This force is given by equation (6.6) as

$$F_M = \frac{GM_M m_a}{r_M^2} = \frac{(6.7 \times 10^{-11}) \times (3.3 \times 10^{23}) \times 0.1}{(2.4 \times 10^6)^2} \approx 0.4\,\text{N}$$

$$F_J = \frac{GM_J m_d}{r_J^2} = \frac{(6.7 \times 10^{-11}) \times (1,9 \times 10^{27}) \times 0.1}{(7.1 \times 10^7)^2} \approx 2.5\,\text{N}$$

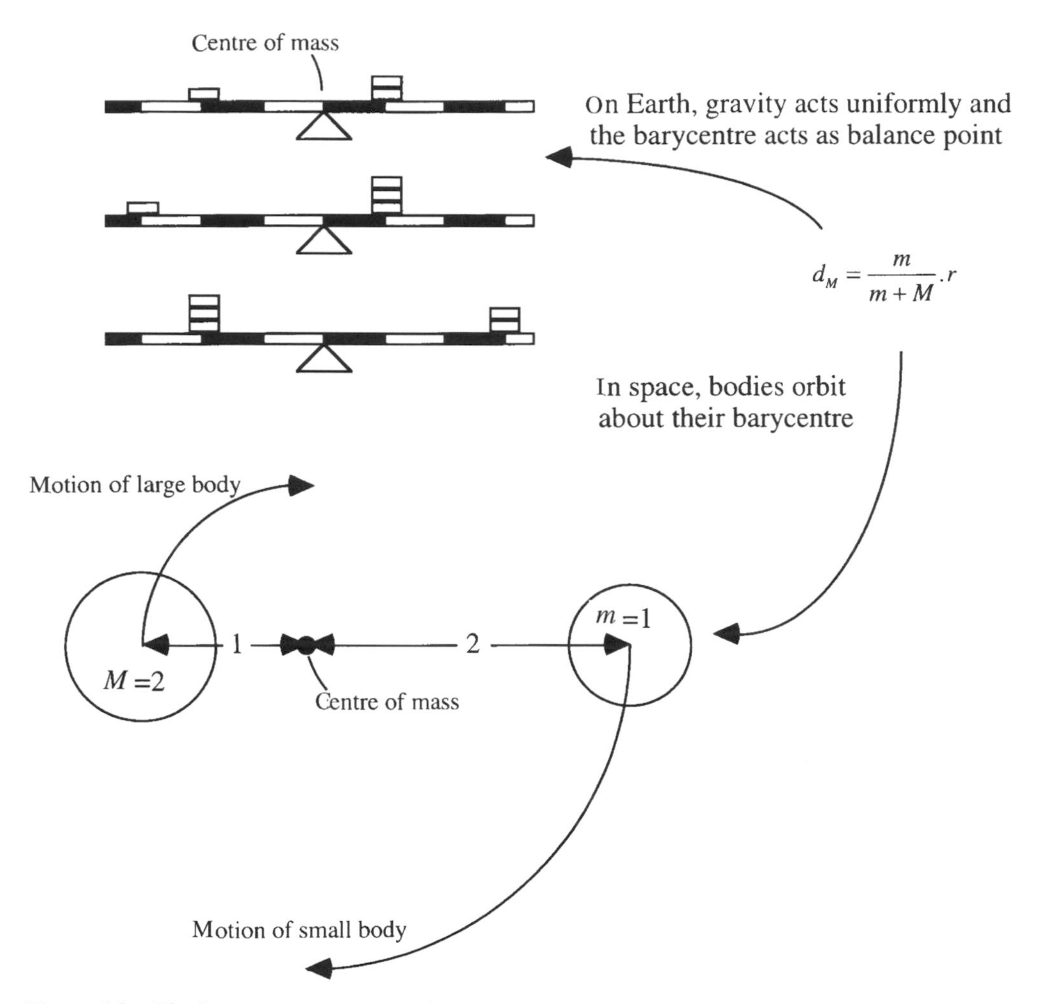

Figure 6.2 The barycentre or centre of mass

The mutual force of attraction between the apple and the planet is thus 0.4 N for Mercury and 2.5 N for Jupiter as compared to 1 N on Earth. The apple will thus feel heaviest on Jupiter and lightest on Mercury though its mass will not have changed. Note that the ratios of forces are not very large (less than ten) despite the fact that the planets vary in mass by factors of almost 10 thousand. This is simply because the force goes up with increasing planetary mass but down with the square of the planetary radius. Jupiter's gravitational influence on surrounding space is nevertheless much greater than Mercury's as planetary radius does not affect the gravitational field outside the body of the planet.

WORKED EXAMPLE 6.5

Q. A planet/moon system consists of a planet of mass 1.485×10^{25} kg and radius 12 000 km and its moon which has a mass of 1.5×10^{23} kg. The centre of mass of the planet is always 1.2×10^6 km from the centre of mass of the moon. Describe the motion of the planet.
A. The distance, d, to the barycentre from the centre of the planet (of mass, M) is given by equation

(6.8):

$$d_M = \left(\frac{m}{m+M}\right)r = \left[\frac{1.5\times10^{23}}{(1.5\times10^{23})+(1.485\times10^{25})}\right]1.2\times10^6 = 12\,000\,\text{km}$$

The binary system rotates about the barycentre which is 12 000 km from the centre of mass of the planet. The planet's radius is also 12 000 km which means that the planet rotates about a point on its own surface. Ignoring the planet's revolution about its parent star and its rotation on its axis, the motion of the planet is therefore as indicated on the diagram.

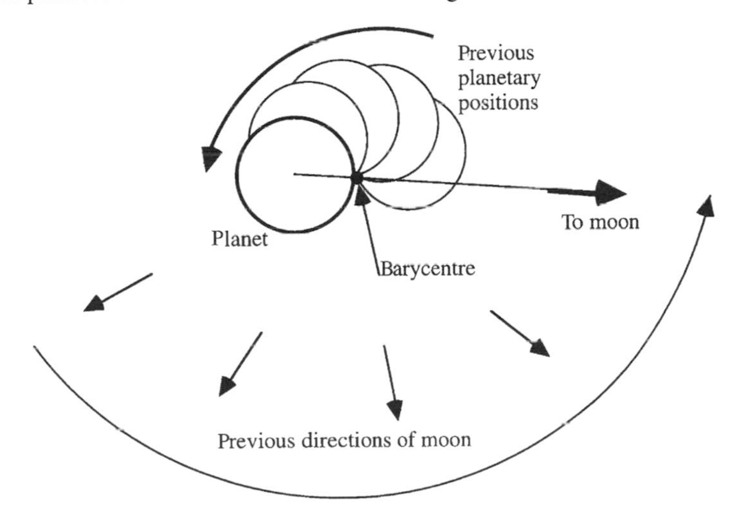

When orbiting bodies have masses within an order of magnitude or two of each other the fixed central body approximation cannot be used and some of the simplifications made in the mathematical derivations above do not hold. In equating centripetal force and gravitational forces (equations (6.4) and (6.6)), the mean distance of attraction remains r but the mean radius of orbit changes to d_M for the large body and d_m for the smaller. Thus:

$$\frac{mv^2}{d_m} = \frac{GMm}{r^2} \quad \text{and} \quad \frac{Mv^2}{d_M} = \frac{GMm}{r^2}$$

By substituting for d_m or d_M, as given by equations (6.8), an expression for the orbital period in terms of the separation of the bodies is obtained. Either route leads to the same final expression but here substitution for d_M is made after v is expressed in terms of d_M:

$$\frac{Mv^2}{d_M} = \frac{M(2\pi d_M/T)^2}{d_M} = \frac{4\pi^2 M d_M}{T^2} = \frac{GMm}{r^2}$$

$$\Rightarrow \left[\frac{4\pi^2 M}{T^2}\right]\left[\frac{mr}{(m+M)}\right] = \frac{GMm}{r^2}$$

$$\Rightarrow T^2 = \frac{4\pi^2}{G(M+m)}r^3 \qquad (6.9a)$$

Expressions for T in terms of d_m or d_M must be obtained independently by substituting for r appropriately:

$$\frac{Mv^2}{d_M} = GMm\left(\frac{m}{d_M(M+m)}\right)^2$$

$$\downarrow$$

$$v^2 = \frac{Gm^3}{d_M(M+m)^2} = (2\pi d_M/T)^2$$

$$\downarrow$$

$$T^2 = \left[\frac{4\pi^2(M+m)^2}{Gm^3}\right] d_M^3 \tag{6.9b}$$

$$\frac{mv^2}{d_m} = GMm\left(\frac{N}{d_m(M+m)}\right)^2$$

$$\downarrow$$

$$v^2 = \frac{GM^3}{d_m(M+m)^2} = (2\pi d_m/T)^2$$

$$\downarrow$$

$$T^2 = \left[\frac{4\pi^2(M+m)^2}{GM^3}\right] d_m^3 \tag{6.9c}$$

Equations (6.9b) and (6.9c) could alternatively be derived by simply substituting for r in equation (6.9a) from equation (6.8a) but the route given above gives a little more insight into the physical processes that dictate the orbital characteristics. The three expressions relate to observations in different ways. Equation (6.9a) refers to the case where a pair of bodies can be seen orbiting about each other such that the separation of the bodies and their period of orbit can be determined but the resolution of the observation is too poor to pinpoint the position of the centre of mass. Such an observation provides a single value of T and a single value of r so that only the sum of masses can be determined. Closer observation allows the precise orbit of both bodies to be found so that d can be found for each body. As T must be the same for each body (to maintain the fixed position of the centre of mass relative to the system) it becomes possible to find the individual masses of the bodies.

A good orbital system on which to employ equations (6.9) is that of the Earth and Moon. External observation of the pair would make it easy to calculate the masses of both the Earth and the Moon. Current observational accuracy for the Pluto–Charon system allows equation (6.9a) to be used accurately but measurements of d remain too poor to use equations (6.9b) or (6.9c) with confidence, meaning that the sum of the masses of Pluto and Charon is accurately known but that the individual contributions are subject to some uncertainty. Note that in Worked Example 6.3, concerning Jupiter and Io, it was stated that nothing could be said about the mass of the orbiting body from the information given. Equations (6.9) allow the mass of an orbiting body to be determined if M and m are within a couple of orders of magnitude of each other. This is not the case for Io and Jupiter so that the perturbation of Jupiter's orbit by Io is tiny. Furthermore, the large number of other moons that orbit Jupiter also influence its motion, making the effect of one orbiting body hard to decouple from the sum.

A further example of orbiting pairs in the Solar System is that of asteroids which often orbit the Sun in pairs. As they have similar masses such a pair will revolve about a point close to the centre of their separation. Furthermore, as the mass of asteroids is relatively small, it is necessary for them to orbit in close proximity to each other at a high rate of revolution. The pairs thus rapidly tumble about each other as they make their way, much more slowly, around the Sun. The centre of mass moves smoothly about the Sun but the orbits of the individual asteroids are more complicated, being composed of the superposition of an orbit about both the Sun and a partner asteroid.

Artificial Satellites

Armed with a knowledge of how orbiting bodies behave and the technology of the twentieth century, it has been possible to send artificial satellites into space and into orbit about the Earth. As an example of a task that is performed by an orbiting satellite consider the motion required of a television relay station in space. So that live coverage of an event on one side of the world is not periodically interrupted at the opposite end, it is necessary to hold the satellite in a constant position relative to the Earth's surface. This is known as a geostationary orbit. As the Earth rotates once on its axis every 24 hours then the satellite must orbit with the same period and direction. It is therefore a simple matter to calculate the distance from the Earth's centre at which the satellite must orbit using equation (6.7). M is 6×10^{24} kg, the mass of the Earth. The distance from the centre of the Earth, r, is calculated to be 42 300 km so that the satellite must orbit about 5.6 Earth radii above the surface. Satellites orbiting inside this distance would revolve more quickly than the required 15° per hour and those outside would revolve too slowly (see Figure 6.3).

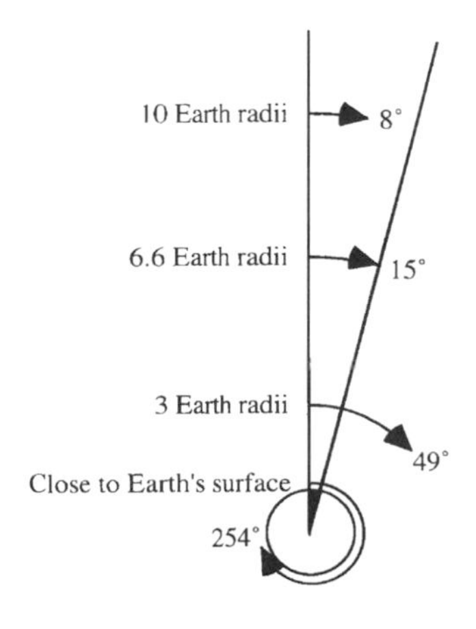

Figure 6.3 The angle through which satellites travel during one hour as a function of their distance from the Earth. The satellite at 6.6 Earth radii from the Earth's centre revolves at the same rate as the Earth rotates on its axis (indicated). This is therefore the correct position for a geostationary satellite

WORKED EXAMPLE 6.6

Q. Leeds United's board of directors have decided to commission a spy satellite to learn more about Manchester United's training methods. Ellandsat I is to skim the surface of the Earth (a sphere of mass 6.0×10^{24} kg and radius 6380 km) at a height of 50 km and take a polar route along Manchester's meridian.
(a) How often will the satellite pass over the training ground and at what speed will the satellite pass over?
(b) If the satellite wishes to pick out Manchester players standing 1 m apart in their traditional red colours using a normal optical telescope, how large would the aperture have to be? Ignore atmospheric effects (there's no atmosphere at Old Trafford anyway!).
(c) Why is it not possible to have a satellite in geostationary orbit directly above Manchester?
A. (a) The time period of the satellite's orbit is given by Kepler's third law, expressed by equation (6.7), where r is the radius of the satellite's orbit which is equal to the Earth's radius plus 50 km (= 6380 + 50 = 6430 km):

$$T^2 = \left(\frac{4\pi^2}{GM}\right) r^3 = \left[\frac{4\pi^2}{(6.7 \times 10^{-11}) \times (6.0 \times 10^{24})}\right] (6.43 \times 10^6)^3 = 2.6 \times 10^7$$

$$\therefore T = 5100 \text{ s}$$

The satellite's speed is given by the circumference of its orbit (distance) divided by its period (time). The orbital circumference is ($2\pi \times 6430$ =) 40 000 km and the period is 5100 s so the speed is 7.8 km s^{-1} (or 28 000 km h^{-1}). This means the satellite will only pass over the training ground once every 85 minutes and it will pass by very rapidly. For instance, if the telescope can observe the training ground for a distance of 100 km in its orbit, only 13 s of viewing time would be available per orbit. A training session is unlikely to last much longer than a few hours and so very little (less than a minute's viewing per day) would be gained through employing such a spying technique. The rotation of the Earth would cause further difficulties.
(b) Using precisely the technique of Worked Example 2.1 where R is given by the ratio of the separation of the two players (1 m) to the distance from which the observation is made (50 000 m), which is (1/50 000 =) 2×10^{-5}, and λ is given by the wavelength of red light which is about 650 nm:

$$D = \frac{1.22\lambda}{R} = \frac{1.22 \times (650 \times 10^{-9})}{2 \times 10^{-5}} = 0.04 \text{ m}$$

Only a 4 cm aperture would be required and this is well within reason (if you're going to the trouble of building a satellite to put it in).
(c) Ignoring the Earth's motion through space, Manchester's motion is circular due to the Earth's rotation on its axis. The centre of the circle is on the Earth's spin axis but is north of the geometrical centre of the Earth. To keep a satellite in circular motion, centripetal force must be applied by the Earth's gravitational field and this acts towards the geometrical centre of the Earth. A satellite is thus unable to orbit on the required path above Manchester. All satellites must travel along great circles (centred on the geometrical centre of the Earth) and the only great circle that allows geostationary motion is above the equator. All television relay stations and similar satellites are thus directly above the equator. As the separation of such satellites from the Earth's surface is 5.6 Earth radii, such a satellite is able to access almost an entire hemisphere (north to south) from its position.

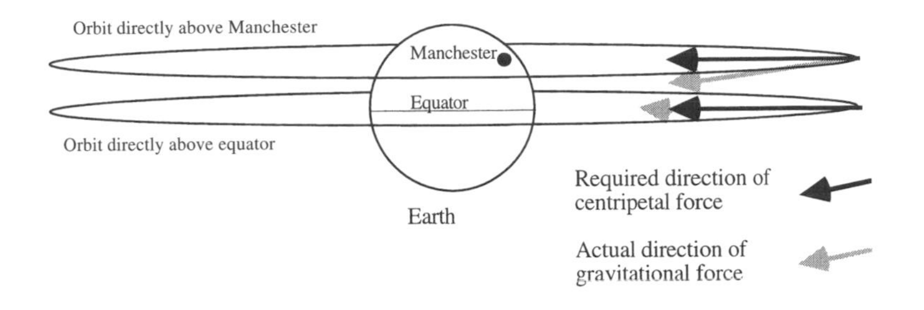

Acceleration Due to Gravity

The strength of the force of gravity at the planetary surface is an important parameter and determines the extent of events such as secondary cratering and the distance that volcanic debris can travel. On the Earth it has important consequences for life. Newton's second law of motion states that when an overall force acts on an object, the object accelerates in the direction of the force. The magnitude of the acceleration, a, is given by the force divided by the mass of the object:

$$a = \frac{F}{m} \tag{6.10}$$

Consider an object at a planetary surface. Its acceleration due to gravity, g, is given by

$$g = \frac{F_g}{m} = \frac{GMm}{r_p^2 m}$$

$$\Rightarrow \quad g = \frac{GM}{r_p^2} \tag{6.11}$$

where r_p is the planetary radius. Note that g is determined only by the planetary parameters. Any two objects will accelerate at the same rate when dropped close to the planet's surface although atmospheric resistance may make this observation difficult with very light objects such as feathers.

Substituting values for the Earth into equation (6.11) yields a value for the acceleration due to gravity of $9.8\ \mathrm{m\,s^{-2}}$. Neglecting air resistance, the speed of any object dropped close to the surface of the Earth (say, within a few kilometres) will increase at the rate of $9.8\ \mathrm{m\,s^{-1}}$ every second. Using Newton's second law provides an easier way to calculate the gravitational force on an object, known as weight, as it is now simply given by the product of g and the object's mass:

$$F_g = mg \tag{6.12}$$

For a 100 g apple this is, as calculated using equation (6.12) (or (6.6)), close to 1 N. It is important to note that weight varies from planet to planet whereas mass is a constant regardless of the gravitational field (neglecting relativistic effects beyond the scope of this book).

WORKED EXAMPLE 6.7

Q. What is the acceleration due to gravity on the planetary surfaces of Mercury (mass, 3.3×10^{23} kg; radius 2400 km) and Jupiter (mass, 1.9×10^{27} kg; radius 71 000 km)? From this information, what is the weight of a 100 g apple placed on their surfaces?
A. This problem can simply be solved by substituting the correct numbers into equation (6.11):

$$g_\mathrm{M} = \frac{GM_\mathrm{M}}{r_\mathrm{M}^2} = \frac{(6.7 \times 10^{-11}) \times (3.3 \times 10^{23})}{(2.4 \times 10^6)^2} \approx 4\,\mathrm{m\,s^{-2}}$$

$$g_\mathrm{J} = \frac{GM_\mathrm{J}}{r_\mathrm{J}^2} = \frac{(6.7 \times 10^{-11}) \times (1.9 \times 10^{27})}{(7.1 \times 10^7)^2} \approx 25\,\mathrm{m\,s^{-2}}$$

Thus, the accelerations due to gravity on Mercury and Venus are $4\,\mathrm{m\,s^{-2}}$ and $25\,\mathrm{m\,s^{-2}}$ respectively. The force due to gravity (weight) on an apple on the planetary surfaces is found by multiplying the apple's mass by the acceleration due to gravity at that point according to equation (6.12). The apple therefore weighs 0.4 N on Mercury's surface and 2.5 N on Venus's surface. These results are identical to those obtained in Worked Example 6.4. Careful inspection reveals that the calculations are essentially the same. The weight of an object is simply the gravitational force that it experiences at any given position. Note that the units of acceleration are $\mathrm{m\,s^{-2}}$ but this is dimensionally equivalent to $\mathrm{N\,kg^{-1}}$ so that the latter units, that are perhaps more obvious in the context of equation (6.12), may also be used.

WORKED EXAMPLE 6.8

Q. Consider Mercury to be a sphere of mass 3.32×10^{23} kg and radius 2430 km orbiting the Sun (mass 1.99×10^{30} kg) in a circular orbit of radius 57.9 million km.
(a) What is the acceleration due to gravity on Mercury's surface due to Mercury's mass alone?
(b) What is the acceleration due to the Sun's gravitational field on Mercury?
(c) Two 100 kg men are standing on Mercury, one with the Sun at his zenith and the other on diametrically the opposite side of the planet. How much does each man weigh?
A. (a) From Worked Example 6.7 (but using three significant figures rather than two) the acceleration due to gravity on Mercury's surface is $3.75\,\mathrm{m\,s^{-2}}$.
(b) Ignoring the influence of Mercury's gravitational field, the acceleration due to the Sun's mass at the distance of Mercury's orbit can be calculated exactly as in Worked Example 6.7, where d_M is the distance of Mercury from the Sun:

$$g_\mathrm{S} = \frac{GM_\mathrm{S}}{d_\mathrm{M}^2} = \frac{(6.67 \times 10^{-11}) \times (1.99 \times 10^{30})}{(5.79 \times 10^{10})^2} = 0.0396\,\mathrm{m\,s^{-2}}$$

So an object in free space, at Mercury's mean distance from the Sun would fall towards the Sun, accelerating at about $0.04\,\mathrm{m\,s^{-2}}$.
(c) Weight is given by equation (6.12). In this case, the complication is that two gravitational fields act on the two men. In the diagram the man on the left has the Sun overhead so that its force appears to be pulling him away from Mercury whereas the opposite is true for the man on the right (there is no gravitational shielding by Mercury and the two men are almost equidistant from the Sun). The

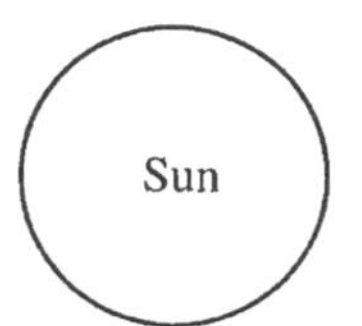

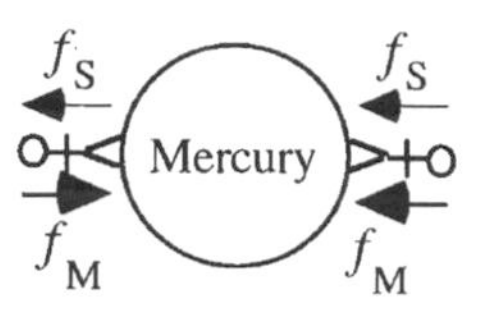

force due to the Sun on each man is therefore given by, $F_S = mg_S$, which is $(100 \times 0.0396 =)$ 3.96 N. The force due to Mercury pulls both men towards the centre of the planet and is given by $F_M = mg_M$, which is $(100 \times 3.75 =)$ 375 N. It may seem therefore that the man on the left weighs $(375 - 4 =)$ 371 N and the man on the right weighs $(375 + 4 =)$ 379 N. This ignores the fact that Mercury itself is also moving under the Sun's gravitational influence so that Mercury is also accelerating towards the Sun at 0.0396 m s^{-2} (that is, its tangential movement is curved into an orbit though its (directionless) speed is constant). This cancels out the influence of F_S on the men so that they both weigh 375 N. An alternative way of thinking of this is that both men and Mercury are in orbit about the Sun. In much the same way that men in Earth-orbiting satellites are weightless (as all bodies fall towards the Earth at the same rate that the Earth's surface curves away) then the men on Mercury are weightless under the Sun's influence. Their weight is determined only by Mercury's gravitational pull (as they are static with respect to Mercury). Small effects due to the difference in distance of the men from the Sun will cause extremely small differences in weight. The weight of the men would only be the 371 N and 375 N calculated above if Mercury were to be held static in space by some additional, external force.

Escape Velocity

It is instructive to consider what is required to completely escape from the gravitational field of a body. In order to move an object from a planetary surface (at r_p) to a point that can be considered outside the planet's gravitational grip (at infinity) it is necessary to do work against gravity. That means that energy must be supplied to the object to propel it in a direction that the gravitational force opposes. The work done in moving an object to oppose a constant force is given by the product of the force and the distance moved. When the force varies with distance, calculus is required to complete the calculation. In essence, the energy required to move a very short distance, dr, over which the force does not vary, is calculated. The sum of these steps is then calculated by integration. In the case of an object of mass, m, being moved from the planetary surface to a place of zero gravitation the work required, W, is given by

$$W = \int_{r_p}^{\infty} F \, dr = \int_{r_p}^{\infty} \frac{GMm}{r^2} dr$$

$$\therefore \quad W = GMm \left[\frac{-1}{r} \right]_{r_p}^{\infty}$$

$$\Rightarrow \quad W = \frac{GMm}{r_p} \tag{6.13}$$

The energy required is proportional to the masses of the planet and the object and inversely proportional to the planetary radius.

A useful concept for characterising the strength of a gravitational field is escape velocity. This refers to the vertical speed at which a projectile has to be launched, without further energy input during its journey, to escape the gravitational field. This can be simply calculated from equation (6.13) as the energy that the projectile uses to escape must be equal to its initial energy of motion, known as kinetic energy. The projectile's kinetic energy is given by $mv^2/2$. Though v will decrease as the projectile is pulled back to the planet it is only necessary to consider the case where the projectile begins its journey with a vertical velocity equal to the escape velocity, v_e. The projectile will come to rest at infinity having just escaped the planet's gravity. The energy required is thus equal to the kinetic energy lost by the projectile:

$$\frac{1}{2}mv_e^2 = \frac{GMm}{r_p}$$

$$\Rightarrow \quad v_e = \sqrt{\frac{2GM}{r_p}} \tag{6.14}$$

The escape velocity is thus a function of planetary parameters only and can be written in terms of the acceleration due to gravity by substituting from equation (6.11) thus:

$$v_e = \sqrt{2gr_P} \tag{6.15}$$

The escape velocity is merely a notional figure. Objects such as rockets that do escape the Earth's gravitational field do not have to attain speeds in excess of $11.2\,\mathrm{km\,s^{-1}}$, the Earth's escape velocity. This is because they are powered throughout their journey. The energy required according to equation (6.13) is provided by the continuous thrust of the rocket's engines so that there is no minimum speed in effect. Nevertheless, the concept of escape velocity becomes quite useful when considering planetary atmospheres later in the book.

WORKED EXAMPLE 6.9

Q. What are the escape velocities from Mercury and Jupiter?
A. The accelerations due to gravity on Mercury and Jupiter have been calculated in Worked Examples 6.7 and 6.8 as being $3.8\,\mathrm{m\,s^{-2}}$ and $25\,\mathrm{m\,s^{-2}}$ respectively. Applying equation (6.15) for each planet gives

$$v_{e_M} = \sqrt{2gr_M} = \sqrt{2 \times 3.8 \times (2.4 \times 10^6)} = 4300\,\mathrm{m\,s^{-1}}$$

$$v_{e_J} = \sqrt{2gr_J} = \sqrt{2 \times 25 \times (7.1 \times 10^7)} = 60\,000\,\mathrm{m\,s^{-1}}$$

So the escape velocity for Mercury, at $4.3\,\mathrm{km\,s^{-1}}$, is about 14 times smaller than that for Jupiter, at $60\,\mathrm{km\,s^{-1}}$. The Earth falls in between these values at $11\,\mathrm{km\,s^{-1}}$. Note that larger planets have higher escape velocities though not directly in proportion to their radius. If all planets had the same density then their mass would be proportional to the cube of their radius so that, substituting into equation (6.14), the dependence of escape velocity on radius would be exactly in proportion to radius. The larger planets tend to have lower densities, however, and so escape velocity tends to have a weaker than linear dependence on radius.

Ocean Tides

Now that the basic ideas of gravitation have been elucidated it becomes possible to explain many of the observed phenomena described in Chapter 5 and one everyday occurrence that may not, at first glance, appear to be connected with gravity.

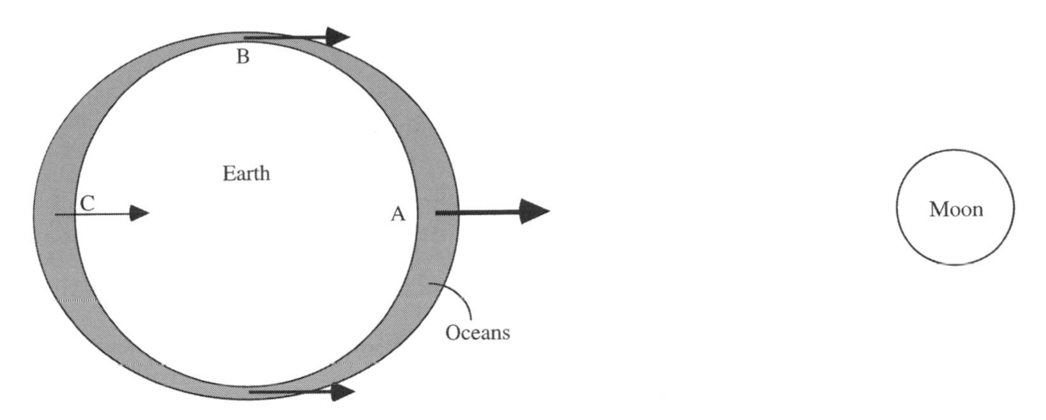

Figure 6.4 The bulge of the Earth's oceans (exaggerated) due to the tidal force of the Moon (see text)

The ocean's tides follow the movement of the Moon in the sky, there being two sequences of high and low tides in the time it takes for the Moon to complete a lap of the sky, about 25 hours. This is not coincidental. Consider Figure 6.4. The acceleration due to the Moon's gravity experienced by the water in position A is greater than that in position B which is in turn greater than that in position C. This is due to the variation in distance of these positions from the Moon (equation (6.11)). As water is a fluid then it is able to move freely so that water in position A drops towards the Moon most quickly while water in position C gets left behind. This forms two bulges of water, one facing the Moon and one on the opposite side of the Earth. In these positions there is a high tide and in position B there is a low tide. As the Earth spins on its axis (and the Moon moves about in its orbit with a 27 times longer period) every longitude on Earth will experience two high tides and two low tides every 25 hours.

The view of ocean tides being influenced only by the Moon is somewhat simplistic. For instance, frictional resistance to oceanic motion prevents the bulges from pointing directly towards and away from the Moon. Local morphology also plays a role, as does the Coriolis force which has a direct influence on the ocean's motion due to the Earth's rotation on its axis.

A further question to ask about tides is, why does the Moon have a stronger influence on tides than the Sun? The force of gravity exerted by the Sun on the Earth is several hundred times greater than that due to the Moon. This is not the important parameter, however. What is important is the difference in force as a function of distance. Consider the rate of change of force, F_t, on a unit mass ($m = 1$) due to the gravitational field of a body of mass, M. This is given by

$$F_t = \frac{dF_g}{dr} = \frac{d}{dr}\left(\frac{GM}{r^2}\right)$$

$$F_t = -\frac{2GM}{r^3} \qquad (6.16)$$

Substituting the values for the Moon and the Sun gives values of $1.7 \times 10^{-13}\ \mathrm{N\,kg^{-1}\,m^{-1}}$ and $7.9 \times 10^{-14}\ \mathrm{N\,kg^{-1}\,m^{-1}}$ respectively. These numbers may appear very small but it should be remembered that 1 $\mathrm{km^3}$ of sea water has a mass of 10^{12} kg and that the diameter of the Earth is more than 10^7 m so that the total tidal forces are appreciable. Second, it is noticeable that the figure for the Moon is a little more than twice that due to the Sun. The Moon thus dominates the ocean's tides on Earth but the Sun's contribution is not negligible. For this reason tides are particularly strong twice per synodic period (29.5 days) at full moon and new moon when the Sun, Moon and Earth are aligned so that the tidal bulges created by the Sun and the Moon are in the same places at the same time. These tides are known as spring tides. When the tidal forces act perpendicularly, at half moon, the tides are smaller and known as neap tides.

Synchronous Orbits

It has been mentioned that there is some frictional resistance that acts against the tidal motion of the water. The motion of any object, even through the air, has to overcome some resistance due to surrounding matter. To overcome a resistive force work has to be done and energy expended. It is a fundamental physical principle that energy must be conserved and so where does the energy to keep moving the oceans against frictional resistance come from? The answer is from the Earth's rotational energy. If the Earth spins more slowly it loses energy. Energy is being lost at a very slow rate. The Earth's rotation period is only increasing by about 1 ms per century. At the same time, according to precise laser measurements, the Moon is moving away from the Earth at a rate of about 4 cm per year while according to Kepler's third law this results in the Moon's orbital period increasing[1]. Eventually the tidal forces will couple the Earth's rotation to the Moon's revolution synchronously so that the two periods will be the same, calculated to be about 55 days.

As the Earth's gravitational field is so much larger than the Moon's, synchronous coupling has already occurred for the Moon's rotational and revolutionary periods. This explains why the same side of the Moon is always tilted towards the Earth. Such a resonance is thought to have become entrenched during the early history of the Solar System when the planets were hotter and softer. Under the tidal action of the Earth the more dense components of the Moon's interior moved within the spherical body. By the time the resonance was reached the Moon's mass distribution had been shifted along an axis pointing towards the Earth and the molten rocks had cooled. This axis became gravitationally held so that it points towards the Earth at all times. Further evidence for this scenario is obtained by comparing the appearance of the facing and opposite surfaces of the Moon. The facing hemisphere has a large number of seas that are solidified lava flows whereas the reverse side has none (see Figure 6.5). The relatively low-density crust is thinner on the side facing the Earth and has thus been ruptured by hotter material from the Moon's interior at certain stages of the Moon's history. The thicker crust on the reverse face has prevented this happening there.

[1] The change in the Moon's orbital period is required to conserve angular momentum (see next chapter) but results in increasing the Moon's orbital energy, also found from the slowing of the Earth's spin.

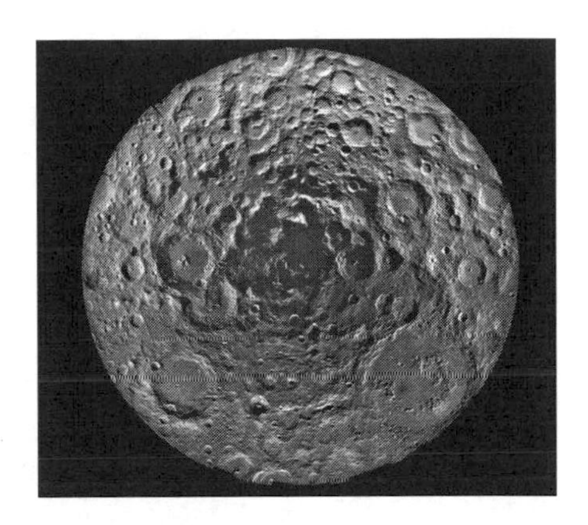

Figure 6.5 A small region of the Moon (centred on the south pole) where the crustal thickness has been sufficient to withstand laval flooding. This region, as for the far side of the Moon, is therefore free of seas but very heavily cratered (NASA, Jet Propulsion Laboratory and National Space Science Data Center)

WORKED EXAMPLE 6.10

Q. What is the ratio of the Earth's (mass, 6.0×10^{24} kg) gravitational force on the Moon (distance from Earth, 380 000 km) to the Sun's (mass, 2.0×10^{30} kg; distance from Earth, 150 million km) gravitational force on the Moon? Why is the Moon trapped in a synchronous orbit about the Earth rather than about the Sun?

A. To calculate the gravitational forces of the Earth and the Sun on the Moon simply requires substitution of data into equation (6.6) for the two cases, followed by division of the two to find the ratio

$$F_{\text{E-M}} : F_{\text{S-M}} = \frac{GM_{\text{E}}m_{\text{M}}}{r_{\text{E-M}}^2} : \frac{GM_{\text{S}}m_{\text{M}}}{r_{\text{S-M}}^2} = \frac{M_{\text{E}}r_{\text{S-M}}^2}{M_{\text{S}}r_{\text{E-M}}^2} = \frac{(6.0 \times 10^{24}) \times (1.5 \times 10^{11})^2}{(2.0 \times 10^{30}) \times (3.8 \times 10^{8})^2} = 0.47$$

Thus the Sun exerts about twice as much gravitational force than the Earth on the Moon. The Moon orbits the Earth and both orbit the Sun and so there is no contradiction between the result of this calculation and the motions of the Solar System. The Moon is trapped in synchronous orbit about the Earth despite its weaker gravitational pull because such an orbit is the result of tidal action, given by equation (6.16). Comparing the ratios of tidal forces can be done in a similar way to that for gravitational forces above:

$$F_{t_{\text{E-M}}} : F_{t_{\text{S-M}}} = \frac{-2GM_{\text{E}}}{r_{\text{E-M}}^3} : \frac{-2GM_{\text{S}}}{r_{\text{S-M}}^3} = \frac{M_{\text{E}}r_{\text{S-M}}^3}{M_{\text{S}}r_{\text{E-M}}^3} = \frac{(6.0 \times 10^{24}) \times (1.5 \times 10^{11})^3}{(2.0 \times 10^{30}) \times (3.8 \times 10^{8})^3} = 190$$

The tidal forces on the Moon due to the Earth are nearly 200 times greater than those due to the Sun and so the Moon became synchronously trapped in the Earth's orbit rather than in the Sun's. The similarity between the two calculations in this problem should be noted. The final ratio expressions (before data substitution) are identical except for the power of the distances involved. Gravitational force is inversely proportional to the square of the separation of two bodies whereas tidal force is inversely proportional to the third power of the separation.

Many other moons in the Solar System orbit their planet in synchronous orbits. Very small moons such as Mars' Phobos and Deimos are unlikely to have evolved in the same way as our Moon but are very small and irregularly shaped (see Figure 6.6) thus making it very easy for them to become trapped in position by tidal action. Larger moons such as Jupiter's Galilean moons were probably trapped in the same way as the Moon. Io experiences particularly strong tidal forces as its distance from Jupiter is similar to the

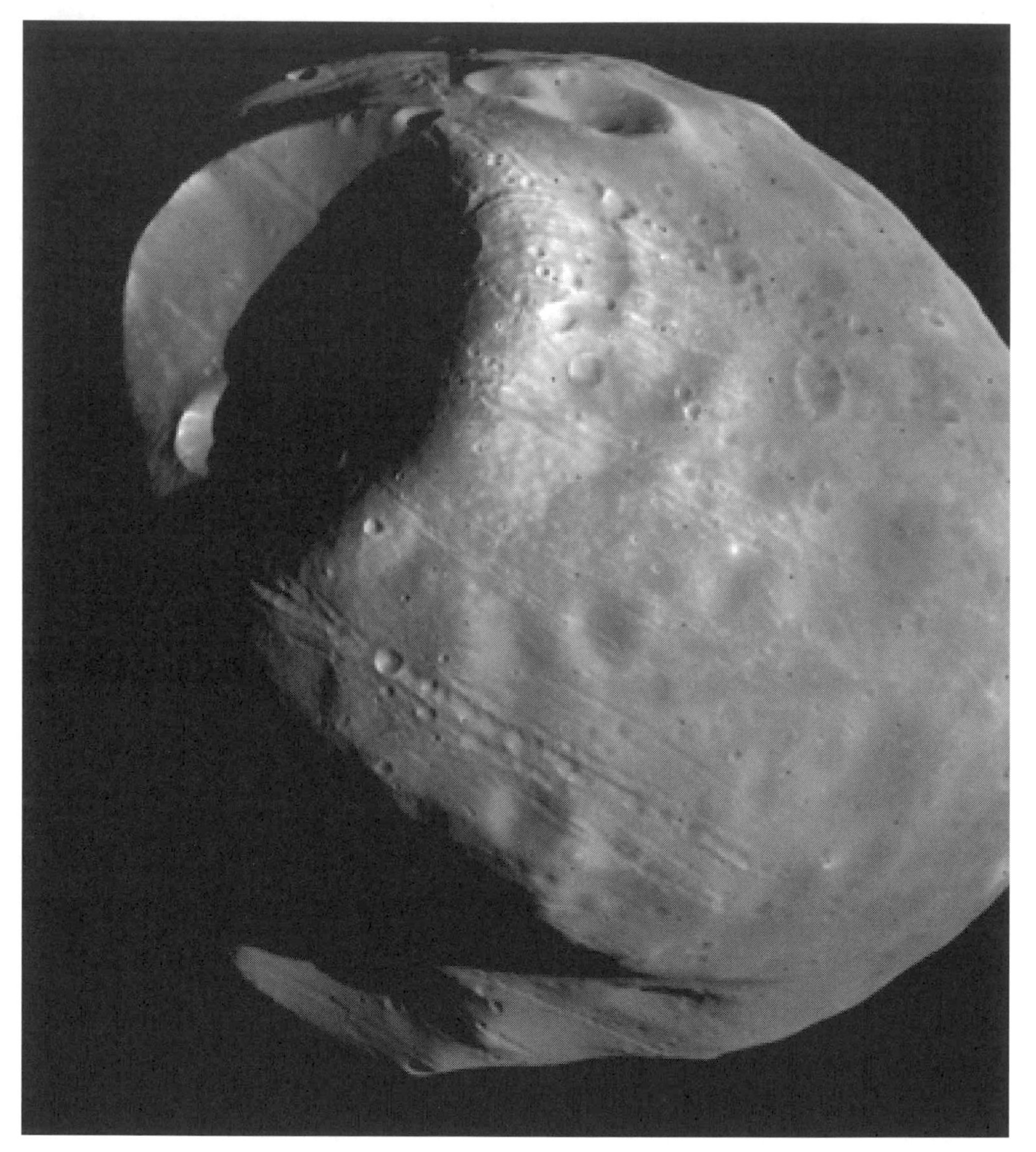

Figure 6.6 Mars' tiny and irregular moon, Phobos, showing the Stickney crater which is 10 km in diameter. Phobos is trapped into a synchronous orbit by the tidal force of Mars (NASA, Jet Propulsion Laboratory and National Space Science Data Center)

Earth–Moon separation but Jupiter's mass is more than 300 times that of the Earth's (see Figure 6.7). Such is the strength of this action that, when added to the gravitational buffeting from the nearby moon Europa and smaller interactions with Ganymede and Callisto, the structure of Io is squeezed and distorted. This causes heating which is released via Io's large volcanoes which spring up from parts of the moon's crust that have been weakened by the stresses of the gravitational pummelling.

The large difference in size between the Jovian planets and their moons means that the moons can quickly be brought into a synchronous orbit whereas the action of the moon

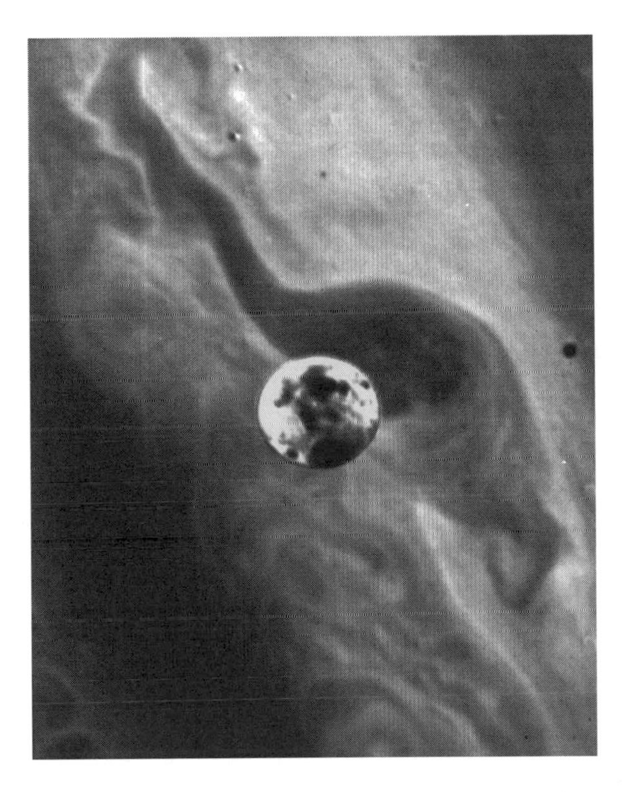

Figure 6.7 Io passing in front of a small part of Jupiter's surface. Io is constantly buffeted in Jupiter's gravitational field, the resulting tidal forces causing volcanism on Io (NASA, Jet Propulsion Laboratory and National Space Science Data Center)

on the planet is negligible. Evidence for this comes from Triton which revolves about Neptune in a synchronous orbit despite the fact that it revolves in the opposite sense to Neptune's rotation on its axis. At the edge of the Solar System lies the antithesis of this scenario, the Pluto–Charon system that consists of the smallest planet which has relatively the largest moon. The eventual fate of the Earth–Moon system, that of both bodies moving in synchronous orbit about the other, has already been reached by Pluto and Charon (see Figure 6.8).

The tidal force due to the Sun on the outlying planets is negligible according to equation (6.16). However, its closest planet, Mercury, has been trapped into a resonant orbit. Mercury is an unusual case as its orbit is relatively eccentric and the ratio between rotational and sidereal periods is not unity but 2:3. This means that Mercury is aligned along the same axis every time it reaches its position of closest approach (perihelion) where tidal forces are greatest (Figure 6.9). The planet has completed one and a half rotations after one orbit so that the axis that is aligned with the Sun is oriented in the opposite direction after each orbit. It is thought that this axis represents the alignment of the mass asymmetry of Mercury as is the case of the axis of the Moon that points permanently towards the Earth. A consequence of Mercury's unusual orbit is that, as it has only one and a half sidereal days per sidereal period, it has only half a Mercurian solar day per sidereal period. The Mercurian solar day is therefore 176 times longer than its

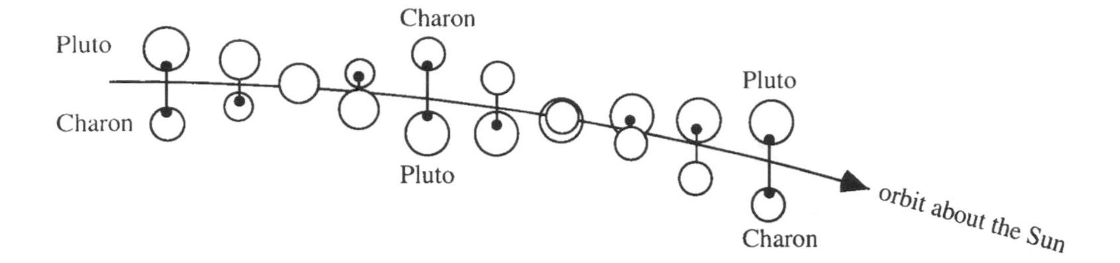

Figure 6.8 The orbits of Pluto and Charon about the Sun. Both bodies orbit about their barycentre, a position contained within neither body and indicated by the arrowed line. The rate of rotation of the Pluto–Charon system is actually more rapid relative to the pair's revolution about the Sun than indicated here. Note that after one quarter of a revolution the bodies will rotate about each other in the same direction as their motion about the Sun

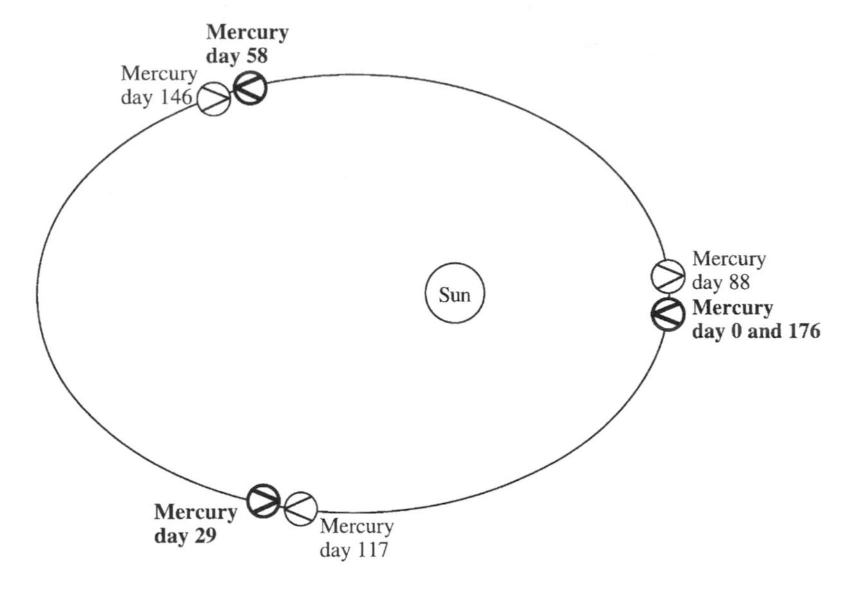

Figure 6.9 The alignment of Mercury's mass distribution axis with the Sun at perihelion. The mass distribution axis is represented by the arrow on Mercury's symbol. Due to the 3:2 ratio of sidereal period to axial spin period the mass distribution axis is always aligned to the Sun at perihelion, alternately in opposite directions

terrestrial equivalent. In the 88 consecutive Earth days during which the nearby Sun is in Mercury's sky the surface is heated dramatically, to 700 K (over 400°C). In the very long night that follows the surface is able to radiate this heat away, unfettered by an atmosphere, to reach a temperature of 100 K (almost −200°C).

Gravity and Planetary Ring Systems

A large number of interesting gravitational effects can be observed by studying Saturn's ring system. As the rings are composed of many small objects orbiting Saturn then they must obey Kepler's third law so that the outer ring constituents travel more slowly than those of the inner rings. The moons of Saturn are bound by the same law. It is noticeable therefore that Mimas orbits Saturn at a distance about $2^{2/3}$ greater than the position of Cassini's division. According to equation (6.1), this implies a period of revolution two times longer. It is thought that the resonance between Mimas and matter in Cassini's division has caused the emptying of this region of the ring system. The strongest gravitational pull by Mimas is imparted on any ringlet component in this region in the same position every other orbit (see Figure 6.10). This causes the rock's orbit to become increasingly distorted until it collides with a section of the ring system closer to or further from Saturn, thus causing its speed to change and consequently its orbital position. This part of the ring system empties of matter. Several other integer ratio resonances with moons have created smaller gaps in the ring system.

Other moons interact with the ring system in different ways. Shepherd satellites are so named because their gravitational influence is to keep the ring constituents held in a certain orbit. A good example of this is the orbit of the moons Pandora and Prometheus. They orbit just outside and just inside the orbit of Saturn's relatively sparse and narrow F ring. Prometheus, being the innermost body in the system, travels at slightly the faster speed and therefore accelerates the constituents of the ring as it passes, causing them to be swept into larger orbits. Pandora, orbiting outside the F ring, moves slightly more slowly so that it decelerates the ring material and pushes it back into a tighter orbit. The opposite actions of the two together keeps the ringlets in place. Such behaviour is seen in other planetary ring systems but mainly in the case of thin, less substantial rings. A nice example, Uranus's moons Cordelia and Ophelia, is shown in Figure 6.11. The outermost of Saturn's dense rings (those easily visible from Earth) is shepherded by two small moons, Pan and Atlas, that are both thought to act in the same way as Pandora, preventing the tidal action of the parent planet from pushing the ring material further out. It is certainly significant that Pan and Atlas are Saturn's innermost moons and that the substantial rings end at their orbit but there is still some way to go before the full complexities of the gravitational dynamics of ring systems is fully understood.

Gravity and Asteroid Belt

The asteroid belt is also swept into bands by gravitational interactions. A number of regions within the extensive belt correspond to orbits with periods that are simple ratios of Jupiter's orbital period. These regions are cleared of asteroids in the same way that the Cassini division in Saturn's ring system is formed, due to resonances with a larger orbiting body. There are a number of so-called Kirkwood gaps in the asteroid belt but also a few resonances that result in an increase in asteroid numbers in particular orbits. The mathematics of such orbits is complicated and attempts at modelling can really only satisfactorily be attempted using chaos theory.

Two clusters of asteroids orbit at the same distance from the Sun as Jupiter (and according to Kepler's third law also have the same orbital period). They are known as the Trojan asteroids and the groups (not rings) orbit 60° in front of and 60° behind Jupiter at positions of orbital stability known as the Lagrangian points. The mathematics that shows why these positions are stable is complicated but does not require the application of chaos theory as is the case for analysis of the Kirkwood gaps.

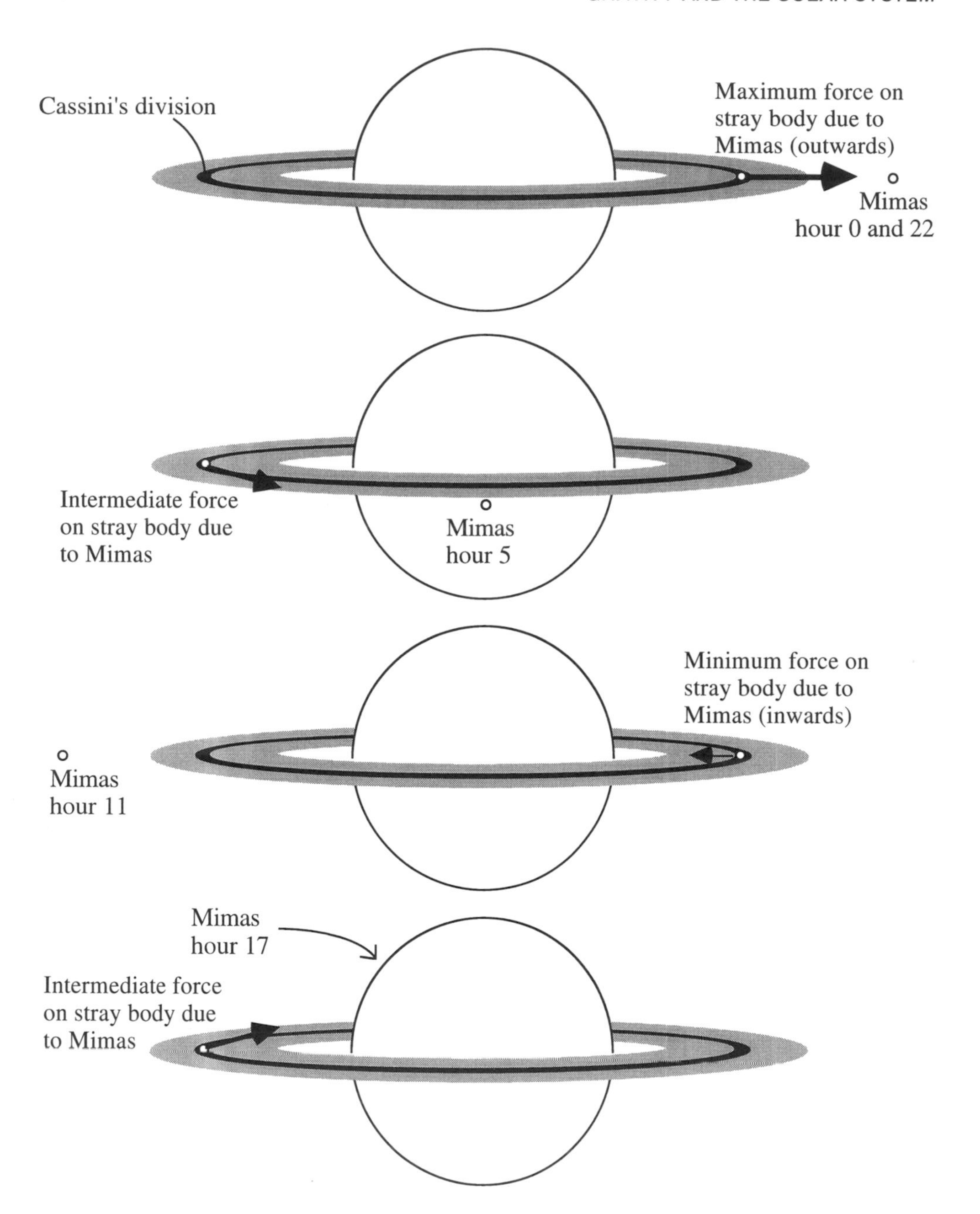

Figure 6.10 The rocking motion of Mimas on a stray body that falls into Cassini's division. The gravitational force that the body experiences due to Mimas in the rightmost position indicated is alternately maximum (outwards) and minimum (inwards). The result of these periodic gravitational kicks is to alter the orbit of the stray body by increasing amounts, in the same way that a swing increases its oscillation if it is pushed periodically at the correct moments. This resonance effect eventually causes the body to collide with material in a surrounding ring and thus to leave Cassini's division

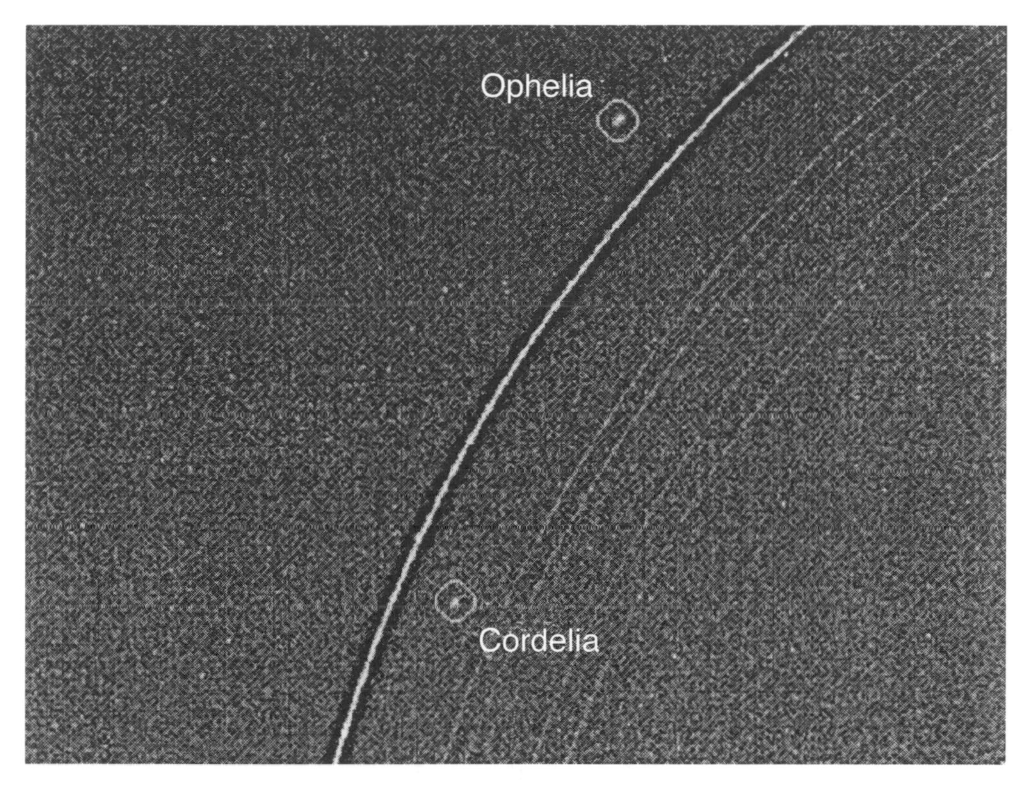

Figure 6.11 The shepherd satellites of Uranus's epsilon ring, known as Cordelia and Ophelia (NASA, Jet Propulsion Laboratory and National Space Science Data Center)

WORKED EXAMPLE 6.11

Q. At what distance from the Sun would an asteroid orbit that has a period one half that of Jupiter?
A. The simplest way to answer this is to consider Kepler's third law (equation (6.1)). If $r^3 \propto T^2$ and T_J is double T_a then

$$r_a^3 = \frac{T_a^2}{T_J^2} r_J^3 = \left(\frac{T_a}{T_J}\right)^2 r_J^3 = r_J^3/4$$

$$\therefore r_a = r_J \sqrt[3]{4} = 0.63 r_J$$

The orbital radius of Jupiter is 780 million km so that the asteroid orbits 490 million km from the Sun. This is well outside the orbit of Mars and close to the centre of the asteroid belt, just outside the orbit of Ceres. The asteroid would be unlikely to remain in this orbit, however, as the resonance with Jupiter is likely to knock the asteroid into an alternative orbit to create a Kirkwood gap.

The Roche Limit

A final concept that should be included in a discussion of gravity in the Solar System that also has relevance to Saturn's rings is that of the Roche limit. This is the distance from a planet at which a body is just able to retain its shape without being shattered by tidal forces. This distance varies with the constitution and geometry of the smaller body. An example of the observation of an object entering within a planetary Roche limit was the collision of Comet Levy–Shoemaker with Jupiter in 1995. As the comet approached Jupiter it was sheared by the massive planet's tidal forces. The impact therefore consisted of many pieces of cometary material colliding with Jupiter right across its face. The Roche limit determines how close the nearest planet to the Sun is able to orbit and the closest approach that moons can make to planets. It is therefore no surprise that the four planets to have ring systems, probably composed of material that has been unable to coalesce into a moon due to being within the Roche limit, are the four planets with the largest gravitational fields. How the material got there in the first place is still open to debate. However, such matters are the subject of the next chapter where the cosmogony of the Solar System will be discussed.

Questions

The universal gravitational constant, G, is 6.7×10^{-11} N m^{-2} kg^{-2}.

Problems

1 Assume the Earth, Venus and Mars move in circular orbits about the Sun with radii of 150 million km, 108 million km and 228 million km respectively. The masses of the Earth, the Sun, Venus and Mars are 6.0×10^{24} kg, 2.0×10^{30} kg, 4.9×10^{24} kg and 6.4×10^{23} kg respectively.
(a) What force does the Sun exert on the Earth?
(b) What force does the Earth exert on the Sun?
(c) What is the maximum force exerted by Venus on the Earth? How does Venus appear as telescopically viewed from Earth at this time?
(d) What is the minimum force exerted by Earth on Venus? How does Venus appear as telescopically viewed from Earth at this time?
(e) What is the minimum force exerted by Mars on the Earth? How does Mars appear as telescopically viewed from Earth at this time?
(f) What is the maximum force exerted by Earth on Mars? How does Mars appear as telescopically viewed from Earth at this time?
(g) Discuss the effects of the gravitational pulls of Venus and Mars on Earth's orbit.

2 The existence of Neptune was predicted through studying wobbles in the orbit of Uranus. What are the maximum and minimum gravitational forces applied to Neptune (mass, 1.02×10^{26} kg; distance from Sun, 4.50×10^{9} km) by Uranus (mass, 8.68×10^{25} kg; distance from Sun, 2.87×10^{9} km)?

3 The Earth's mean distance from the Sun is 150 million km. Use Kepler's third law to calculate the mean distance between the following planets and the Sun from the periods of revolution given in parentheses, using no further information.
(a) Mercury (0.24 years)
(b) Jupiter (12 years)
(c) Pluto (250 years)

4 A geostationary satellite travels in an orbit of 6.6 planetary radii about the Earth's centre. What would the orbital radius of geostationary satellites be about the following planets (in terms of their planetary radii)?
(a) Mars (mass, 6.4×10^{23} kg; radius, 3400 km; orbital period, 24 h 37min)
(b) Jupiter (mass, 1.9×10^{27} kg; radius, 71 000 km; orbital period, 9h 51min)
(c) Venus (mass, 4.9×10^{24} kg; radius, 6100 km; orbital period, 243 days)

5 What is the ratio of the acceleration due to gravity on the surface of Venus to that on Mars given only that Venus has a diameter 1.78 times greater and a mass 7.59 times greater?

6 Where is the centre of mass of the Earth–Moon system given that their masses are 6.0×10^{24} kg and 7.2×10^{22} kg respectively? Their separation is 384 000 km and the Earth's radius is 6400 km. Draw a sketch that shows how the Earth moves in its orbit about the Sun, including this perturbation.

7 Given that Titan orbits Saturn at a mean distance of 1.2 million km and takes 15.9 days to complete one revolution, what is the mass of Saturn?

Teasers

8 What is the acceleration due to gravity experienced by a man of mass 80 kg standing on the surface of Venus (mass, 4.9×10^{24} kg; radius, 6000 km)?

Exercises

9 The sidereal period relating to the Moon's orbit about the Earth is 27.3 days whilc thc synodic period is 29.5 days. How often do neap tides take place on the Earth and what is the phase of the Moon at these times?

10 How are the Kirkwood gaps and Cassini's division related?

7 The Origin of the Solar System

The previous two chapters have given a basic description of what the Solar System looks like and the fundamental force, gravity, that holds it together. This information is sufficient to piece together a history of the Solar System since its formation almost 5 billion years ago. It is based on observations of the Solar System as it is today and on our understanding of the influence of gravity on matter. It would be meaningless to state that gravity played a more important role in the formation of the Solar System than the other fundamental forces of nature because all are essential protagonists, but gravity is involved throughout the succession of processes. In what follows, a middle line between the nominated theories is taken. No single version of these theories is completely established as yet. Little that is written here is controversial as the most important ideas are almost universally accepted and it is the details that remain to be agreed upon. This chapter restricts its discussion to the eight largest planets, the asteroids and comets. The formation of planetary moons and Pluto is discussed later in the book.

Local Space Before the Solar System

The date of formation of the Solar System was about 5 billion years ago, several billions of years after the Big Bang, and so it is reasonable to imagine that space looked then very similar to how it does now. As discussed in Chapter 5, most of space is devoid of condensed matter in the form of stars or planets. This does not mean that it is entirely empty. Space is filled with a varying density and composition of independent atoms, ions[1] and molecules as well as occasional coagulations of different molecules that can be thought of as dust particles. Hydrogen and helium dominate the composition of the interstellar medium. Usually there are only a few hundred thousand atoms, ions or molecules per cubic metre ($\sim 10^{20}$ times less than the Earth's atmosphere) and even fewer dust particles but in some places the density increases considerably. Such regions are known as nebulae. The root of this word is 'mist' and is applied to many diffuse objects in

[1] An ion is an atom or molecule that has an overall electrical charge due to an imbalance between the number of electrons and protons that constitute the particle. In hot regions, atoms and molecules often lose electrons to become positively charged cations. Under certain other circumstances, atoms or molecules may capture extra electrons to become negatively charged anions (see Appendix 2).

the night sky in order to describe appearance rather than to label a specific type of body.

Of interest here are dark nebulae that are more informatively known as molecular clouds. Such regions of space have dimensions measured in tens of light years and their composition is once again dominated by molecular hydrogen though hundreds of different, mainly organic molecules[2] are present. These regions have particle densities hundreds of times greater than is typical for interstellar space though this remains, initially at least, billions of times less than the sparsest vacuum that the pump of a household vacuum cleaner can create. Nevertheless, the enormous size of such regions means that they are kept cold by the insulating effect of their outer layers. This works in a similar way to atmospheric absorption on the Earth, discussed in Chapter 2, whereby molecules absorb the radiation of nearby stars thereby preventing the energy from reaching further in. As there are so many different molecules in the cloud most regions of the electromagnetic spectrum are absorbed and the temperature of the bulk of the molecular cloud is therefore only about 10 K. At such low temperatures there is little activity.

The Collapse of the Solar Nebula

It was stated in the previous chapter that gravity acts from the centre of mass of a body. Figure 7.1 shows why this is the case for a molecular cloud. Though the force on a single molecule due to a second single molecule elsewhere in the cloud is minuscule, such is the number of molecules in this enormous volume that the sum of all forces can be significant. A particle towards the edge of the cloud experiences forces that act mainly towards the centre. A molecule at the centre of the cloud is pulled in all directions equally and thus experiences no overall force. The resulting motion of the cloud is therefore to slowly contract towards the centre. In fact molecular clouds are not of uniform density and several gravitational contraction centres develop. One such sub-cloud, known as the Solar Nebula, went on to form the Solar System. As the cloud continued to collapse two effects began to manifest themselves.

In Chapter 6 it was shown that energy is required to remove an object from the surface of a planet to a place effectively outside the gravitational field. Applying this argument in reverse leads to the conclusion that the motion of an object drawn towards the centre of a gravitational field must release energy. A simple example is of an apple falling from a tree. As the apple accelerates towards the Earth it gains kinetic energy. A molecule being pulled towards the centre of an increasingly dense ball of gas also speeds up. The increased kinetic energy of all molecules is shared throughout the ensemble as collisions take place thus causing energy to be passed from molecule to molecule. By definition, a gas that has faster-moving particles is a hotter gas. As the cloud collapses it heats up. The speed of collapse gradually escalates as the gravitational attractive forces increase with decreasing particle separation.

The second significant effect that occurs during the collapse of the cloud is that its rotational speed increases. The usual analogy for this effect is of a spinning ice skater. If a certain rate of rotation is induced while the skater has their arms outstretched then it can

[2] An organic molecule is a collection of atoms, including one or more carbon atoms, that are electrically bound together to form a single particle. Organic molecules that consist of carbon and hydrogen atoms, such as methane, ethane, ethene and ethyne, are known as hydrocarbons. Organic molecules that contain carbon, oxygen and hydrogen, such as carbohydrates, are the basis of all known lifeforms.

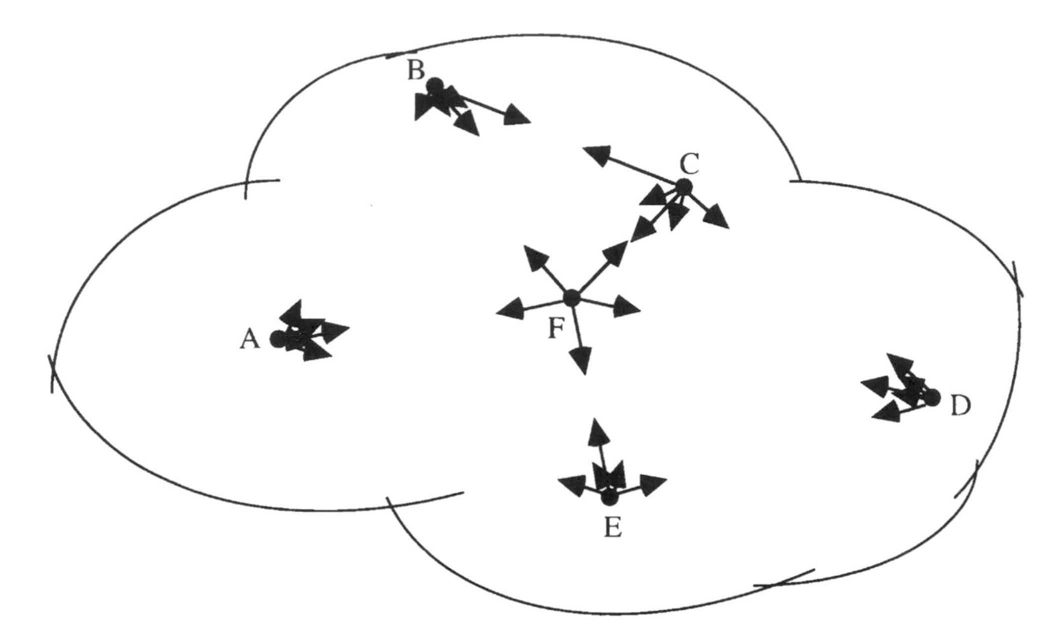

Figure 7.1 The forces on particles in a molecular cloud. Six particles are chosen to represent the enormous number of particles in a large molecular cloud. The direction and magnitude of the gravitational forces due to each of the other molecules is indicated by the arrows. The resultant force on the outer molecules is inwards whereas the resultant force on the central molecule (F) is close to zero, that is it experiences approximately equal forces in all directions

be increased by moving the arms towards the body. This is an effect of the principle of conservation of angular momentum. For a single molecule slowly drifting about the centre of a molecular cloud its angular momentum is given by the product of its transverse speed and its separation from the centre. As it moves towards the centre this quantity must be conserved. As the distance decreases then the rotational speed must increase. Again angular momentum becomes shared throughout the ensemble through molecular collisions. The result is that as the cloud collapses its speed of rotation increases. This concept is illustrated in Figure 7.2.

Once the Solar Nebula had collapsed from an entity on the parsec scale (10^{13} km) to a ball of gas on the Solar System scale (10^{8} km) its temperature had risen by a factor of more than a hundred and its rotation period had reduced from several million years to a matter of days or less. At about this stage the ball of gas that had slowly formed itself into a sphere began to become oblate about the spin axis. The centripetal force required by gas molecules to maintain their circular motion was harder to provide for faster-moving molecules at the equator than for those near the poles and, with further contraction, the ball became lenticular (Figure 7.3). Matter then began to spew out of the equatorial region of the Solar Nebula to form a disk (Figure 7.4).

Separation into Star and Disk

In the case of the solar nebula the disk of matter began to form when the radius of the central ball had reached about 20 million km. This is about 30 times the size of the

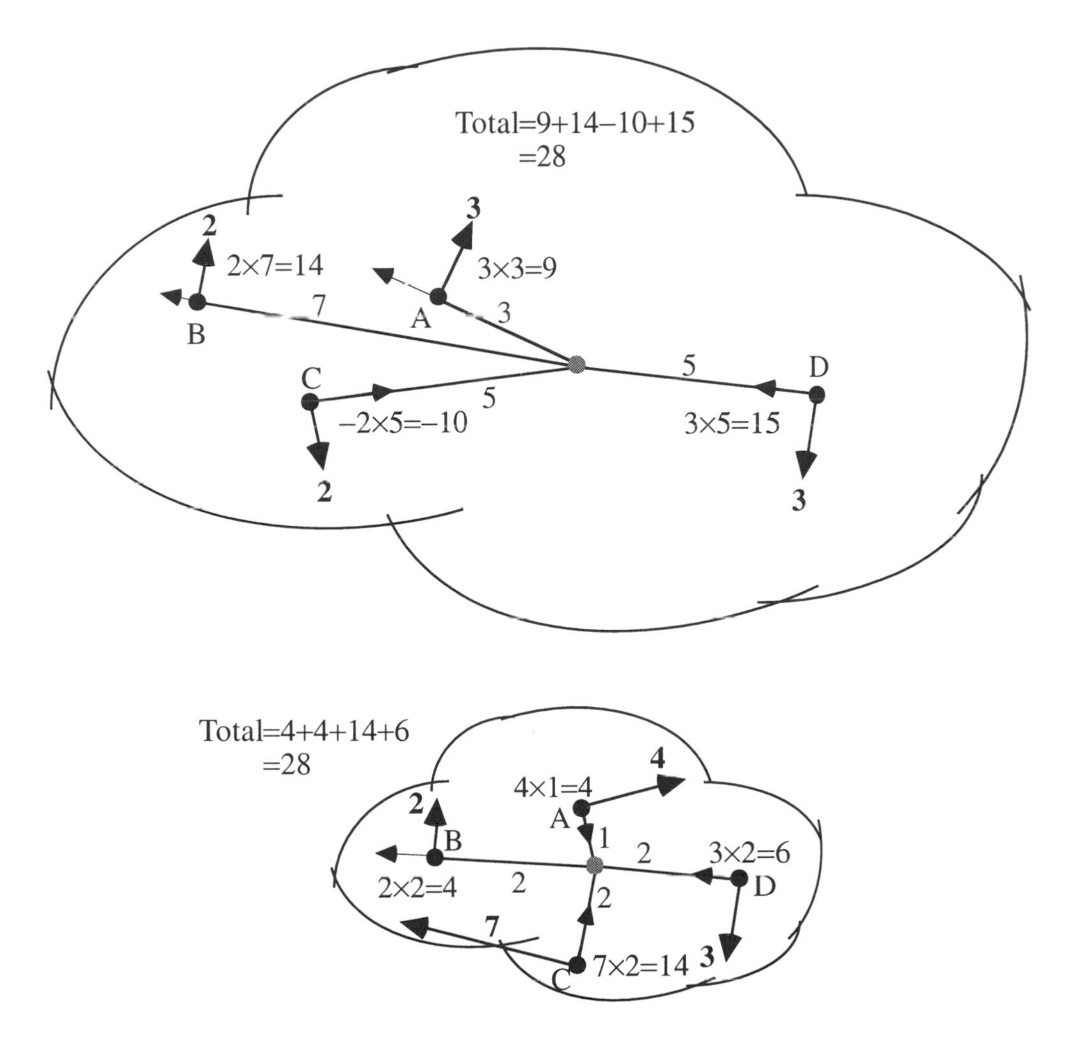

Figure 7.2 Conservation of angular momentum in a collapsing molecular cloud. Four particles are chosen to represent the enormous number of particles in a large molecular cloud. The motion of each particle is represented by two arrows, one indicating lateral motion relative to the cloud's centre of mass and one indicating radial motion. The lateral velocity is the component that is important for conservation of angular momentum and its value is indicated in bold. The distance of the particle from the centre is also indicated. The angular momentum of each particle is then calculated as being the product of the two indicated parameters. The convention adopted here is that motion is considered positive in a clockwise direction and negative in an anticlockwise direction. As the cloud collapses the total angular momentum of all particles is conserved though it becomes redistributed among the particles during collisions. Note the increased lateral speeds in the collapsed cloud

present-day Sun and represents a size of about one third of Mercury's orbit. At this stage the story diverges into two parts. The central ball of matter, now heated to a temperature of about 3500 K continued to shrink, eventually creating conditions in the interior sufficient to allow nuclear reactions to begin. The new outward thermodynamic forces thus generated could then resist further gravitational contraction until an equilibrium

Figure 7.3 A protoplanetary disk surrounding a forming star in the Orion nebula. The Solar Nebula probably once resembled this structure (reproduced by permission of Space Telescope Science Institute)

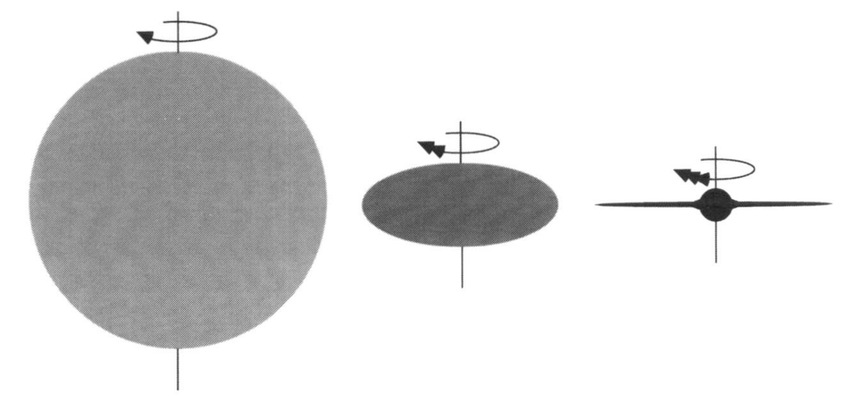

Figure 7.4 The gravitational collapse of the Solar Nebula. As the cloud collapses it first becomes lenticular and then forms into a nucleus and disk. The rotational speed increases throughout the collapse

between inward and outward forces was reached. In this way, the newly formed star slowly stabilised to become the Sun. This part of the story will be dealt with in detail later in the book. Here the body at the centre of the disk will be merely treated as a source of heat and, initially, of matter while an explanation of the establishment of the planets from the disk is given.

As the matter began to stream out of the equatorial region of the nucleus of the Solar Nebula it took with it some of the angular momentum so that, because this quantity must be conserved, the nucleus slowed in its rotation. However, the nucleus had not finished contracting and so there was an opposing effect causing the nucleus to continue rotating more quickly and, at the same time, to distribute more material into the disk. By the time the nucleus had become a star a large part of its mass (probably enough to form a second star) should have been ejected into the disk. There is no evidence of this in the modern Solar System. The Sun contains 99.9% of the mass of the Solar System and, though some mass has escaped the region altogether during the last 5 billion years, it is clear that the vast majority of the mass was retained by the nucleus.

What slowed down the rotation of the nucleus of the solar nebula sufficiently to stop its leakage of mass to its surrounding disk? The best clue comes from analysing the distribution of angular momentum in the Solar System as it is now. The spin of the Sun on its axis contributes just 0.5% of the total. Almost all of the angular momentum of the Solar Nebula, in the form of its rotation, has been transferred to a small fraction of its mass, mainly in the form of the planetary orbits about the Sun. Some process that redistributes the angular momentum is required to explain this.

The rotating Solar Nebula did not consist of condensed matter so that continuous variations in the speeds of different parts of the gas were possible. This gas was very hot. The particles in hot gases move very quickly and collide with each other. Such collisions can cause atoms and molecules to lose electrons thus creating what is known as a plasma, consisting mostly of electrically charged particles (ions and free electrons). The motion of an electrically charged particle creates a magnetic field. The nucleus of the Solar Nebula thus had associated with it a magnetic field. The matter close to the nucleus in the disk was also hot and partially ionised. It thus interacted with the magnetic field, moving with it. The temperature of the disk decreased with distance from the centre so that more and more electrically neutral molecules would have been found further out. As the charged particles within the disk were forced around by the magnetic field then the uncharged particles, unaffected by such an electromagnetic force, would have collided with the charged particles. The charged particles thus dragged the uncharged particles with them and in so doing increase the angular momentum of the disk at the expense of the nucleus.

If a particle in the disk of the Solar Nebula increased its angular momentum it would have moved away from its centre of rotation, the nucleus, in much the same way that the Earth and Moon are moving apart as the Earth's rotation slows in order to conserve the total angular momentum. The important difference is that the magnetic force that operated in the Solar Nebula was not tidal so that there was no requirement that an equilibrium would eventually be reached. The nucleus continued to contract thus increasing its rotational speed (albeit at a reduced rate due to angular momentum transfer) while the matter spiralling into the disk slowed as it orbited further and further from the centre (according to Kepler's laws). In this way the whole of the Sun's angular momentum could be transferred to the disk if the disk continued to contain a significant number of charged particles. Rough calculations show that the transfer of angular momentum required to match the values in the Solar System observed today would only require a few thousand years to be complete. At this point the nucleus would be spinning quite slowly and therefore ejecting charged particles to the disk at a much-reduced rate.

Condensation and Accretion

Once the mass and angular momentum were distributed more or less in their present proportions by the mechanisms described above only one more major step of evolution was required to produce the Solar System as it is today as. The orbiting nebula disk coagulated into a series of bodies composed of condensed matter. There were three steps involved in this, two or more of which may have acted during the same eras.

The first process to act was the one that operates on the smallest scale, that is, to join atoms or molecules into collections of atoms or molecules. This process is known as condensation and acts in the same way that water condenses in a cloud to form raindrops.

The temperature is of critical importance to the condensation of different materials. Atmospheric water condenses into clouds and clouds into precipitation depending most strongly on the local temperature. The same was true in the disk of the Solar Nebula except that there were many chemical species present, all of which condense under different conditions. Of course, the temperature of the Solar Nebula decreased with distance from the principal source of heat at the centre. As the core had not finished contracting at this time its energy output would have been greater than the stabilised Sun today and so temperatures would have been higher. Close to the nucleus nothing would have condensed because the effective temperature would have been above 2000 K.

Close to Mercury's present orbit the temperature would have dropped to around 2000 K. This is still above the condensation temperature of most materials if the heat source is isotropic. In the case of all the heat being radiated from a single direction the effective temperature for a solid body drops by a factor of up to 30%. This is because only one side of the body is heated while the opposite side is able to radiate the energy away. Good conductors condense more easily than poor conductors as they are better able to pass thermal energy through their volume (from the sunward to the shaded side). At the temperatures present in the inner solar nebula refractory materials such as metals and minerals were able to condense. As metals are much better thermal conductors than rocks they were the most likely materials to condense close to the nucleus.

It should not be forgotten that during the early condensation stages material from the core was still moving outwards. Most of the material was simply gaseous hydrogen and helium but its volume was substantial and relatively dense close to the nucleus. These volatile materials could therefore easily sweep small condensates out with them until the gas density began to drop further out in the disk. Condensed matter would have to reach sizes of the order of metres to become immovable by the gas flow, a statistically unlikely but not impossible occurrence. Metals would condense from inside Mercury's present orbit outwards while rocks would start to form a little further from the centre but both could be carried steadily outward by the motion of the disk. By the present orbit of Mars the temperature would have dropped to 1000 K so that this region was the principal domain for the condensation of metals and rocks. Much further out the temperature dropped sufficiently for more volatile materials such as water, methane and ammonia to condense. Throughout the disk of the solar nebula small bodies of varying composition began to form. Eventually these bodies were bound to start colliding with each other.

The second process of planetary formation is known as accretion. This is simply a fancy word for sticking together after a collision. In the early stages of accretion such collisions can be relatively low energy events. Newly condensed matter in the early stages of the birth of the Solar System would have been quite soft and so collisions would have simply consisted of two small bodies thudding into each other with a small relative speed, coalescing to form a larger body, perhaps with an irregular shape. Gravity caused the condensing grains to be attracted to the plane of the disk and this restriction to an almost two-dimensional region caused the collision rate to be enhanced.

As the bodies grew in mass they began to exert a more considerable gravitational field. Once this occurred it became more difficult for low-energy collisions to take place. The bodies, now known as planetesimals, accelerated towards each other and collided at high speeds releasing their kinetic energy in the form of heat. The planetesimals remained hot and soft thus increasing the probability of accretion.

WORKED EXAMPLE 7.1

Q. Consider two planetesimals orbiting the Sun in the early Solar System at a distance of 0.99 and 1.01 AU with masses of 1% and 4% of the Earth's mass respectively. The Earth's mass and radius are now 6.0×10^{24} kg and 6400 km and its orbital radius is 150 million km.
(a) Calculate the maximum gravitational force on the inner body due to the outer body relative to the gravitational force on the inner body due to the Sun.
(b) Estimate the energy released if the two bodies were to collide.
A. The maximum gravitational force will occur when the two planetesimals are at their closest proximity, that is, 0.02 AU. To calculate the gravitational force between the planetesimals and of that of the Sun on the small planetesimal simply requires substitution of data into equation (6.6) for the two cases, followed by division of the two to find the ratio

$$F_{P_1-P_2} : F_{S-P_1} = \frac{GM_{P_1}m_{P_2}}{r^2_{P_1-P_2}} : \frac{GM_S m_{P_1}}{r^2_{S-P_1}} = \frac{M_{P_2} r^2_{S-P_1}}{M_S r^2_{P_1-P_1}}$$

$$= \frac{[0.04 \times (6.0 \times 10^{24}) \times (0.99 \times 1.5 \times 10^{11})^2]}{(2.0 \times 10^{30}) \times [0.02(1.5 \times 10^{11})^2]} = 3 \times 10^{-4}$$

The influence on the orbit of the inner planetesimal of the outer planetesimal is very small relative to the Sun's gravitational field but cannot be ignored. A precise calculation of the energy released when the two bodies collide (as they eventually are likely to) is very complicated and so, as a zero-order estimate, the fact that they are orbiting the Sun will be ignored. That is, the calculation will be performed as if the bodies are situated in free space and are gradually attracted to each other until they collide. To do this a similar calculation to that used to derive equation (6.12) is required. The smaller body can be considered to be drawn into the larger body's gravitational field thus releasing energy (equal to the work done that would be required to perform the reverse action). The energy released during a small movement is equal to the product of the force between the bodies and the distance moved through. As the force varies as a function of separation the small movements must be summed by integration from their initial separation to the final separation. Note that the final separation is not zero but the separation of the two centres of mass. If the bodies have about the same density as the Earth then their radii will scale with the cube root of their masses. The radii are therefore ($\sqrt[3]{0.01} =$) 0.22 and ($\sqrt[3]{0.04} =$) 0.34 of the Earth's radii or 1400 km and 2200 km. Actually, the Earth's present density is likely to be somewhat larger than that of the planetesimals that formed it due to increased gravitational contraction in the Earth's much larger mass so the planetesimal radii will be taken to be 2000 km and 3000 km. The calculation can now proceed as for equation (6.12):

$$W = \int_{r_{S-P_2}-r_{S-P_1}}^{r_{P_1}+r_{P_2}} F\mathrm{d}r = \int_{r_{S-P_2}-r_{S-P_1}}^{r_{P_1}+r_{P_2}} \frac{Gm_1m_2}{r^2}\mathrm{d}r$$

$$= Gm_1m_2\left[\frac{-1}{r}\right]_{r_{S-P_2}-r_{S-P_1}}^{r_{P_1}+r_{P_2}} = Gm_1m_2\left\{\left[\frac{-1}{r_{P_1}+r_{P_2}}\right] - \left[\frac{-1}{r_{S-P_2}-r_{S-P_1}}\right]\right\}$$

$$= (6.7 \times 10^{-11}) \times [0.01 \times (6.0 \times 10^{24})][0.04 \times (6.0 \times 10^{24})]$$

$$\times \left\{\left[\frac{-1}{2 \times 10^6 + 3 \times 10^6}\right] - \left[\frac{-1}{(1.01 - 0.99) \times 1.5 \times 10^{11}}\right]\right\}$$

$= -2 \times 10^{29}$ J (minus sign indicates energy released rather than work done)

An enormous amount of energy is released. Though this calculation is very sketchy it does give some indication of how much energy is released in such collisions. Assuming this energy is used to heat the newly joined body and a round figure estimate of the specific heat capacity (energy required to raise 1 kg of material by 1 K) of the planetesimal is 1 kJ kg^{-1} K^{-1} then the temperature increase of the new body (assuming it fuses into a single body) is given by the energy imparted divided by the product of the body's mass and its specific heat capacity. This value is ($2 \times 10^{29}/(0.05 \times 6 \times 10^{24} \times 1000) =$) 700 K. While this problem is an entirely invented one and the method of calculation is highly simplified, it does demonstrate the intensity of such collisions. The effect of raising the temperature of such a body by many hundreds or a few thousands of degrees will be profound. If the body's temperature had been initially below a few hundred kelvin then vaporisation of all surface volatiles would take place and, though mixtures of rock and metal will not be fully melted below temperatures of around 2000 K, the body will become much softer and will glow red-hot. If the temperature of the planetesimals had initially been 1000 K or more then the whole of the new body would become molten liquid, still held together by gravity and thus quickly able to reform into a spherical shape.

The forming bodies became larger and larger but fewer and fewer until single protoplanets came to dominate their part of what was the disk of the solar nebula. Rival bodies would have been successively sucked into the protoplanets by their ever-increasing gravitational field. The rate at which collisions took place would necessarily slow considerably as the number of orbiting bodies decreased. To reduce the planetesimal population to a few thousand would have taken only a few tens of thousands of years but the whole process of planet formation may have taken more than 100 million years.

The Solar System evolved to be dominated by nine planets spaced in such a way that planetary separation increases steadily with distance from the Sun. Each body has been continuously bombarded by other bodies from all directions resulting in final orbits that reflect an averaging to zero of the perturbing influence of impacts. Thus the orbits are close to being circular, taking with them the vast majority of the former Solar Nebula's angular momentum, all moving in the same direction as the Sun's rotation. Most spin on their axes in the same direction. Final major impacts may have knocked the smallest of the planets, Pluto and Mercury, into a more elliptical orbit while Venus, Pluto and Uranus may have had their spin axes toppled by a similar major collision. Some moons, such as Mimas (Figure 7.5) also show signs of discrete, massive impacts.

Gravitational Differentiation

It has already been shown that a large amount of energy in the form of heat results from the impact of two large bodies. In the early Solar System such impacts were common and so planets remained molten for a long period, well after the newly formed Sun's luminosity reduced and allowed the remaining gas in the Solar Nebula to cool. By now the terrestrial planets were composed of a mixture of metals and rocks due to the accretion process. Metals are denser than rocks and so, under the influence of the planetary gravitational field, metals tended to sink towards the centre of the planet, pushing rocks further out and any more volatile materials to the surface. This process is known as gravitational differentiation and relies on the materials present in the planets being somewhat fluid (not necessarily liquid but at least able to ooze slowly). As collisions became less frequent and their heating influence less important the outer parts of the terrestrial planets could solidify permanently while the inner, metallic cores remained hot and molten due to the extreme pressure at the centre of the bodies. A significant

Figure 7.5 A classical crater. The impact that created the Herschel crater on Mimas almost ripped this moon of Saturn apart (NASA, Jet Propulsion Laboratory and National Space Science Data Center)

contribution was, and still is, made to the heating of planetary interiors by the decay of radioactive nuclei.

The Formation of the Giant Planets

The formation of the terrestrial planets up to the present day has been explained and it might seem to be a simple step to extend the ideas to the outer planets. Once the matter of the disk had been swept sufficiently far from the nucleus the temperature would drop so that more volatile materials such as water, methane and ammonia could condense. The problem that is difficult to explain, however, is that the next two planets from the Sun after the terrestrial planets are actually composed of two of the most volatile elements of all, hydrogen and helium, and it is necessary to travel out as far as Uranus and Neptune to find the first 'icy' planets.

In the early stages of the Solar System the temperature of the disk was such that the region of the Jovian planets was optimum for the condensation of icy materials and this is where the next major stage of planetary formation began. In the original solar nebula there were considerably more compounds such as water, methane and ammonia than of the metals and silicates, for instance. It is not surprising that the Jovian planets are much larger than the terrestrial planets, each of the four largest planets having a mass at least ten times greater than the combined masses of the four inner planets. However, the four most massive planets all contain significant amounts of hydrogen and helium, elements that could have never condensed into droplets in the early Solar System.

The third of the planetary forming processes alluded to much earlier in this chapter had a significant impact on the growth of the giant planets. The protoplanets in the outer part of the Solar Nebula could accumulate large masses due to the significant, additional contribution of icy materials to the rocks and metals to which inner protoplanets were restricted. In moving through the thick cloud of the disk large protoplanets were then able to act as a strong gravitational centre. A second process of gravitational contraction took place but this time the protoplanet acted like a snowball as it orbited the newly formed Sun, mopping up the hydrogen and helium that dominated the Solar Nebula. Once this process became established it would have proceeded quite quickly as the protoplanet became larger and larger thus becoming an increasingly stronger gravitational trap. Jupiter and Saturn reached sufficient sizes to enable runaway nebula contraction to occur whereas Uranus and Neptune, while collecting some hydrogen and helium, never became large enough to allow exponential growth. The cooler temperatures in the outer disk also helped with the contraction but, in the same way that the gravitational collapse of the nucleus of the Solar Nebula caused heating, then so did the contraction of the disk gases into the emerging giant planets. The planets thus became quite hot. This possibly explains why, for instance, in the Galilean moons there is a progressively larger ratio of icy to rocky materials, resulting in decreasing densities in moving away from Jupiter. The heat given off by Jupiter evaporated the icy materials from close-by moons.

The Survival of Small Bodies

As large bodies formed, the process of accretion would not always have been as straightforward as considered so far. Near-misses would inevitably be common. Such events would cause the bodies to be considerably perturbed in their orbits, a process known as gravitational scattering. From the region of the giant planets some planetesimals were sent to one of two cometary stores, the Kuiper disk and the Oort Cloud (see Chapter 5). Some of these comets can be periodically observed as they travel close to Earth along their highly eccentric orbits.

The asteroids are also thought to have been influenced by gravitational scattering. They fall in the region between major aggregation regions and this zone might be expected to be the home of a much larger planet. Ceres is the largest of the asteroids, containing about one third of the total mass of all asteroids but still only has a mass of 1% of the Moon. The small quantity of matter in this region is thought to be due to the proximity of Jupiter that has gravitationally removed most of it while also preventing Ceres from accreting all of the local mass that remains during nearly 5 billion years.

Volatiles on Earth

The final puzzle that will be tackled in this chapter is the appearance of volatiles on the surfaces of the terrestrial planets, particularly water on Earth. There was little possibility for water to condense in the region of the Earth during the early stages of the Solar System. By the time the Earth was cool enough, after the accretional impacts had reduced to a lower level so that the crust could solidify, it is thought that one or two last collisions could have taken place. This time the Earth was hit by a rogue body that had accreted much further from the Sun but which had been swung into an eccentric orbit by a collision near-miss. A single body would have been able to bring with it most of the volatiles on the Earth today. Such a collision could have also taken place earlier in the accretional history of the Earth. In this scenario, the water was trapped within the Earth's body and slowly made its way to the surface (during volcanic events, for instance) over the following billions of years.

This uncertain theory brings to a conclusion the history of the origin of the Solar System. The story has not yet ended and the planets continue to change. In the next two chapters more examples will be given of such evolution as the bodies of the Solar System are examined in greater detail, armed now with a knowledge of the workings of gravity and the history of their formation.

Questions

Problems

1 Consider two planetesimals orbiting the Sun in the early Solar System at a distance of 5.19 and 5.21 AU (close to Jupiter's present position) with masses of 1% (6.0×10^{22} kg) and 4% (2.4×10^{23} kg) of the Earth's mass respectively. Take the density of the planetesimals to be about that of Jupiter's icy moons at around 2000 kg m^{-3}.
(a) Estimate the energy released if the two bodies were to collide.
(b) By how much would the temperature of the bodies increase as a result of the collision if they were each composed of similar materials with an average specific heat capacity of 1 kJ kg^{-1} K^{-1}?

2 Consider two planetesimals orbiting the Sun in the early Solar System at a distance of 0.99 and 1.01 AU with masses of 5% (3.0×10^{23} kg) and 25% (1.5×10^{24} kg) of the Earth's mass respectively. Take the density of the planetesimals to be 4000 kg m^{-3}, a little less than the Earth's current density .
(a) Calculate the maximum gravitational force on the inner body due to the outer body as a proportion of the gravitational force on the inner body due to the Sun.
(b) Estimate the energy released if the two bodies were to collide.

Teasers

3 If two thirds of the Earth's surface is covered by oceans with an average depth of 4 km and all the water on Earth arrived during one collision with a stray planetesimal containing 30% water-ice what would the planetesimal's radial size have been relative to the Earth's current radius of 6400 km?

4 Kepler's third law was derived in Chapter 6 for circular motion. It also holds for elliptical orbits (see Appendix 3) where the orbital radius is replaced by the average distance of the orbiting body from the dominant, central mass. Approximate the orbital period of a comet in a highly elliptical orbit that takes it to 1 light year from the Sun (mass, 2×10^{30} kg).

Exercises

5 Explain why Mercury has its present basic constitution.

6 Explain why Jupiter has its present basic constitution.

7 Explain why Ceres has its present basic constitution.

8 Explain why high-energy collisions between planetesimals during planetary formation were important in determining the final internal structures of the terrestrial planets.

9 Compare the relative orbital radii of Venus, Earth, Mars and the asteroids about the Sun with the relative orbital radii of the Galilean satellites about Jupiter. What might be surmised about the formation of these two sets of bodies as a result of these observations?

10 Explain the three terms condensation, accretion and gravitational contraction, in the context of the formation of the planets of the Solar System.

8 A Closer Look at the Terrestrial Planets

In this chapter the details of the inner four planets and the only significantly sized moon in the region (our Moon) will be examined. Each of these bodies has already been described from an observational point of view in Chapter 5. Here the object is to discuss the underlying processes and phenomena that affect the densest bodies in the Solar System.

Internal Differentiation

That Earth's three closest planetary neighbours are the densest bodies in the Solar System is no surprise given that they are composed of the least volatile material to have condensed from the Solar Nebula; metals and rocks. A glance at Table 5.2 (or Appendix 5) shows that the Earth has the highest density of all (at 5520 kg m^{-3}) but that Mercury and Venus are within 5% of the Earth's value. Mars, at 71% of the Earth's density, comes next. The Moon is a little less dense than Mars, at 3330 kg m^{-3}. Density data is very useful in determining what the interior of a planet is composed of and to deduce the state of different regions of gravitational differentiation.

Planets were heated during their formation by the accretion process and have continued to be warmed by the energy release that accompanies the radioactive decay of nuclei. Consequently, the terrestrial planets have hot interiors and their temperature increases with pressure towards the planetary core.

The Earth is much the easiest of the planets to study because telescopes or rockets are not required to make observations. The science of seismology has provided a wealth of information on the Earth. By definition, seismology is the study of earthquakes but by association the way in which mechanical waves are propagated or reflected by different parts of a planet's interior is also included in the field. A combination of experimentation and observation has led to the conclusion that the Earth is differentiated into three main zones. At the centre, with a radius of about 3500 km, is a region into which the densest materials, the metals, have settled. A mixture of iron and nickel is the main contributor. The core can be divided into three parts. The central third has a temperature of around 6000 K but, due to the extreme pressure in this region, is probably solid. The intermediate part of the core retains a similar temperature but the pressure begins to drop so that the iron and nickel are molten. The outer third is also molten but the temperature reduces with a much stronger gradient to about 4000 K.

Outside the metallic core, the Earth is composed mostly of rocks, much of which is a mineral known as olivine, an iron and magnesium silicate. The next layer of gravitational differentiation of the Earth outside the core is known as the mantle and stretches almost all the way to the Earth's surface. The temperature varies gradually to a low of a little over 1000 K closest to the surface. Under these conditions the rock takes on an intermediate state between solid and liquid that might best be described as plastic. The rock is able to flow but very slowly like very thick honey.

The outer few tens of kilometres of the Earth is known as the crust. The temperature quickly drops to the typical surface temperature of around 300 K and the rock that comprises it is solid. A whole series of books could be (and have been) written about the current knowledge of the Earth's crust but as this is an astronomy book only the most salient points will be included here.

There are two different types of crust. A thin ($\sim$10 km) layer of basalt provides the structure on which the oceans flow whereas the continents are mainly composed of granites with a much greater depth ($\sim$70 km). Both basalt and granite are igneous rocks which implies they are solidified from molten lava. Other rocks, known as sedimentary rocks, form at lower temperatures through the drying and compression of sands, muds and clays. While sedimentary rocks are common on the surface of the planet they do not constitute the bulk of the crust.

The implications that there are two layers of crust are many, particularly as the continental crust does not cover the entire planet. As water is a fluid, it flows with gravity to the lowest possible point and so spends most of its time on the lower oceanic plates. The term 'plate' implies that the crust is not a single structure like an eggshell but rather is more like an armadillo's covering, being composed of separate sections. In between the gaps in the plates heat that builds up in the Earth's centre can be released. Such volcanic action may be sporadic, occasional or regular while being violent or benign and may take place on land or under the ocean. As the Earth's internal structure was formed by gravitational differentiation, the crust must be the least dense region of the Earth and it can be thought of as 'floating' on the mantle. The plates are constantly, though very slowly, moving relative to one another so that a map of the Earth needs to be updated every few million years. As the plates rub up and down against each other they can slowly build up mountain ranges or suddenly slip to cause an earthquake. These phenomena, known as plate tectonics, are thought to be rare throughout the rest of the Solar System though the other planets are not well enough understood to rule this out completely.

A comparison of the interiors of the terrestrial planets is shown in Figure 8.1. Much of the information required to model the interiors derives directly from the size and mass of the planets and this is easily attainable. For instance, the internal structure of Venus is likely to be similar to the Earth as the two bodies are about the same size and have almost the same densities. It is reasonable to assume therefore that internal pressures in both compress the inner regions of the planets similarly and the gravitational differentiation is also likely to be of the same form. This fits with models of Solar System formation as Venus and Earth represent bodies accreted from a similar region of the disk of the Solar Nebula.

As Mercury is a small planet its internal pressures are less than in much larger planets like the Earth. However, Mercury's mass shows that it has almost the same density as the Earth. The implication of this is that it must be largely composed of the densest of planetary constituents, the metals. As it is the closest planet to the Sun this fits nicely with

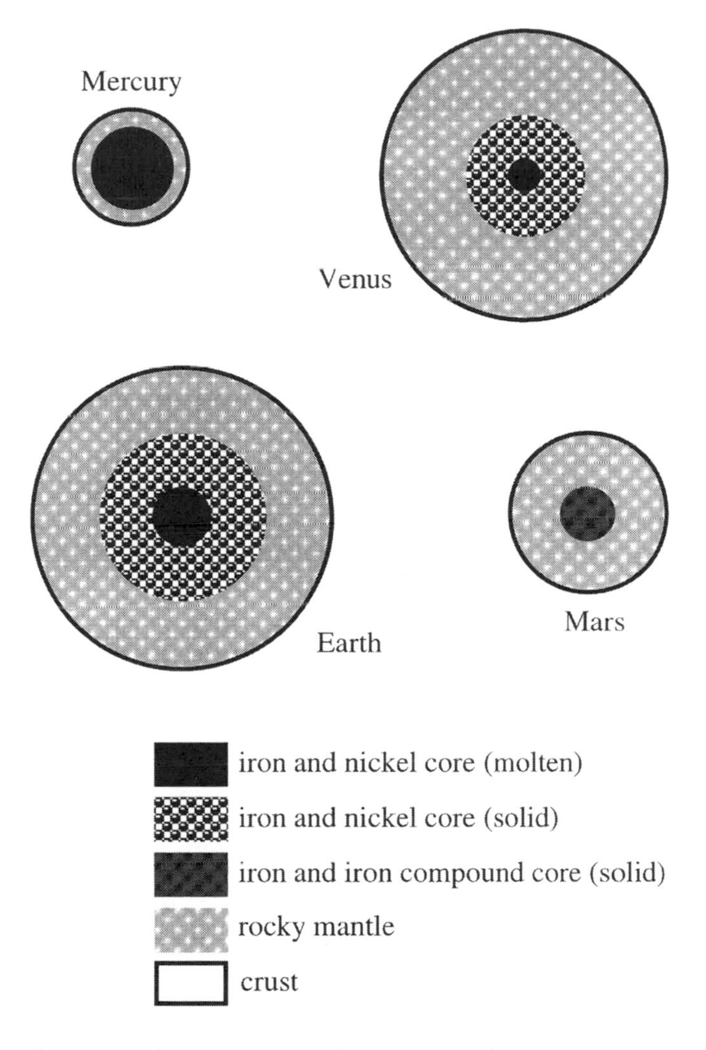

Figure 8.1 The internal differentiation of the terrestrial planets. The diagram is drawn approximately to scale with slightly exaggerated crustal thicknesses

the model of the condensation of the disk of the Solar Nebula. Metals are the materials that condense at the highest temperatures and so bodies that accreted in this region of the Solar Nebula would be expected to be metal rich. Mercury has an iron–nickel core that stretches 80% of the way to its surface. It is possible that this figure was artificially increased by a very large impact in the latter stages of Mercury's growth. This could have stripped Mercury of much of its mantle and sent the remnant planet into its somewhat eccentric orbit observed today.

Mars is of intermediate size between the Earth (and Venus) and Mercury. The internal pressures should also be intermediate and so, assuming that its metal content is substantially below that of Mercury, it is not surprising that the density of Mars should be somewhat lower than the other terrestrial planets. The implication is that

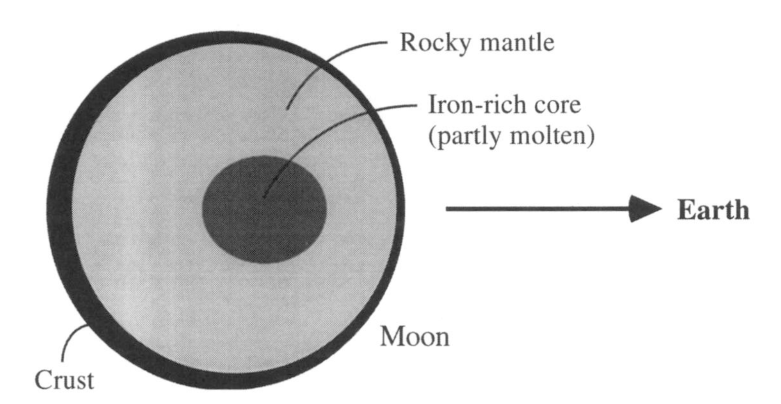

Figure 8.2 The internal differentiation of the Moon. The influence of the Earth's tidal forces on the Moon during formation and cooling have caused non-spherical symmetry in the Moon's mass distribution and for it to become trapped in a synchronous orbit about the Earth. The crust's thickness is exaggerated to show the variation in thickness between those hemispheres that face toward and away from the Earth

the core of Mars is a little smaller and that much of it may not be hot enough to be molten. Again, this corresponds well with the predictions of the model for planetary formation, continuing the trend of decreasing metal proportion with distance from the Sun.

Though the Moon is not a planet it is included in this discussion of gravitational differentiation as it shows similar features to the planets, having a crust, mantle and core. The Moon follows Mercury as being the smallest of the terrestrial bodies discussed so far. Its mass is about 20% of Mercury's but its density shows that its inner constitution is more similar to Mars. With a density of 3330 kg m^{-3} it is unlikely that the Moon has a core that is substantially metallic or that the internal temperature is sufficient to warm the mantle to plasticity. The crust does not therefore float on the mantle and lunar quakes similar to those on Earth do not occur. The crust is interesting in one other way however. That is that the tidal interaction with the Earth, as discussed in Chapter 6, has caused the crust to be thinner on the side facing towards the Earth where the inner, denser regions shifted under the Earth's tidal action while its internal regions were still mobile. This is illustrated in Figure 8.2.

The Moon's properties, in particular its large size relative to the Earth, sparsity of metallic constituents and the similarity in its density to the Earth's mantle indicate the mechanism by which it was formed. Current theories suggest that the Earth and the Moon were once the same body. A collision with a large rival protoplanet during the early stages of the Solar System caused a large chunk of the Earth's mantle to be splintered off. This remained in the Earth's gravitational field and stabilised to become the Moon. The so-called giant impact theory for the Moon's formation has rivals but is currently the most widely accepted.

Volcanism

Beyond simple physical data, there are two other direct indicators of planetary interior; volcanism and magnetic field. Volcanism has already been discussed for the Earth. Its

Figure 8.3 Part of the Caloris Basin (left) and a region of plains (right). The centre of the Caloris Basin is beyond the terminator (the line between light and dark hemispheres) but semicircular folds in the surface indicate its shape, in particular where ranges of volcanic hills were formed as a result of the massive impact that formed the basin (NASA, Jet Propulsion Laboratory and National Space Science Data Center)

study gives some information on the conditions beneath the crust and the nature of the crust itself. The planetary magnetic field gives information on the conducting material, concentrated at the core. However, the magnetic fields of the terrestrial planets are still not well understood and so it is volcanism that will be looked at in greater detail.

There is strong evidence of volcanic activity on all the terrestrial planets though not necessarily of its continuation today. As previously discussed, the planets were much hotter in earlier times, being warmed by more rapid bombardment and higher rates of radioactive decay. Energy has subsequently been released from the planetary interiors and volcanism has played an important role in this. For instance, there is evidence that a single collision on Mercury caused substantial volcanism across the whole of the planetary surface. The result of the collision can still be seen today in the shape of the Caloris Basin, a 'crater' some 1300 km across that is rimmed by 2 km high mountains (see Figure 8.3). Interestingly, there is a region on Mercury's surface, diametrically opposite the Caloris Basin, where the crust is wrinkled and distorted in an unusual way. This area is called the Weird Terrain and is thought to have resulted from the recombination of shock waves from the basin-forming impact on the opposite side of the planet.

Small objects cool more quickly than large ones and so Mercury cooled more quickly than the other terrestrial planets. Consequently, volcanic activity is no longer taking place. That Mercury cooled quickly is evidenced by its wrinkled surface. These wrinkles, known as scarps, were formed by rapid surface shrinkage and cross the surface of the planet running for hundreds of kilometres.

The largest volcanoes in the Solar System can be seen on Mars though all are now extinct. Olympus Mons (Mount Olympus) has a base diameter of 600 km and a height of

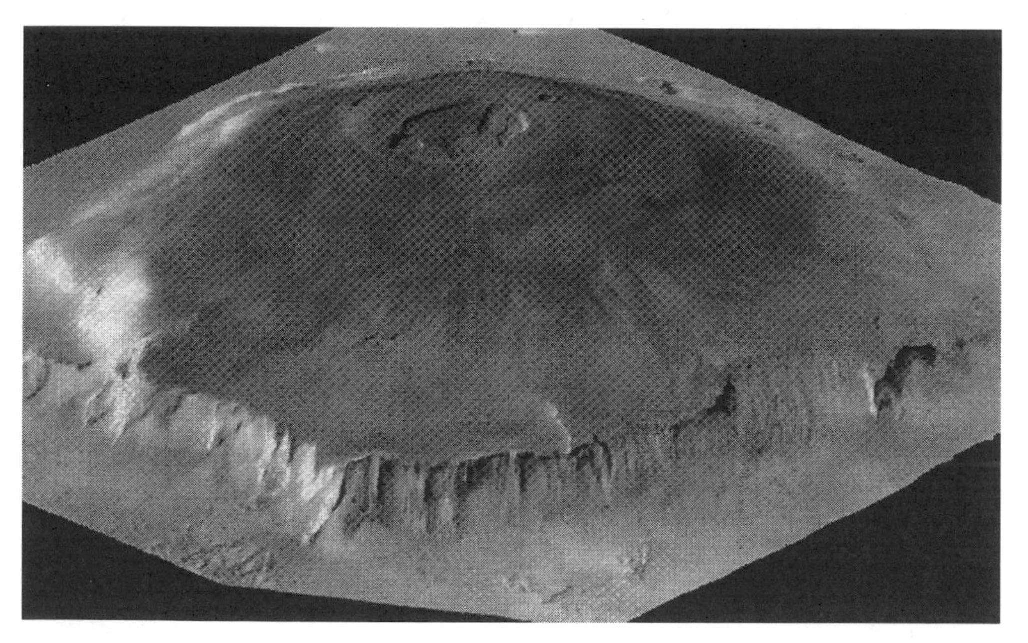

Figure 8.4 The largest volcano in the Solar System, Olympus Mons on Mars. Olympus Mons is 25 km high and has a base diameter of over 600 km, This picture is a computer-fabricated combination of views from several perspectives. (NASA, Jet Propulsion Laboratory and National Space Science Data Center)

Figure 8.5 The volcano Sapas Mons (image centre below horizon) as reconstructed from radar data collected by the Magellan mission. The vertical view is exaggerated by a factor of ten to emphasise the mountains. In fact Sapas Mons is only 1.5 km high with a base diameter of 400 km. Maat Mons (background, on the horizon) has a height of 8 km (NASA, Jet Propulsion Laboratory and National Space Science Data Center)

25 km (see Figure 8.4) making it of a similar area to the whole of England and almost three times higher than Mount Everest. This is particularly striking on a planet much smaller than Earth. The volcanoes of Mars are clustered in one particular region called Tharsis and not distributed in planet-wide chains as on the Earth. This is evidence of a lack of plate tectonics on Mars. Volcanism has also manifested itself on Mars in somewhat less spectacular fashion. Most of the northern hemisphere is considerably depleted in crater impacts relative to the south, showing that lava floods must have covered the northern hemisphere since the early days of the Solar System when cratering impacts, and in particular major collisions, were more frequent. Mars, being the second smallest of the terrestrial planets would, like Mercury, have cooled quite quickly and volcanism ceased there many hundreds of millions of years ago.

Venus is of a similar size to the Earth and orbits at a similar distance from the Sun. Its gravitational differentiation appears to be like the Earth's and so it is reasonable to expect volcanism to show the same characteristics. Indeed there is strong evidence of past volcanism from photographs of volcanic-like rocks strewn on the Venusian surface and from radar images of coned mountains (see Figure 8.5). The largest of these volcanoes is known as Maxwell at about 11 km in height. Like Mars, however, the distribution of volcanoes is not in chains, again suggesting a lack of plate tectonics. Furthermore, the surface height variation on Venus falls into a single distribution rather than the two, representing oceans and continents, on Earth. The crust of Venus is thus quite different from Earth's, being composed of a single shell. Venus's thick clouds and atmosphere have meant that its surface has only recently been uncovered. It still holds many mysteries among which is whether volcanism still continues. There is circumstantial evidence that it does. First, sharp rocks have been photographed by landers. As the atmospheric pressure is very high, weathering is expected to be rapid. Sharp rocks must therefore be relatively new. Second, lightning has been detected close to volcano summits, as might be expected during an eruption or out-gassing event.

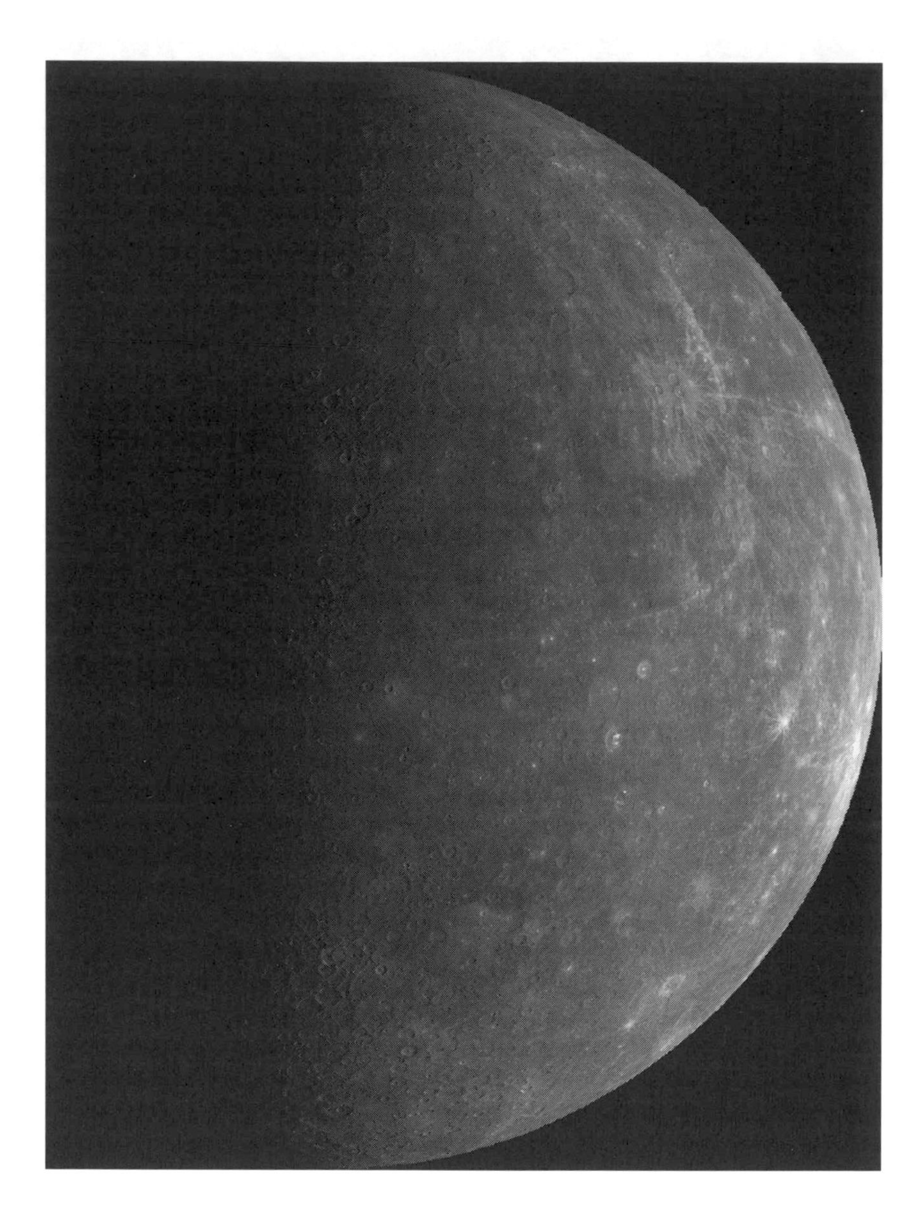

Figure 8.6 A photomosaic of Mercury taken by the Mariner 10 space probe. Note Mercury's dark, crater-covered surface, the large rays that emanate from some of the craters and the extensive scarps that run around the planet (NASA, Jet Propulsion Laboratory and National Space Science Data Center)

Cratering

The nature of volcanic events goes some way towards explaining the appearance of planetary surfaces. Violent ongoing explosions are likely to produce mountains whereas steady seepage of lava can create large plains. The time at which volcanic activity occurred also influences present-day planetary surfaces. As the debris of the Solar System has slowly been soaked up, the size and frequency of collisions has gradually reduced. An ancient surface that still remains must have been subject to neither atmospheric weathering or recent volcanic activity. The best example of this among the terrestrial planets is Mercury, which has a very sparse atmosphere and a very ancient surface. Almost the whole of Mercury's history can thus be read in its surface from enormous impacts such as the collision that formed the Caloris Basin to the myriad smaller cratering events to the scarps that betray Mercury's rapid cooling.

The large number of impact craters on Mercury's surface (see Figure 8.6) is a clear indication that volcanic activity has not taken place for a very long time. A lot of information can also be derived from the size and shape of craters. The impact speed of an incoming object must be at least equal to the planet's escape velocity as this is the speed to which a body at rest will be accelerated by the planet's gravitational field. This is simply the reverse of the logic used to derive equation (6.13). In Mercury's case, v_e is calculated to be a little over 4 km s^{-1} (which is 15 000 km h^{-1}). The local region of the planet is warmed upon impact as much of the invading body's kinetic energy is released suddenly as heat. The precise planetary deformation is determined by the types of rock of which the surface and supporting crust and mantle are composed and the size of the impacting body. Usually, the first response of the surface after impact is to flatten as the local rock is unable to deform to precisely the same shape as the invading body at a fast enough rate. This flattened part of the surface is pressed downwards while the planetary interior resists this compression. The interior acts like a spring, at first being compressed but with ever-increasing reactive force, building until it eventually arrests the compression and pushes back. In this way, material can be re-ejected across large regions of the planet and, for high-speed collisions, all the way back into space. A central mound is often left in the centre of the crater where the ejecta disconnects from the planetary surface. This whole process is illustrated in Figure 8.7. In Mercury's case, secondary craters, caused by the impact of the ejected material, are often relatively close to the primary crater. The ejecta is quickly pulled back down to the surface by the planet's gravity. This is an indication of the fact that Mercury has a relatively high gravitational field (for a small planet), further evidence of a large dense core. Nevertheless, some craters have light streaks of material, known as rays, radiating from them across significant distances.

The Moon's surface is the best place to look for strikingly large rays and, indeed, the Moon's surface is often compared with that of Mercury. It has an ancient surface, little affected by weathering and is also a small body that cooled relatively quickly. There are two important differences, however. The Moon escaped the wrinkling effect that gave Mercury its scarps (possibly because of the Moon's unusual formation mechanism) and the Moon has a much greater covering of lava-filled seas. The reason for the latter effect has been discussed in Chapter 6 in terms of the Moon's thinner crust that has formed in the direction facing Earth due to tidal action. However, it represents a good example of how such events can be dated. The cratering intensity in these seas is lower than elsewhere

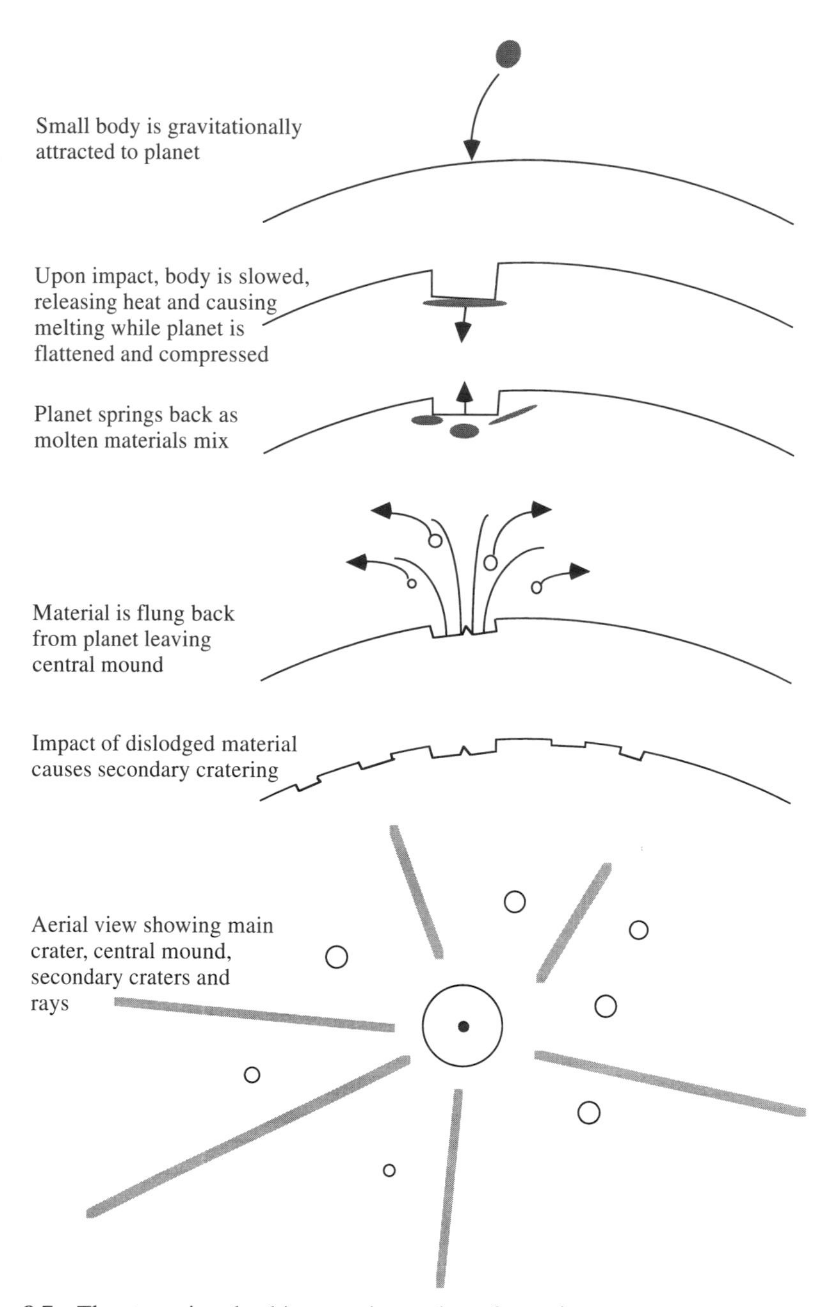

Figure 8.7 The stages involved in cratering and ray formation

(see Figure 8.8) meaning that the lava floods occurred more recently than the period of most frequent cratering events. Current models date the Moon's surface in the region of the seas at around 3 billion years, around one and a half billion years younger than the oldest rocks discovered by the manned missions to the Moon of the 1970s.

The Earth and Venus provide the greatest contrast to the cratering of Mercury and the

Figure 8.8 A sea and a major impact site on the Moon. In the foreground is the sea, Mare Imbrium, a large laval flood plain. Just below the horizon, 400 km away, is the crater Copernicus which is 107 km in diameter. Many of the smaller craters in the foreground are secondary craters created by the Copernicus impact (NASA, Jet Propulsion Laboratory and National Space Science Data Center)

Moon. In particular, the Earth shows little of its history of bombardment. It is in exactly the same region of space as the Moon and must therefore have been involved in just as many impact events per unit area, the strength of which would have been always higher due to its more intense gravitational field. There are only a few impact craters remaining on the Earth, however. The reasons for this are many and include the Earth's continued plate tectonics and volcanism, the atmosphere and associated weather and, perhaps most importantly, the fact that life has evolved. The influence of a rain forest or a herd of wildebeest or, for that matter, a mechanical digger on an impact crater can be quite profound! Venus has not had this helping hand but does have a very dense, unstable and corroding atmosphere as well as volcanism that, like the Earth, means the vast majority of its surface material is less than a few hundred millions of years old. Without the influence of life and running water, however, Venus has still managed to retain many impact craters on its surface (see Figure 8.9).

Unique Features on Mars

As mentioned above, cratering varies strongly in intensity between the hemispheres of Mars, indicating the relative youth of the northern half of the planet's surface. In between the hemispheres, a huge rift valley runs for about 5000 km, sometimes reaching a width of 500 km (see Figure 8.10). Such features form where the crust pulls apart and separates and

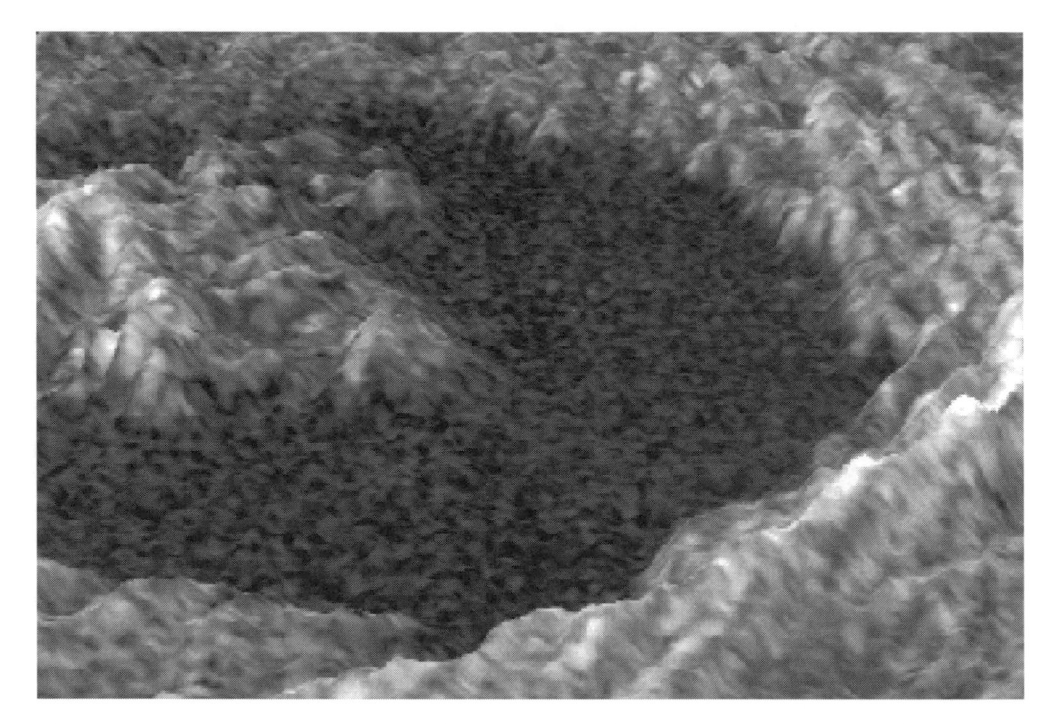

Figure 8.9 An impact crater on the surface of Venus (called Golubkina). Craters on the surface of Venus are both volcanic and impact-related. This 30 km crater shows impact characteristics such as terraced inner walls, formed as the surface collapses, and a small central peak, formed during the rebound of the inner crater floor (NASA, Jet Propulsion Laboratory and National Space Science Data Center)

this represents the only approximation to plate tectonics on Mars. It is close to the Tharsis region where Mars' large volcanoes are situated and this may not be a coincidence.

The Martian surface has many unique features. The most visible of these are its polar caps. At first sight it would be easy to confuse these as being similar to those of the Earth. The Martian caps are most easily distinguished from Earth's by the large change in their size with season. The tilt of Mars on its spin axis is quite similar to the Earth's and so seasons progress in a similar way. The main difference is that the coloration of the Martian caps is not due to water ice. It is due to dry ice (solidified carbon dioxide). In the local winter such a layer becomes thicker as atmospheric carbon dioxide freezes onto the surface. When summer returns, the carbon dioxide sublimes[1] back into the atmosphere as the shrinking pole is directed back towards the Sun. Other regions such as sheltered basins also become white during winter due to dry ice.

A more permanent layer of less visible water ice does underlie the polar caps, however, and it is thought that a great deal of water is frozen into crustal rocks just beneath the surface across the whole planet. The low atmospheric pressure on Mars does not allow water to exist as a liquid. Reports of canals on the surface of Mars from the turn of the century were spurious. There was free-running water at some time in Mars's history as

[1] To sublime means to change state from solid to gas without passing through a liquid state.

Figure 8.10 Photomosaic of Mars taken by the Viking 1 Orbiter. Note the massive canyon, Valles Marineris, running through the centre of the picture. It is about 5000 km long and up to 8 km deep. On the left, the three blotches are the Tharsis volcanoes (including Olympus Mons) (NASA, Jet Propulsion Laboratory and National Space Science Data Center)

can be seen from a large number of channels and regions of collapsed surface. This indicates that Mars must once have had a denser atmosphere and a warmer temperature so that liquid water could stabilise.

Atmospheres

Mars' atmosphere is also responsible for subtle variations in its surface. Though its density is only about 1% of the Earth's this is sufficient to cause occasional, tremendous dust storms that rage for many weeks at a time. Of course, the planet appears very different at these times as the surface is obscured, but more permanent changes also occur as the surface becomes more eroded and more of the rusty dust is created that gives Mars its distinctive red appearance.

The atmosphere on Mars is largely composed of carbon dioxide, the compound that freezes onto the surface of the planet in cold areas. The other 5% of the atmosphere is mainly nitrogen and argon with a little oxygen. Occasional clouds of water-ice crystals can be seen. The density of the atmosphere and the number of clouds are both too low to affect the climatic conditions beyond the dust storms discussed above. The temperature is largely determined by radiation from the Sun and was measured to vary between 190 and 240 K by the Viking lander though more extreme temperatures might be expected at the equator and poles.

It is interesting to contrast the atmospheres of Mars and Venus as they are composed of very similar proportions of gases. The big difference is the density of the gases. The surface pressure on Venus is about 10 000 times that on Mars so that the influence of the Venusian atmosphere on its climate is considerable. It acts like a blanket to keep heat in and to keep the heat distributed across the planet regardless of the time of day. It is significant that the most abundant gas in the Venusian atmosphere is carbon dioxide as it is particularly effective at absorbing energy (heat). As discussed in Chapter 1, the Sun emits energy throughout the electromagnetic spectrum with most of its radiative power being concentrated in the visible and infrared regions. The first barrier to this energy reaching the Venusian surface are the thick clouds that swirl about the planet once every four days. They completely surround the planet and reflect a large proportion of the visible light. They are largely composed of sulphuric acid and other sulphur compounds mixed with a little water. The electromagnetic energy that penetrates the clouds must then travel through the very thick atmosphere. Carbon dioxide has a series of very strong absorption bands in the infrared and so energy absorbed at these wavelengths warms the atmosphere. The remaining light reaches the planetary surface which, like most of the rocky bodies of the Solar System, has a low albedo. Most of the energy that reaches the surface is thus absorbed. So the surface is warmed by two external sources; absorption of penetrating electromagnetic radiation and contact with the atmosphere. The planet itself acts like a black body radiator but its temperature is reduced by a factor of ten compared to the Sun. From Wien's displacement law (equation (1.4)), the energy radiated by the surface of Venus is shifted mainly into the infrared. The infrared-absorbing atmospheric gases thus have a second chance to absorb the energy and heat up further. This is known as the greenhouse effect as analogous effects take place in glass buildings due to the fact that glass, like the Venusian atmosphere, passes visible radiation but not infrared (though atmospheric confinement magnifies the effects in small buildings). The end result is a surface temperature on Venus of over 700 K, warmer even than that on Mercury which is much closer to the Sun.

WORKED EXAMPLE 8.1

Q. Estimate the mean temperature that Venus would maintain if it were an inert, atmosphere-free body but with all other physical parameters unchanged. Venus has a radius of 6000 km, an albedo of 0.72 and is 110 million km from the Sun.

A. The approach to this problem is to consider the influx and output of energy to and from the planet in terms of black-body radiation. The input can be calculated using the exact technique of

Worked Example 1.4. There, the Sun's power output was calculated to be 4×10^{26} W. This power is radiated equally in all directions and so, at the orbit of Venus, the power is spread over a sphere with a diameter of 220 million km. The power density is therefore:

$$\text{Power per unit area} = \frac{\text{Total power}}{\text{Total area}} = \frac{4.0 \times 10^{26}}{\pi(2.2 \times 10^{11})^2} = 2600\ \text{W m}^{-2}$$

The surface area that Venus presents to the Sun is that of a disk with a radius equal to that of Venus (imagine viewing Venus from the Sun). This area is ($\pi \times (6 \times 10^6)^2 =$) 1.1×10^{14} m^2 so that the total power received by Venus is the product of the power per unit area and the presented area at ($2600 \times 1.1 \times 10^{14} =$) 2.9×10^{17} W. The albedo of Venus is 0.72 which means that only 28% of the incident radiation reaches the planet, that is ($0.28 \times 2.9 \times 10^{17} =$) 8×10^{16} W. For Venus to have a steady-state temperature, it must radiate this same amount of energy from its surface. Application of equation (1.3) provides a solution to the unknown temperature, T:

$$P_A = \sigma T^4 \qquad \therefore \text{Total power, } P = \sigma T^4 \times \text{total surface area}$$

$$= \sigma T^4 \times 4\pi r^2 = (5.67 \times 10^{-8}) T^4 \times 4\pi (6 \times 10^6)^2$$

$$= 2.6 \times 10^7\ T^4\ \text{W}$$

$$\text{Power received} = \text{power radiated}$$

$$\therefore\ 8 \times 10^{16} = 2.6 \times 10^7\ T^4 \quad \Rightarrow T = 240\ \text{K}$$

The value of 240 K is somewhat lower than a more sophisticated calculation (that takes into account the variation of albedo with wavelength and the non-perfect behaviour of the planet as a black-body radiator) would give but nevertheless shows that the influence of the atmosphere on the planetary temperature is very significant. It should be noted that internal heating also plays a role in increasing the ambient temperature.

The Earth is also warmed by the greenhouse effect. As far as life is concerned, this is of vital importance. The extra 35 K that may be attributed to greenhouse warming means that most of the water on Earth is liquid. Some of it evaporates from the oceans to join the atmosphere and to form clouds. Dry air is composed of 78% nitrogen, 21% oxygen, 1% argon and only 0.03% carbon dioxide. So where does the heating of the greenhouse effect come from? In the Earth's case it is the water vapour, that can make up as much as 4% of air in humid conditions, that performs the role of carbon dioxide. There is a balance therefore between the surface temperature of the Earth and the proportion of water in the atmosphere. If the average humidity were to increase a little then the temperature of the atmosphere would also rise as greenhouse warming would increase. This would cause greater evaporation of the oceans leading again to increased temperature and so on. Such a scenario is an example of a phenomenon known as positive feedback and leads to runaway instabilities. Eventually it is possible that Venus-like conditions could be established on the Earth as the oceans evaporate completely to produce a dense atmosphere of gaseous water. Small increases in the Earth's surface temperature over the past century are thought to attest to increases in carbon-based greenhouse gases due to the burning of fossil fuels and the destruction of carbon-dioxide-removing vegetation. Other greenhouse gases, such as sulphur dioxide, are also increasing due to industrial pollution.

The Earth's atmosphere is being disturbed in other ways too. In the Earth's upper atmosphere, the interaction of short wavelength radiation from the Sun with normal oxygen molecules (O_2) produces ozone (O_3). This prevents radiation damaging to life from reaching the Earth's surface. Some man-made compounds, halogenated hydrocarbons, are disturbing this process and allowing larger quantities of ultraviolet radiation to penetrate to the Earth's surface.

A final important region of the Earth's atmosphere is the ionosphere, a layer of ionised gas 80 to 300 km above the surface. The atmospheric molecules in this region absorb ultraviolet and X-ray radiation from the Sun and in so doing lose one or two of their electrons. The remaining electrically charged gas reflects electromagnetic radiation in a certain part of the radio spectrum. This is useful for communication between two places that do not have line of sight. To send a message between the two you just bounce it off the ionosphere!

Neither Mercury nor the Moon have significant atmospheres, a further similarity between these bodies. It is quite easy to see why this is the case, especially for Mercury. Though Mercury has a relatively high gravitational field, due to its large metallic core, it is strong only by the standards of a small body. As it is a small body it possesses a weak gravitational field in absolute terms. This implies a small escape velocity. Atmospheric particles are thus easily able to attain enough thermal energy to reach escape velocity as the planet is very close to the Sun. Furthermore, the particles that make up Mercury's transient atmosphere are small and thus more easily accelerated to high speeds. Some of the atmosphere is hydrogen and helium that is trapped from particles ejected by the Sun into space, the solar wind. More helium is derived from radioactive α-decay of various elements within Mercury's solid body. When the temperature on Mercury reaches its highest, at around 700 K during the heat of the very long day, sodium and potassium are released by surface rocks to bolster the atmosphere. Oxygen is also present but it is not clear from where it comes. Despite the various sources of particles, the atmospheric pressure on Mercury remains a tiny fraction of the Earth's.

The Moon's atmospheric density is also tiny. The temperature on the Moon is significantly less than it is on Mercury but, as the Moon is smaller and less dense, the escape velocity for the Moon is also lower. All that remain are a few particles that include neon and helium.

This chapter has sought to connect the physical properties of the major terrestrial bodies to both the theories of Solar System formation and evolution and to their present appearance. The size of the planet has been shown to be of critical importance. Large planets are able to retain their heat for longer periods and have therefore continued to evolve for longer. The Earth's surface continues to evolve through volcanism and plate tectonics while Venus, being slightly smaller, seems to have a slightly older surface. Most of the Martian surface has an intermediate age of around 2 billion years while the smallest bodies, Mercury and the Moon, have remained unchanged, except for an occasional crater-forming impact, for about 3 billion years. The planet's size is also of critical importance to the atmospheric density. Venus and the Earth, being the largest planets, are able to hold the thickest atmospheres, Mars is once again intermediate while Mercury and the Moon have virtually no atmosphere. It is clear, therefore, that size does matter!

Questions

Problem

1 Estimate the mean temperature that the Earth would maintain if it were an inert, atmosphere-free body but with all other physical parameters unchanged. The Earth has a radius of 6400 km, an albedo of 0.39 and is 150 million km from the Sun.

Teaser

2 The rotational period of Venus is more than four times that of Mercury but the former planet's temperature is considerably more constant across its surface than the latter. Why?

Exercises

3 Mercury is a small planet that orbits close to the Sun. Discuss the implications of this statement for the planet's atmosphere.

4 (a) The Martian atmosphere was first detected using star occultations. Explain how this technique works.
(b) The Martian atmosphere is composed of 95% carbon dioxide but Mars has no significant heating due to the greenhouse effect. Explain why not.

5 Why is only one side of the Moon visible from Earth? What is the name of the effect that causes small parts of the other side to become visible?

6 What evidence is there for volcanic activity on each of the terrestrial planets?

7 Are the following statements about Mercury true or false?
(a) It is difficult to observe due to its small size, low albedo and proximity in the sky to the Sun.
(b) It has a very sparse atmosphere
(c) Temperatures fall to below −150°C (120 K) on the surface.

8 Using examples, describe what information can be obtained about the age of a planet's surface and about its gravitational field through the study of its surface cratering.

9 Compare and contrast the internal differentiation of the terrestrial planets.

10 (a) Discuss the difficulties in sending a probe to make scientific measurements on the surface of Venus.
(b) Discuss techniques that may be used in order to find a suitable landing place for such a probe.

9 A Closer Look at the Jovian Planets

The asteroids fall between the orbits of Mars and Jupiter and mark the boundary between sets of planets that are fundamentally different in nature. The previous chapter discussed the differences between the terrestrial planets and showed these to be significant. However, variation in the terrestrial planets is essentially qualitative, being due to a number of influences manifesting themselves to differing extents. Jupiter, Saturn, Uranus and Neptune have little in common with the terrestrial planets, being enormously larger, almost entirely fluid, composed of different compounds, surrounded by rings and much colder. The outermost planet, Pluto, is exceptional in many ways and though little is known about it, Pluto seems to have more in common with the larger moons of the Jovian planets than any of the other planets. The four giant planets may themselves be divided into two pairs. Jupiter and Saturn soaked up large quantities of the hydrogen and helium from the Solar Nebula through gravitational contraction and have volumes in the region of 1000 Earths. Further out, Uranus and Neptune contain much less hydrogen and helium but a greater proportion of the compounds of intermediate volatility, such as water, aggregated mainly through accretion. Their volumes are equivalent to about 60 Earths. The overall density of all four giant planets are, however, considerably lower than the Earth's so that their masses are only roughly equivalent to 300, 100, 15 and 17 Earths (from Jupiter to Neptune).

Internal Heating, Rotation and Appearance

It is instructive to start a discussion of the giant planets by relating their physical appearance to simple calculations that can be made to determine the magnitude of their internal source of heat. In the case of the fluid planets, most of the heat released internally is due to continued planetary contraction and differentiation that releases gravitational potential energy as heat. It is easy to calculate the temperature that a large body of a known composition should have when heated by a single distant source (the Sun). Jupiter, Saturn and Neptune all have temperatures greater than predicted and must all therefore be releasing gravitational potential energy. It would appear that the contraction of Uranus has ceased. For the other three giant planets the descending order of energy release per unit area is Jupiter, Saturn, Neptune. By coincidence, this is the order of distance from the Sun and it turns out that each of these three planets releases a few times more internal energy than is received from the Sun. Of course, Neptune receives much less energy from the Sun as it is much further away than Jupiter while Saturn represents the intermediate case. A quick glance at the appearance of the four giant planets shows the influence of energy outflow upon their atmospheres. The order of apparent surface inhomogeneity, once again in descending order, is Jupiter, Saturn, Neptune, Uranus. It would appear therefore that internal energy release drives the 'weather'.

WORKED EXAMPLE 9.1

Q. Estimate the mean temperature that Jupiter would maintain if it were an inert, atmosphere-free body but with all other physical parameters unchanged. Jupiter has a radius of 70 000 km, an albedo of 0.70 and is 780 million km from the Sun.
A. This problem is identical to Worked Example 8.1 but with different data. At the orbit of Jupiter, the Sun's power is spread over a sphere with a diameter of 1.56 billion km. The power density is therefore:

$$\text{Power per unit area} = \frac{\text{Total power}}{\text{Total area}} = \frac{4.0 \times 10^{26}}{\pi(1.56 \times 10^{12})^2} = 52\,\mathrm{W\,m^{-2}}$$

The surface area that Jupiter presents to the Sun is that of a disk with a radius equal to that of Jupiter. This area is ($\pi\,(7.0 \times 10^7)^2 =$) $1.5 \times 10^{16}\,\mathrm{m^2}$ so that the total power received by Jupiter is ($52 \times 1.5 \times 10^{16} =$) 7.8×10^{17} W. The albedo of Jupiter is 0.70 which means that only 30% of the incident radiation reaches the planet, that is ($0.30 \times 7.8 \times 10^{17} =$) 2.3×10^{17} W. For Jupiter to have a steady-state temperature, it must radiate this same amount of energy from its surface. Application of equation (1.3) provides a solution to the unknown temperature:

$$P_A = \sigma T^4 \quad \therefore \text{ Total power, } P = \sigma T^4 \times \text{total surface area}$$

$$= \sigma T^4 \times 4\pi r^2 = (5.67 \times 10^{-8})T^4 \times 4\pi(7.0 \times 10^7)^2$$

$$= 3.5 \times 10^9\ T^4\ \mathrm{W}$$

$$\text{Power received} = \text{power radiated}$$

$$\therefore\ 2.3 \times 10^{17} = 3.5 \times 10^9\ T^4 \ \Rightarrow T = 90\,\mathrm{K}$$

The value of 90 K is a little lower than the surface temperature which varies between about 110 and 150 K on Jupiter. The extra heat that raises the temperature is supplied by the contraction of the planet. If the average surface temperature of Jupiter is taken to be 130 K then the total power radiated (which must be equal to that received from both internal and external sources) can be calculated from above as being

$$3.5 \times 10^9\ T^4 = 3.5 \times 10^9 \times 130^4 = 1.0 \times 10^{18}\ \mathrm{W}$$

As 2.3×10^{17} W are received from the Sun then ($1.0 \times 10^{18} - 2.3 \times 10^{17} =$) 7.7×10^{17} W must originate from gravitational contraction. It should be remembered that these calculations falsely assume no wavelength dependence on albedo so that a more sophisticated calculation will give somewhat different values.

Variation in the 'weather' on the giant planets can be thought of as the level of atmospheric turbulence. It has already been mentioned that the extent of the atmospheres of Jupiter and Saturn are difficult to define as the planet is composed largely of hydrogen and helium throughout with the mixture changing state as the temperature and pressure vary. The same applies to Uranus and Neptune though there is a more profound change in composition at the gas–liquid interface. It is convenient therefore to define the atmosphere as being the whole of the shell of gas that surrounds the liquid part of the planet. For all four giant planets this region is composed mainly of hydrogen and helium. The addition of methane gives Neptune and Uranus their blue–green colours by absorbing red light and reflecting blue. A larger mixture of impurity gases, including methane, ethane, ammonia, ethyne, water, carbon monoxide and phosphine, cause the yellows, browns and reds of Saturn and Jupiter. It is this coloration that allows the intensity of atmospheric disturbances to be seen.

The final ingredient that determines the dynamics of the gaseous part of the giant planets is the rapid rate of planetary rotation. Jupiter and Saturn rotate on their axes in about 10 hours whereas Uranus and Neptune take about 17 hours. For sizeable planets like Uranus and Neptune this sweeps the outer atmosphere around at very high speeds and so for the much larger Jupiter and Saturn the speeds become enormous. The speed of circulation of the atmosphere can be measured by following individual disturbances as they rotate around the planet. Is this the planetary rotation speed, though? Venus has clouds that rotate in four days while the planet below takes sixty times longer. For the giant planets, the rotation speed of the bulk is equal to that of the atmosphere. Confirmation of this comes from magnetic field measurements. As the magnetic fields of the planets are not perfectly aligned with their spin axes it is possible to measure their rotation rate from outside the condensed matter with which they are associated (Figure 9.1). For Saturn such a measurement is difficult as the spin and magnetic field axes are separated by less than 1° but for Jupiter the angle increases to 10°. The equivalent angles for Uranus and Neptune are 59° and 47°. The physical mechanism that underlies such large angles is not understood.

The patterns on the surface of the giant planets can now be explained but it is worth briefly reviewing the facts first. Jupiter has the most spectacular surface, swathed in coloured bands and swirling ovals, the structure of which can persist for centuries or longer. Saturn's banding is more subtle and smoother. Local disturbances that occur tend to last for only a few weeks or months. Neptune's features are still more subtle, its colour varying only slightly in shades of blue. Uranus is almost featureless. The face that points towards the Sun is somewhat reddened, probably due to the presence of ethyne in the atmosphere but there is no banding or local disturbance.

A common explanation can be given for the appearance of all the giant planets. Turbulence in the atmosphere is caused by the release of energy from the centre of the planet. This energy heats the atmospheric gas, causing it to expand, become less dense and thus to rise. Cooler gas rushes in to fill the gap the warm gas leaves. This process is known as convection. Similar effects, on a much smaller scale, occur on Earth. For instance, on a sunny day by the sea it is common to experience an off-shore breeze. This occurs as the darker land absorbs more of the Sun's heat than the reflective sea. The air above the land thus warms up and tends to rise while the cooler air from above the sea moves across to equalise the pressure thus causing the breeze. As the warm gases rise through the atmospheres of the giant planets, rapid rotation causes the gas stream to be wound into a band that circles the planet. Thus the bands that cause the planet to be striped parallel to the equator (perpendicular to the spin axis) represent regions of rising and falling gas. The light regions are called zones and represent cool regions in which initially warm gases have been forced up above the mean surface of the upper atmosphere. The darker bands, known as belts, are areas where the air is falling back towards

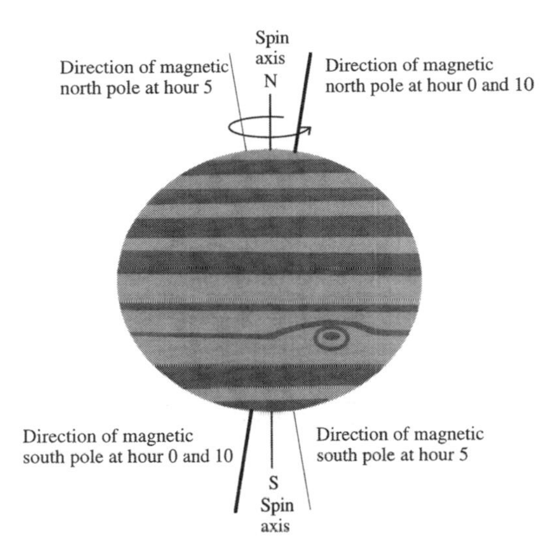

Figure 9.1 The effects of Jupiter's spin. As Jupiter's magnetic field is not aligned with its spin axis (as in many planets) then the field is swirled around in space every ten hours. Such is the rate of rotation that Jupiter is oblate perpendicular to its spin axis and convective pockets of gas are whirled into bands

the planet and once again being warmed. The zones are thus a few tens of kilometres above the belts. The relative brightness of zones and belts is not a blackbody radiation phenomenon as their temperature is low and they are seen only through reflected light. (Figure 9.2)

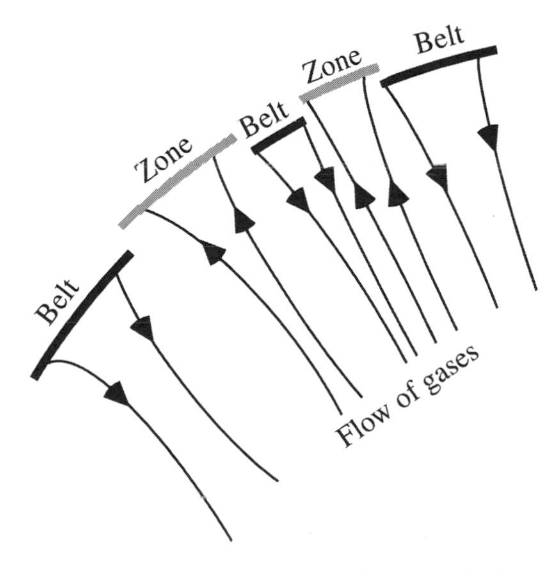

Figure 9.2 Convection currents in the outer parts of Jupiter (similar to Saturn and Neptune). Columns of gas rise, taking heat with them, cool at the surface and then sink again. Due to Jupiter's rapid rotation, the columns of gas are swirled into bands, known as zones in the case of the lighter regions where gases have just reached the surface and belts where the gases are falling back

In between the zones and belts enormous disturbances can be seen. Precisely how these enormous storms are able to persist for very long periods of time is not known but their basic nature is understood. They are equivalent to cyclones and anticyclones on the Earth, the former being a region of low pressure into which neighbouring gases spiral and the latter a region of high pressure out of which gases spiral. The direction of the spiral is determined by the sense of spin of the planet and the hemisphere that the storm is in. The same effect causes the direction in which water spirals down plugholes to be different in the Earth's northern and southern hemispheres and is known as the Coriolis force. Cyclones and anticyclones rotate in opposite directions and reverse direction between hemispheres. It is therefore simple to diagnose the type of disturbance that causes any given storm by simple observation. The Great Red Spot (Figure 9.3), for instance, is an

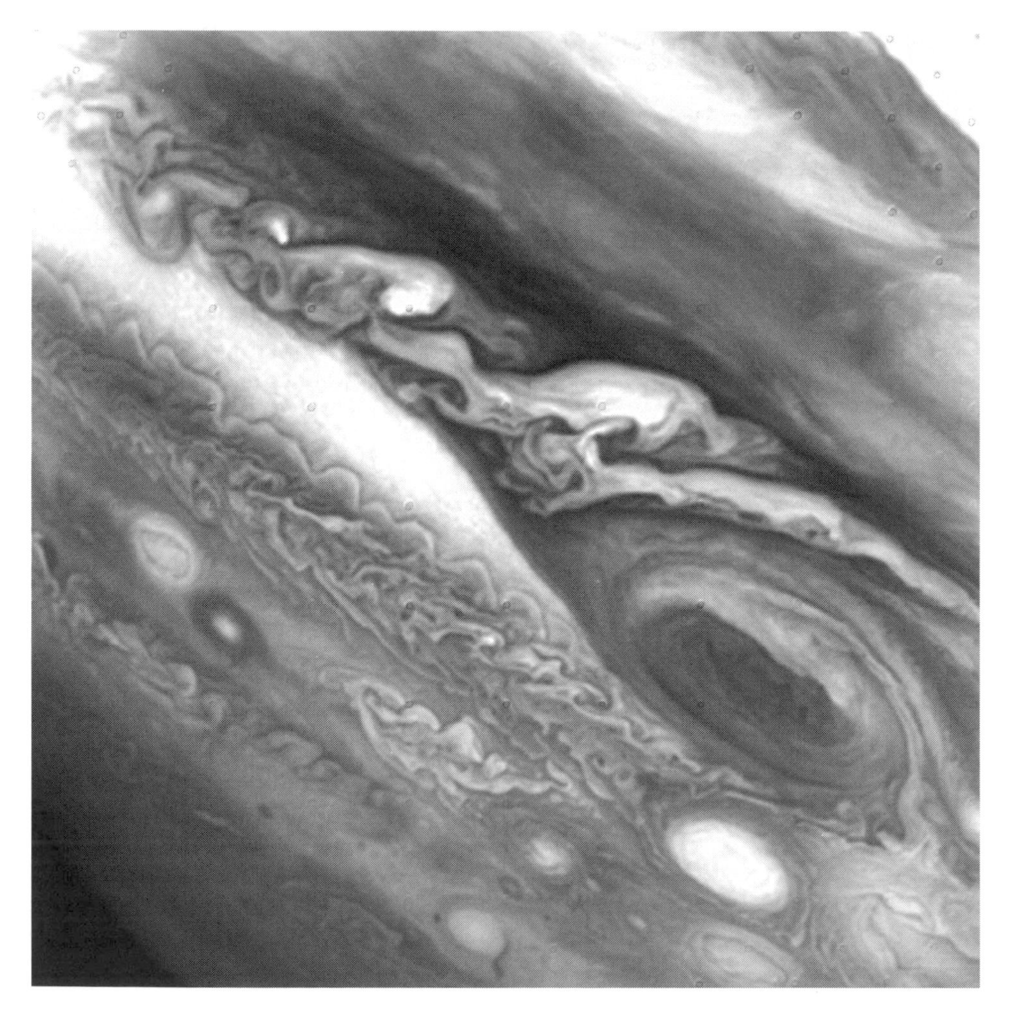

Figure 9.3 Close-up view of the anti-cyclonic storm known as the Great Red Spot that has been raging for centuries. The White Oval (below) and surrounding turbulence are also shown. The frame is 60 000 km across (NASA, Jet Propulsion Laboratory and National Space Science Data Center)

anticyclone, a storm that is larger than the entire Earth causing winds to blow at speeds upwards of 300 km h^{-1}. Lower in the atmosphere the wind speed increases further and can exceed 600 km h^{-1}.

The colour of any particular disturbance is thought to be due to the prevailing conditions within it. It should be re-emphasised that planets are observed by the light they reflect as they are too cold to radiate significantly as a black body in the visible part of the spectrum. Different coloration is therefore almost always due to a chemical effect. In the case of the giant planets, different conditions support the existence of differing quantities of variously coloured gases. The planets therefore differ in colour both between themselves and across individual surfaces. Once again, greater energy outflow creates greater inhomogeneity of surface conditions so that the descending order of colour variation is Jupiter, Saturn, Neptune, Uranus.

The fluid state of the giant planets manifests itself in two particularly interesting ways. First, fluidity removes the restriction that a planet must rotate on its axis at the same rate at all latitudes. On Jupiter and Saturn different bands rotate at different speeds thus gradually slipping in position relative to each other. There is a variation of about 5 minutes between the rotation period at the equator and poles, the rate at the equator being more rapid. It is the rapid rotation of the planets that leads to a second effect associated with fluidity. The outer parts of Jupiter close to the equator are being whirled around at over 40 000 km h^{-1}. Closer to the poles the speed drops considerably. This is due to the much smaller radius of rotation rather than the much smaller effect of differential rotation or the quite separate concept of wind speed. The effect of the difference in transverse speeds with varying latitude is to cause the planet to be oblate, that is, to have a greater circumference around the equator (4.5×10^5 km) than around the poles (4.2×10^5 km) (Figure 9.1). This is reminiscent of the early stages of the formation of the Solar System itself as the centre of the Solar Nebula began to spin rapidly and then became oblate.

Planetary Interiors

The interiors of the giant planets have been analysed in detail and coherent models have been developed that describe their gravitational differentiation. For the giant planets this refers not to a variation in the type of material that accumulates at a particular distance from the centre of the planet but to the state of the matter in any given region. There are similarities between the giant planets but greater clarification is obtained by treating them sequentially.

Jupiter can be thought of as an enormous ball of hydrogen with some helium mixed in. The contributions of the other components, other than at the core which remains somewhat mysterious, are too small to affect Jupiter's fundamental constitution. It is first important to note that Jupiter's average density is significantly less than that of the terrestrial planets, being about one quarter that of the Earth. At first sight this may seem obvious as Jupiter is composed of the lightest of the elements rather than of metals and rock but Jupiter's enormous gravitational field that is constantly squeezing itself and thus increasing the density should also be remembered. It is Jupiter's enormous gravitational field that allows it to maintain such a dense atmosphere of very small particles that would rapidly escape from many of the other planets. It is also this gravitational field that causes the pressure and density of the hydrogen to increase towards the centre and thus to change state.

About 1000 km beneath Jupiter's cloud tops the gas pressure reaches 5600 times the Earth's sea-level atmospheric pressure (written as 5600 bar) and the temperature has climbed from the external value of 130 K to about 2000 K. These are the conditions that are required to make the molecular hydrogen of the atmosphere adopt a liquid state. Moving towards the centre of the planet the temperature, pressure and density of the liquid continue to increase until a pressure of a few million bars and a temperature of over 10 000 K is reached. Under these almost unimaginable conditions the hydrogen changes state again and becomes an electrical conductor. The state is somewhat like a metal as the hydrogen nuclei are so closely crushed together that the electrons (see Appendix 3) are no longer localised in the region of parent nuclei but are associated with the whole body. Electrons may thus move relatively freely which is another way of saying that the state is conducting. This region behaves like a metal and hydrogen in this state is known as metallic hydrogen. Close to Jupiter's core the temperature is thought to be a few tens of thousands of kelvin and the pressure around 100 million bars. The details of the solid core that represents the initial body around which the planet began to gather and then gravitationally contract can currently only be guessed.

Saturn's body is very similar to Jupiter's in many ways but as it is about three times less massive, the conditions that its gravitational field can generate are less extreme. It is interesting to note that, despite its smaller mass, Saturn is only a little smaller in size than Jupiter so that its average density is only half that of Jupiter and almost exactly one eighth that of the Earth. The conditions in Saturn's centre are sufficient to cause the metallic state of hydrogen to exist and yet the planet's overall average density is less than that of a glass of water on Earth! The physical conditions required for metallic hydrogen to exist in Saturn are the same as for Jupiter and anywhere in nature. As temperature and pressure increase more slowly under Saturn's weaker gravitational field then the proportion of the planet that is composed of metallic hydrogen is reduced from that of Jupiter, though it is still significant. The temperature at the core is probably in the region of 15 000 K.

The gravitational differentiation of Uranus and Neptune may be considered together. Again, much of the detail of the model for these planetary interiors is based on mathematical guesswork. Good observations of Uranus and Neptune only became possible when space probes visited there relatively recently and so the mathematical modelling continues. The planets are considerably smaller than Jupiter and Saturn and so the temperatures and pressures do not reach such large values. Despite this, Uranus has a density similar to that of Jupiter and Neptune's is somewhat higher. This is a definite indication that Uranus and Neptune are composed of a larger proportion of the intermediate volatility class of compound such as water, methane and ammonia and that they have, proportionally, a more substantial rocky core. It is conceivable that conditions are extreme enough close to the core to allow water to take on its conducting state in much the same way that hydrogen does in Jupiter and Saturn.

A major indicator of the state of matter inside a planet is the magnetic field that extends into the surrounding space. Jupiter has a very strong magnetic field (that traps passing charged particles into large orbits extending millions of kilometres into space) and this is evidence for the large quantity of electrically conducting hydrogen in Jupiter's body. The Earth has a similar, but smaller, magnetic field due to its molten iron core. Saturn, Uranus and Neptune all have significant magnetic fields though all are smaller than Jupiter's.

Ring Systems

The four giant planets also have the common characteristic that they are surrounded by ring systems. The appearance (Chapter 5) and dynamics (Chapter 6) of the rings have been discussed previously. There is little to add here other than to give a general comparison of the systems. The first point to note is that the vast majority of the material that comprises the ring systems is within the respective Roche limit of the parent planet. Thus, the rings of Neptune and Uranus are much closer to the planetary surface than those of Saturn and Jupiter. Furthermore, the rings of Neptune and Uranus (Figure 9.4) are banded into a small number, four and eleven respectively, of quite narrow rings whereas Saturn and Jupiter possess much more extensive systems. Saturn's rings are much the most impressive, being composed of large (measurable in metres) pieces of

Figure 9.4 The rings of Uranus are very dark so that Voyager 2 cameras required a time exposure to take this photograph. The streaks represent the motion of background stars during the exposure. The frame covers a distance of about 10 000 km (NASA, Jet Propulsion Laboratory and National Space Science Data Center)

Figure 9.5 A view of Saturn in which Cassini's division is clearly visible as a dark band in the ring structure. Saturn's southern limb can be seen through the rings indicating their non-continuous nature (NASA, Jet Propulsion Laboratory and National Space Science Data Center)

material with high albedo (see Figure 9.5). The biggest contrast with Saturn in terms of ring component size is provided by Jupiter which has rings that are composed of tiny particles, each only about 10^{-5} m across. Though the mechanics of the rings are beginning to be understood it is still not clear why the particle sizes should vary so much from planet to planet. It could be that Jupiter's much greater tidal forces have almost completely prevented grain accretion within the Roche limit though it is probable that other processes are also important.

The formation of ring systems is still not understood though the connection with the Roche limit is clearly important. Material in these regions is unable to accrete into moons but the question is, where did this material come from in the first place? The two possible sources are from the planet itself or through the gravitational capture of a passing body that is subsequently ripped apart by tidal forces. The difficulty with the latter model is that only certain types of body with certain entry trajectories would behave in this way. Most would crash into the planet, possibly in many pieces, or simply be sling-shot into a different orbit. The former model is therefore favoured but there remains the possibility that both processes make contributions. An intermediate idea in which a collision dislodges material from the planet, in much the same way that the Moon was formed, has also not been fully discounted.

The Moons of the Giant Planets

The idea that the rings of the giant planets are composed of planetary outflow material is most favoured and also fits with current ideas on the formation of many of the moons in this part of the Solar System. The analogy between the spin of the giant planets and that of the Solar Nebula in the early stages of formation has already been drawn. It is thought that similar equatorial matter outflow to form a disk surrounding the giant planets probably took place during their formation. The processes of condensation and accretion then formed the rings and moons depending on separation from the planet. Of course, the scale on which these processes operated was very different from that of the formation of the planets themselves but the fundamentals are the same. The Galilean moons, in particular, fit this theory as their separations, at 0.4, 0.7, 1.1 and 1.9 million km, have a very similar pattern to that of the planetary orbits about the Sun and there is a steady variation in the nature of the moons with separation from Jupiter. One difficulty in confirming the planetary disk theory is that the orbits of the moons have fully evolved. That is, unlike planetary orbits about the Sun, moon orbits about the planets have almost all been trapped into synchronicity thus obscuring information on the initial state of the system.

Jupiter's orbital system represents the best evidence for the planetary disk theory. Within the Roche limit lies the ring system composed of very small particles. There then lie four very small moons, two of which appear to act as shepherding satellites for the rings. The Galilean moons, Io, Europa, Ganymede and Callisto follow. Though the radii of the moons are all between 1500 and 2700 km, their average densities decrease steadily from 3550 $kg\,m^{-3}$ for Io to 1830 $kg\,m^{-3}$ for Callisto. This fits neatly with the planetary disk theory as outer moons should accumulate more volatiles and these are less dense compounds. Due to the small size of the moons, none form through gravitational contraction of the nebula to form fluid moons but the innermost are able to accumulate any rocky or metallic substances while the water, ammonia and methane compounds tend to condense further away. While the raw numbers thus fit the theory quite well it should not be forgotten that Io is heated by the tidal action of Jupiter and may thus have lost formerly accumulated volatiles with time. The same applies to Europa but to a lesser extent. Beyond the Galilean moons there lie four small, low-density moons that orbit in very similar paths just beyond 11 million km and a second group of four about 23 million km from Jupiter. It is thought that these groups did not form with Jupiter, its rings and other moons but that they are bodies that have been captured, perhaps from the nearby asteroid belt. Good evidence that this is the case is derived from the fact that the outer four moons orbit Jupiter in the opposite direction to the others (and the vast majority of bodies in the Solar System). This would be extremely unlikely to evolve from a planetary disk that would swirl exclusively in one direction.

The moon systems of the other planets do not fit the disk model quite as well as for Jupiter but do not conflict with the paradigm sufficiently to defeat it. Most of the moons fit into 'regular' or 'irregular' groupings based on their orbital characteristics and these are taken to signify moons that respectively are or are not formed from a planetary disk. Some examples are obvious, such as Phoebe that orbits about Saturn out of the equatorial plane in a retrograde direction more than three times further from the planet than the next closest moon and is thus clearly an irregular, externally formed body. Other examples are not quite so easy to classify. For instance, Triton is easily Neptune's largest moon, suggesting that it should have evolved from the planetary disk but its motion is retrograde, suggesting the opposite. Though such examples are difficult to classify, the fact that the total amount of matter in the moons of Neptune and Uranus is considerably smaller than that in the moons of Saturn and Jupiter is further supportive evidence for the planetary disk theory; the smaller planets would be expected to have had smaller outflows to create their satellite nurseries.

Much as the terrestrial planets have differing time scales for their evolution, and in particular the evolution of their surfaces, then so too do the moons of the giant planets. The boundary between large and intermediate-sized moons is easy to distinguish. Six moons of the giant planets have radii between 1300 km and 2700 km. The next largest moon is Uranus's Titania with a radius of 790 km. The border between small and intermediate is somewhat arbitrary but moons start to become aspherical when their radius drops below about 200 km, thus below this size the bodies resemble large boulders rather than moons. This smallest class generally orbit close to the rings of the giant planets, often as shepherd satellites, or very far from the parent planet, usually falling into the 'irregular' classification. Bodies in the small category will not be discussed any further. Not surprisingly, it is the large moons that show the most planetary-like behaviour.

Only Jupiter possesses more than one of the large category moons and so it is, once again, the Galilean moons that are most appropriate for study. The density relationship to orbital radius has already been discussed but similar patterns emerge when considering, for instance, surface evolution. Io's surface is clearly still evolving, covered as it is by enormous, tidally driven volcanoes (Figure 9.6). Europa's surface is thought to have frozen over only about 30 million years ago and still has the appearance of broken sea ice on the Earth (see Figure 9.7). It is thought that beneath Europa's ice there could be salt water oceans with a depth of about 100 km. These oceans could be heated by tidal action on the core and it is just possible that life could exist underwater in the same way that life survives near volcanic outflows deep in oceans on the Earth. Indeed, it is also thought that 'ice volcanoes' may exist on Europa. The next moon out, Ganymede, has a much older surface as evidenced by substantial cratering. However, parallel ridges on its surface, similar to features observed on moving glaciers on Earth (as well as other evidence) suggest that Ganymede was active at some time in its past (Figure 9.8). Finally, Callisto is also heavily cratered and shows little evidence of any activity in its past (see Figure 9.9). Thus there is a clear progression in the surface ages of the Galilean moons. Io continues to evolve, Europa has only recently frozen over, Ganymede shows evidence of ancient activity while Callisto has remained inert for billions of years. Such a progression is caused by a combination of the decrease in gravitational tidal forces with distance from Jupiter and the formation process that determined the initial composition of the moons.

The same factors that determine surface dynamics also influence the internal differentiation of the moons. It appears, for instance, that Io has a large iron core in much the same way as the Sun's nearest neighbour, Mercury. In Jupiter's case, however, the distribution of the small amounts of metal among its orbiting bodies is more skewed. Io seems to have taken the lion's share leaving mainly rocky and icy material for the outer moons. Nevertheless, all the Galilean moons are thought to be gravitationally differentiated with rocky materials forming a core around which the less dense ice lies. Ganymede's ice layer is thought to be about 800 km thick, for instance. This ice need not be static but may adopt a plastic form similar to the mantle of the terrestrial planets. Thus slow-moving convection currents allow the moons to discard their internal heat. This is the theory that underlies Europa's postulated ice volcanoes wherein heat release is not gentle but sudden, as is the case in Io's more familiar volcanoes.

In Io's case, the volume of the outflow of matter is sufficient to support a very weak

Figure 9.6 The unique surface of Io. Io has several volcanic regions but an otherwise smooth surface covered by frequently refreshed lava flows (NASA, Jet Propulsion Laboratory and National Space Science Data Center)

Figure 9.7 The surface of Europa. The icy surface is considerably fractured but otherwise featureless, indicating that it is relatively modern. It may overlie an ocean of water (NASA, Jet Propulsion Laboratory and National Space Science Data Center)

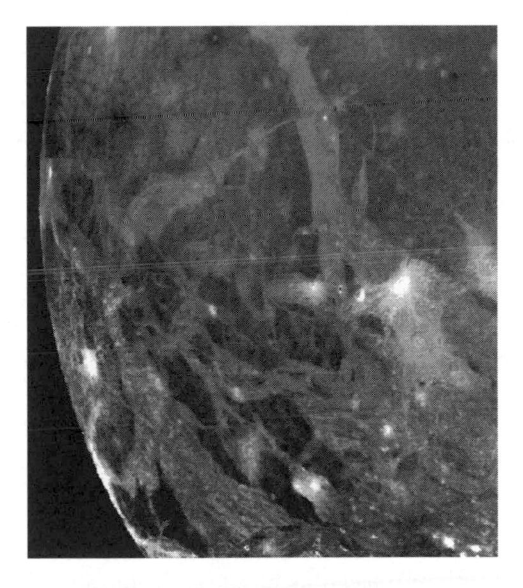

Figure 9.8 Ganymede's surface shows signs of its history. Striations indicate ancient ice volcanism while cratering indicates impacts over a considerable period. As for all such bodies there are no mountains as ices are not rigid enough to be self-supporting over long periods (NASA, Jet Propulsion Laboratory and National Space Science Data Center)

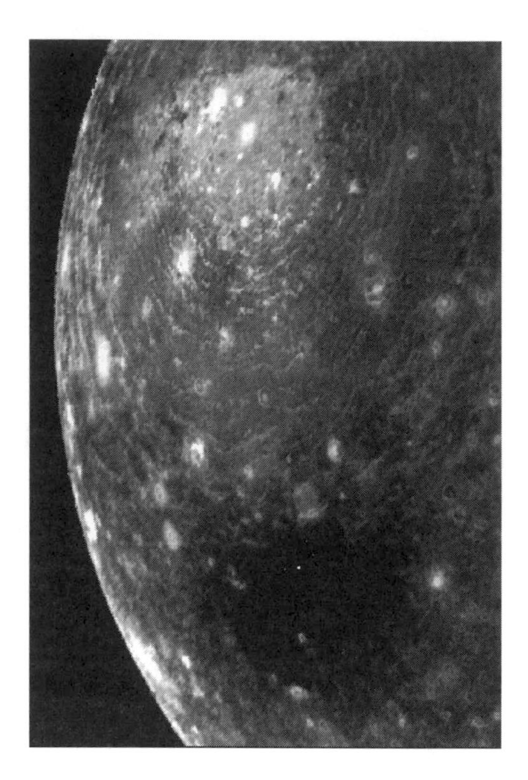

Figure 9.9 The Valhalla basin is an impact site that has created a multi-ring effect on the surface of Callisto stretching over more than 2000 km. The large number of impact craters attests to the great age of the surface. Impacts on the 'icy' bodies in the outer Solar System are somewhat gentler processes than those on the terrestrial bodies due to differences in the mechanical properties of ices and rocks (NASA, Jet Propulsion Laboratory and National Space Science Data Center)

atmosphere, similar in density to Mercury's though composed largely of sulphur dioxide. Only two other moons are known to have an atmosphere, Neptune's Triton and Saturn's Titan. Both of the latter atmospheres are dominated by nitrogen with significant quantities of methane but Titan's has much the greater density. Titan has a surface atmospheric pressure that is 50% greater than Earth's despite having a surface temperature that is three times lower. The surface of Titan can only be speculated upon at this time because the organic molecules present in Titan's atmosphere—methane, ethane, ethyne and ethene, among others— cause the mixture to be opaque. From Titan's density it is expected that the planet is composed of an approximately equal mixture of rocky and icy materials, probably internally differentiated. It is further thought likely that a combination of surface ice and atmospheric gases may have liquefied to form substantial organic oceans on the moon's surface.

Such is the difference in mass of Jupiter and Saturn that different mechanics would have been involved in the birth of their satellite systems from a planetary disk. In Saturn's case 95% of the disk's mass was gathered by a single moon. Nevertheless, Titan is surrounded by a host of other moons, four of which have radii greater than 500 km: Tethys, Dione, Rhea and Iapetus. These and the smaller moons are all heavily cratered balls of ice with one exception. Enceladus has a radius of only 250 km but orbits Saturn five times closer to the parent planet than does Titan. The surface of Enceladus is less cratered than other moons and has a series of grooves crossing it. Both features are thought to be indicative of tidal heating of the moon, leading to ice volcanism in a similar manner to Jupiter's Europa.

The moons of Uranus are similar in size and nature to those of Saturn except for the absence of a dominant contributor. Miranda, Ariel, Umbriel, Titania and Oberon are the five largest moons, varying in radius from 200 to 800 km. Again, the innermost moon, Miranda, shows signs of tidal heating at some time during its history (see Figure 9.10) while the three outermost moons are heavily cratered showing their surfaces to be ancient. Ariel is situated between Miranda and the outer three larger moons and, as might be expected, has intermediate properties including evidence of large flows of water ice.

Neptune's innermost six moons range in radius between 30 and 100 km, becoming larger with distance from the central planet. Their separations increase at a similar rate indicating that they have each accumulated the matter in their region of an ancient planetary disk. There then follows a large gap from 120 000 to 350 000 km to much the

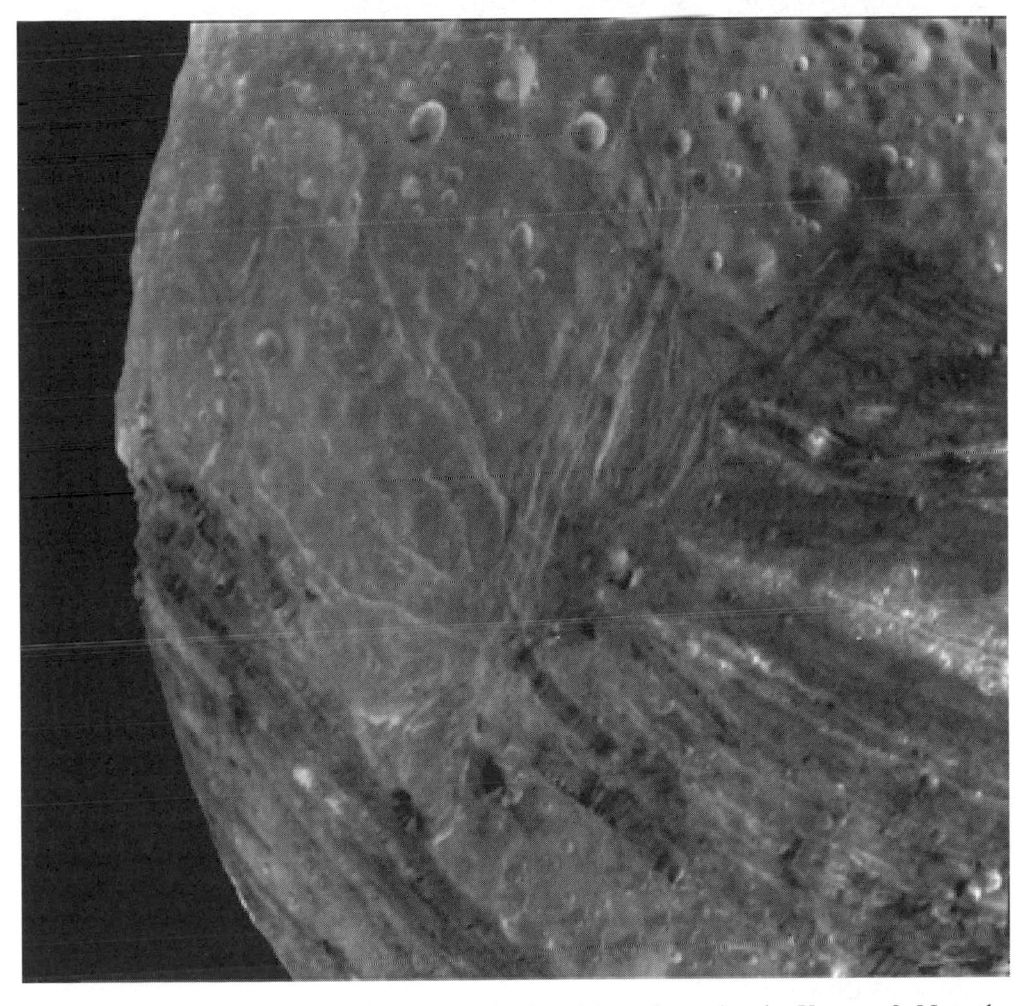

Figure 9.10 High-resolution image of part of Miranda's surface taken by Voyager 2. Note the range of surface features that include grooves with depths of several kilometres. The image is of a region about 240 km across (NASA, Jet Propulsion Laboratory and National Space Science Data Center)

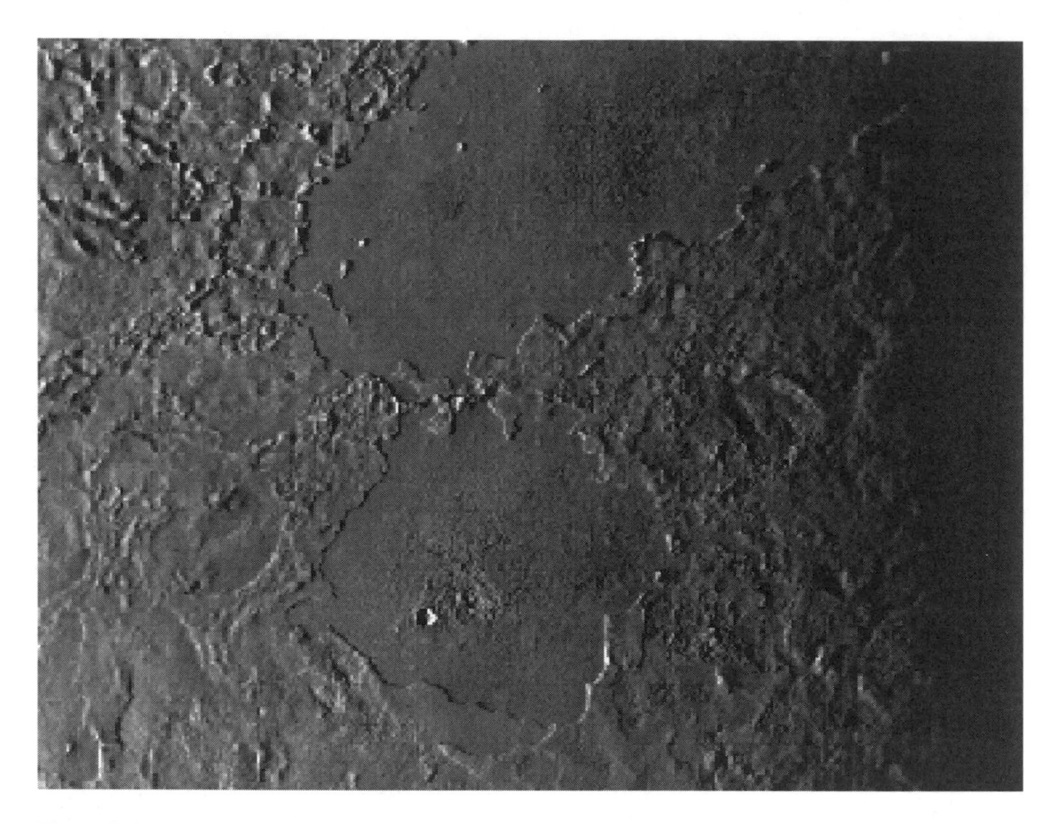

Figure 9.11 Features on the surface of Triton resemble frozen lakes and suggest that Triton once had open liquid on its surface. It is now completely frozen over though nitrogen slowly evaporates from the surface to provide a weak atmosphere (NASA, Jet Propulsion Laboratory and National Space Science Data Center)

largest of the Neptunian moons, Triton which has a radius of 1350 km. At 5.5 million km follows Nereid which is much smaller (170 km) and orbits Neptune in a very eccentric orbit ($e = 0.76$, see Appendix 3). A simple model for the evolution of such a system might be to assume that Triton (Figure 9.11) is separated from its fellow planetary disk satellites due to its large size that has allowed it to dominate a large region by gravitationally capturing smaller nearby bodies in the usual accretive manner. Nereid would then be an irregular moon captured from elsewhere. A further piece of data casts doubt on this theory. That is, that Triton orbits in a retrograde direction. Thus the vast majority of angular momentum associated with Neptune's moon system has the opposite sense to most of the rest of the bodies of the Solar System. It would therefore seem likely that Triton was not formed from Neptune's planetary disk but that it has been captured by a chance encounter. In so doing it is possible that it threw Nereid out of a regular orbit.

Pluto and Charon

Beyond the giant planets lies the tiny Pluto–Charon system. Despite recent imaging of the pair by the Hubble Space Telescope and earlier data obtained during mutual eclipsing

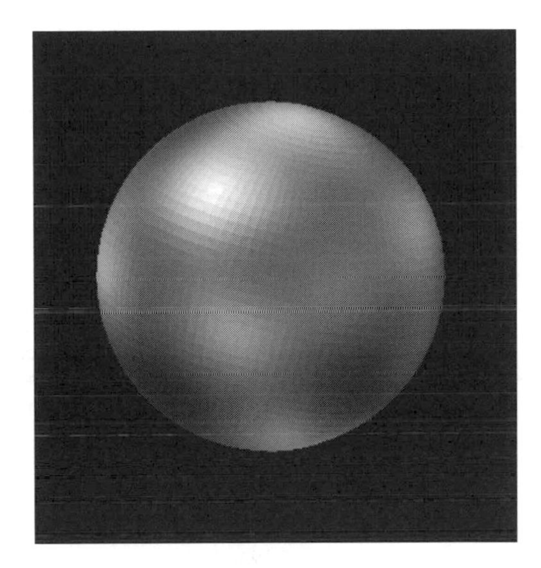

Figure 9.12 A computer enhanced image of Pluto taken by the Hubble Space Telescope. Reasons for the mottled appearance are still being speculated upon (NASA, Jet Propulsion Laboratory and National Space Science Data Center)

events, details of the system remain sketchy (see Figure 9.12). Calculations suggest that it is extremely unlikely that the two bodies evolved together but that they became locked in orbit about each other after a major collision. These calculations are consistent with the system's unusual orbital and rotational characteristics and the apparent make-up of the two bodies. At present it is thought that Pluto has a higher rock content than Charon. Both are believed to be internally differentiated and the heating required for such a process may have been obtained from their mutual collision (see Worked Example 7.1) or from the higher bombardment rate by smaller bodies during the first few hundred million years of the Solar System. Nuclear disintegration would have also made a contribution. The differentiation of the bodies is thought to be quite simple, consisting of a rocky core covered by a thick layer of ices, similar to many of the moons of the giant planets. Pluto seems to have a large proportion of methane while Charon's ice is dominated by water.

Pluto has an atmosphere that has been detected using occultation. It is likely to be composed of methane that sublimes from the surface in much the same way that the water contribution in Mars' atmosphere does. This analogy can be extended in that there is also likely to be seasonal variation. In Pluto's case this is not caused so much by the angle between its spin axis and the ecliptic but by the variation in its distance from the Sun. Perihelion is at 29.7 AU and aphelion is at 49.3 AU and this leads to a variation in light energy per unit area of nearly a factor of three. Thus it is thought that the perihelion atmosphere is likely to be significantly denser than that at aphelion. In the latter case the atmosphere will lie on the planet as surface frost.

It is interesting to note the apparent similarities between Pluto, Charon and Neptune's moon, Triton. With radii of 1140 and 590 km respectively, Pluto and Charon have a similar size to Triton. They also have a similar mean density to Triton. Why should there be only three bodies similar to Pluto, Charon and Triton at the edge of the Solar System and how did Pluto and Charon manage to find each other to collide in the midst of such wide-open space while Triton found Neptune to be trapped in its orbit? The answer is that they are probably not the only ones. Many bodies of similar size are thought to have been gravitationally scattered, mainly by the giant planets, to the Oort cloud and the Kuiper disk (see Chapters 5 and 7). Only when observation techniques have been greatly improved or space probes have been sent to the extreme edges of the Solar System can such theories as well as the many remaining mysteries of the Pluto–Charon system be resolved.

Questions

Problem

1 (a) Estimate the mean temperature that Saturn would maintain if it were an inert, atmosphere-free body but with all other physical parameters unchanged. Saturn has a radius of 58 000 km, an albedo of 0.75 (assume no wavelength dependence) and is 1.4 billion km from the Sun.
(b) Estimate the proportion of Saturn's thermal energy that is derived from gravitational contraction given that Saturn's actual surface temperature is 97 K.

Teasers

2 Why can Jupiter be labelled a 'failed star'?

3 The orbital periods of the inner three Galilean moons, Io, Europa and Callisto, are in the ratio 1 : 2 : 4. What effect might this have on Io?

4 What is the connection between Pluto, Charon, Triton and the Kuiper disk?

Exercises

5 Describe the visual appearance of Jupiter, explaining the causes of all the main features.

6 Explain how Jupiter's constitution varies, from its outer atmosphere to its central core.

7 Compare and contrast the internal differentiation of Mercury and Saturn.

8 Explain the connection between the gravitational contraction of the giant planets and their external appearance.

9 How does the rapid rotation of Jupiter affect its appearance?

10 Explain the connection between the Roche limit and the extent of the ring systems of the giant planets.

10 The Sun

At the centre of the Solar System lies the body that gives our local region of space its name. The Sun dominates its realm, its mass of 2×10^{30} kg being 99.9% of the Solar System's total and its power output of 4×10^{26} W being more than 99.9% of that released within the whole of the Solar System. The Sun is a sphere with a diameter of about 1.4 million km, meaning that its average density is only a little greater than liquid water on the Earth's surface at 1.4 kg per litre (1400 kg m^{-3} in SI units). Such a density is very similar to Jupiter's and it would be easy to make the mistake of considering them to be similar objects. Both are the result of the gravitational contraction of the Solar Nebula so that both are principally composed of hydrogen and helium. Both are very large compared to the Earth and the other terrestrial planets but it is size that once again counts. While Jupiter is a planet, the Sun is a star.

What qualifies the Sun to be a star? The answer to this lies in scale. The Sun represents the vast majority of a sizeable part of a contracted molecular cloud. Jupiter represents the majority of a small proportion of the Solar Nebula's matter that the Sun could not hold on to. The mass ratio is a little over 1000:1 and this drives quite different processes. During the formation of both bodies large amounts of energy were released due to gravitational contraction. The Sun's larger mass provided greater heating and higher internal pressures. Thus the early Sun contained at its centre very hot, highly compressed gas greatly beyond the scale of even Jupiter's unimaginable conditions. So extreme were the conditions that nuclear reactions began to take place in the Sun. The core of the Solar Nebula thus became a star.

The Sun, upon the establishment of nuclear reactions in its centre, had two opposing forces acting upon it. Gravity continued to act to compress the star while the energy output from nuclear reactions heated the gas further thus creating an outward pressure that resisted contraction. Eventually an equilibrium between expansive and compressive forces was reached and the Sun has remained largely unchanged in the 5 billion years that have since passed.

In the same way that the planets are internally differentiated into concentric shells composed either of differing compounds or similar compounds in different states then so is the Sun (see Figure 10.1). Gravity determines the differentiation of the planets but the Sun has the added complication of having an enormous nuclear furnace at its centre. While gravity plays an important role in the differentiation of stars, particularly as they age, the Sun's concentric shells are best defined by the thermal processes that dominate their behaviour. As calculated in Chapter 1, the surface temperature of the Sun is around

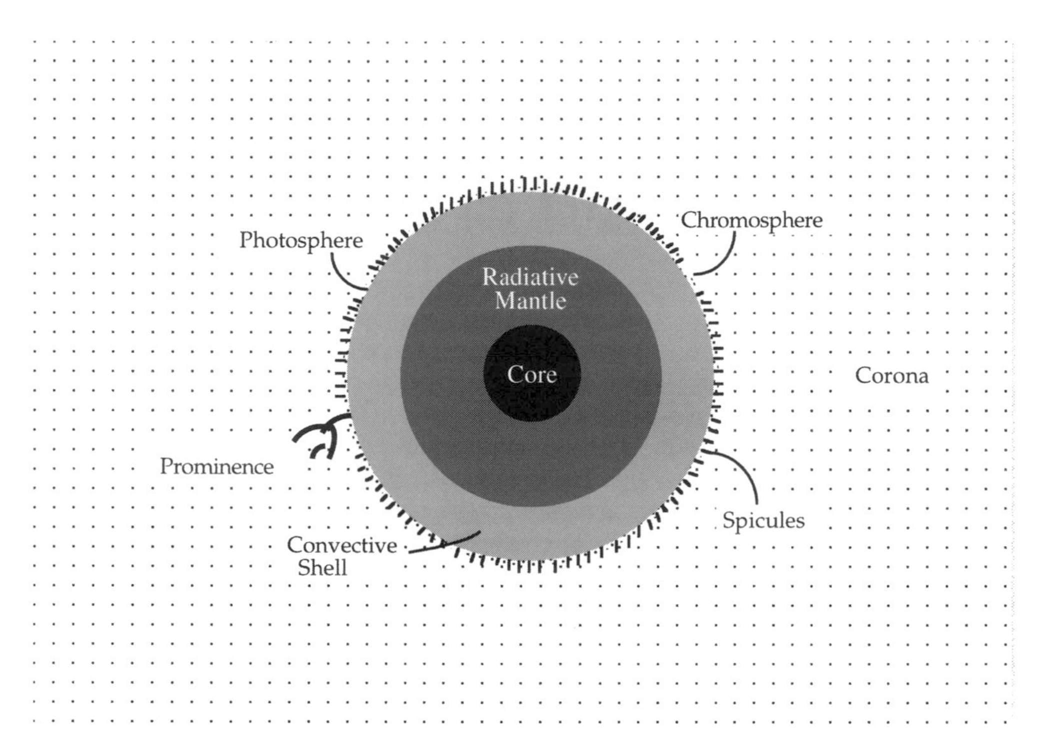

Figure 10.1 A cross-section through the Sun. The schematic diagram is roughly to scale except for the size of the spicules which are exaggerated and the corona which is considerably less homogeneous

6000 K. Moving towards the centre the temperature increases to around 1.5×10^7 K. Nowhere in the Sun's volume is the temperature cool enough for atoms to hold on to all of their electrons (see Appendix 2). At the surface, atoms may only lose a few electrons to the volume of the gas but at the centre nuclei and electrons are entirely liberated. Such a soup of electrically charged particles, in both cases, is known as a plasma.

The Sun's Core and Energy Generation

The plasma at the core is composed of positively charged nuclei and negatively charged electrons (see Appendix 2). The electric force operates in a similar way to gravity whereby the force between two charged particles is proportional to each of their charges, Q and q, (cf mass for gravity), and inversely proportional to the square of their separation, r:

$$F = \frac{1}{4\pi\varepsilon} \frac{Qq}{r^2} \tag{10.1}$$

where ε, the permittivity, is a constant that depends on the medium in which the interaction is taking place. As Q and q can be positive or negative (unlike gravity, where mass is always positive), then a pair of positive charges repel each other as do a pair of negative charges. Positive and negative charges attract each other. Atoms and molecules, at low temperatures, contain equal numbers of positive charges (associated with protons) and negative charges (associated with electrons) but as the temperature increases a small number of electrons can be thermally excited so that they can escape the electrical attraction of the proton-containing nucleus. This is analogous to the escape from a gravitational field of thermally excited particles from planetary atmospheres. In the Sun's plasma electrons and ions continue to attract each other and periodically recombine, releasing energy as light. Ions continue to repel each other. However, in the extreme conditions at the centre of the Sun, the electron-stripped nuclei can obtain so much kinetic energy (i.e. be moving so fast) that the electrical repulsion barrier can be overcome during a collision and the nuclei fuse to create a single, larger nucleus. This is a nuclear fusion reaction. Such is the temperature and density of the plasma in the Sun's core that these reactions occur at a very high rate.

Each nuclear fusion reaction releases a small amount of energy, the quantity of which depends on the nuclei that fuse (see Figure 10.2). In nuclear reactions the mass of the products is a little less than the mass of the reactants. The mass loss associated with the

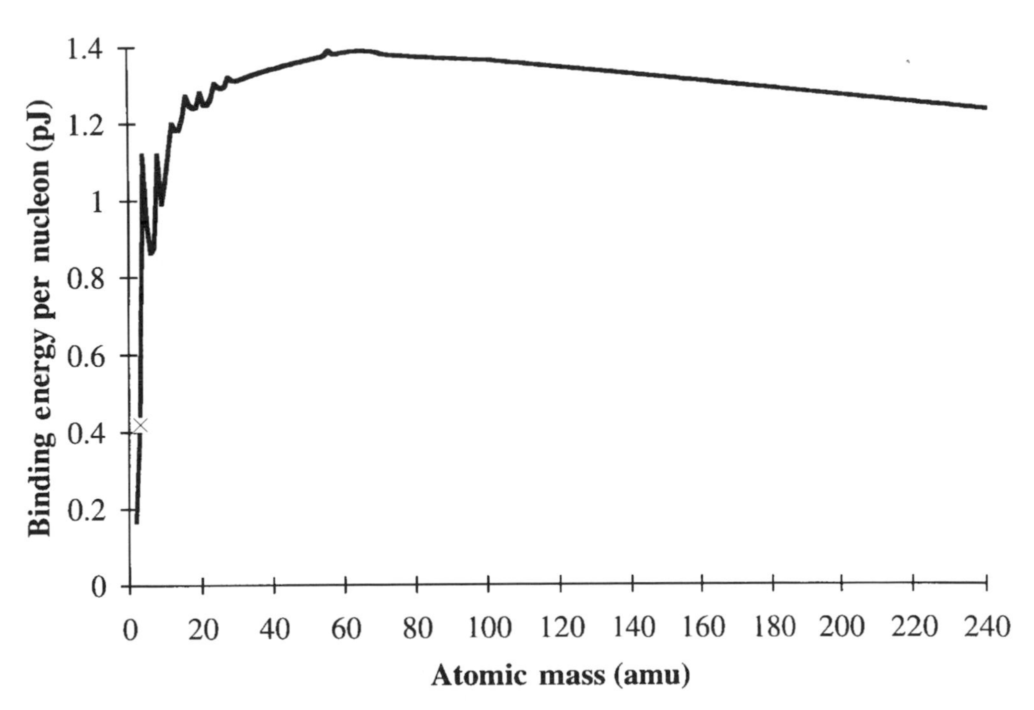

Figure 10.2 The binding energy per nucleon for naturally occurring nuclei. In fusion reactions two small nuclei bind together to produce a larger nucleus with a larger binding energy per nucleon. The energy difference between reactants and products is released during the reaction through the conversion of mass. In fission reactions a large nucleus is split into two smaller nuclei, again increasing the overall binding energy per nucleon and releasing energy proportional to the mass defect. The data point for 3 mass units is marked by a cross for clarity

reactants reappears in the form of energy according to Einstein's famous relationship:

$$E = mc^2 \tag{10.2}$$

Here, E represents the energy produced through the transformation of a mass, m. The constant of proportionality is the square of the speed of light. This relationship can be derived quite simply from first principles but this involves the incorporation of special relativity theory and so will be omitted from this text.

WORKED EXAMPLE 10.1

Q. How much energy is released per gram of transformed mass?
A. Remembering that the SI unit of mass is the kilogram, so that $1\ \mathrm{g} = 10^{-3}\ \mathrm{kg}$, the energy released is calculated by substituting this mass into equation (10.2):

$$E = mc^2 = 10^{-3}(3 \times 10^8)^2 = 9 \times 10^{13}\ \mathrm{J}$$

Almost 10^{14} J of energy, a few hours' worth of power output of a typical power station, is released through the transformation of just one gram of mass.

The principal nuclear transformation that takes place in the Sun's core is the conversion of hydrogen to helium. The most common hydrogen nucleus consists of a single proton whereas helium nuclei are most often composed of two protons and two neutrons. As protons and neutrons have similar masses but different electrical charges (one electron charge of positive sign and zero respectively) four hydrogen nuclei are required to produce one helium nucleus and two of these protons must change into neutrons in the process. To achieve this, the transformation proceeds in stages, each of which involves the fusion of two nuclei. Before examining the chain of events that leads to the production of helium nuclei it is worth mentioning the nomenclature that is used to symbolise nuclear reactions. Reactants are placed on the left of an arrow and products are put on the right thus:

$$^{x}\mathrm{E} + {}^{y}\mathrm{F} \longrightarrow {}^{x+y}\mathrm{G} + \mathrm{s} + \mathrm{t} + \ldots \tag{10.3}$$

Here elements with chemical symbols E and F fuse to make a larger nucleus with symbol G. The chemical symbol used is determined by the number of protons in the nucleus; for instance, one proton means the nucleus is hydrogen, H; two indicates helium, He, and so on. The superscript indicates the total number of nucleons, that is, protons plus neutrons. The number of neutrons associated with a particular number of protons may vary. For instance, hydrogen may have zero, one or two neutrons in its nucleus but only one proton. These three isotopes are represented as $^{1}\mathrm{H}$, $^{2}\mathrm{H}$ and $^{3}\mathrm{H}$, the latter two also having the special names of deuterium and tritium respectively. Equation (10.3) shows that the number of nucleons is conserved during a reaction. Nevertheless, the average mass per nucleon decreases slightly so that the total mass changes. A little mass may be taken out of the nucleus by the production of new particles and the rest is converted to energy, much of which is often carried away by a photon of electromagnetic radiation. The new particles, including the photons, are represented by a special series of symbols represented by s and t in equation (10.3).

By far the most common nucleus in the Sun is $^1\mathrm{H}$, a single proton. It is the fusion of two protons that starts the sequence that leads to the production of $^4\mathrm{He}$. This initial reaction is written thus:

$$^1\mathrm{H} + {}^1\mathrm{H} \rightarrow {}^2\mathrm{H} + e^+ + \nu \qquad (10.4)$$

The nucleus created is composed of a proton and a neutron so that one proton has been converted to a neutron. To preserve charge neutrality a new particle must be created. This is an electron in all but the sign of its charge. The positron (e^+) has the same magnitude of charge and mass as an electron but is positively charged. The plasma contains an enormous quantity of electrons and the positron is certain to be rapidly electrically attracted to one. When such matter and antimatter particles collide, the particles annihilate thus converting more mass to energy, released as two X-ray photons. A second particle called a neutrino, ν, is also created during reaction (10.4). Neutrinos have no mass (or very little depending on which theory one believes) and are extremely weakly interacting, making them very hard to detect and thus study. The energy created by the fusion of two protons is about 6×10^{-14} J. In other words, more than 10 million million such reactions need to occur to create a single Joule of energy.

The second stage of the transformation occurs when the product of the first reaction, a deuterium nucleus, fuses with a third proton:

$$^1\mathrm{H} + {}^2\mathrm{H} \rightarrow {}^3\mathrm{He} + \gamma \qquad (10.5)$$

This reaction is much simpler than the first as the proton is simply 'absorbed' to create a new nucleus consisting of two protons and one neutron. A gamma-ray photon, γ, carries away much of the energy created through mass transformation. The second stage creates more than ten times as much energy as the first at 9×10^{-13} J.

The final stage occurs when two $^3\mathrm{He}$ nuclei fuse:

$$^3\mathrm{He} + {}^3\mathrm{He} \longrightarrow {}^4\mathrm{He} + {}^1\mathrm{H} + {}^1\mathrm{H} + \gamma \qquad (10.6)$$

Once again nucleon transformations are not required. A stable helium nucleus has been produced. At the same time, two protons are regenerated and more gamma radiation carries away much of the energy that replaces the mass transformed. The total energy created by this reaction is a little over 2×10^{-12} J, the most productive stage of all.

The three stages together are known as the proton–proton chain and can be summarised as a single reaction. To create a helium nucleus two stage one reactions must precede two stage two reactions before the $^3\mathrm{He}$ products fuse in the final stage. Taking into account the fact that the positrons created in stage one are quickly annihilated by electrons to produce two gamma-rays each, the overall reaction can be written as

$$4\,{}^1\mathrm{H} + 2e^- \longrightarrow {}^4\mathrm{He} + 2\nu + 7\gamma \qquad (10.7)$$

The mass of a proton is 1.6726×10^{-27} kg while an electron mass is 1830 times smaller. The mass of the reactants thus adds up to 6.6922×10^{-27} kg. The helium nucleus has a mass of 6.644×10^{-27} kg. A mass defect of 4.8×10^{-29} kg has thus been converted to 4.3×10^{-12} J of energy.

WORKED EXAMPLE 10.2

Q. The Sun radiates energy at a rate of 4×10^{26} W. The Sun's total mass is 2×10^{30} kg. (a) What proportion of the Sun's mass is converted per second? (b) How many proton–proton chains are completed per second (assuming no other nuclear reactions are taking place)?

A. (a) The total transformed mass per second, m_s, can be calculated by rearranging equation (10.2), remembering that the power produced, P, is equal to the energy released per second so that

$$m_s = \frac{P}{c^2} = \frac{4 \times 10^{26}}{(3 \times 10^8)^2} = 4.4 \times 10^9 \text{ kg s}^{-1}$$

The mass transformed is thus more than 4 billion kg s^{-1} but this is just a tiny proportion of the Sun's total mass at ($4.4 \times 10^9/2 \times 10^{30} =$) 2.2×10^{-21} s^{-1}.

(b) If one proton–proton chain transforms 4.8×10^{-29} kg, then to transform 4.4×10^9 kg, ($4.4 \times 10^9/4.8 \times 10^{-29} \approx$) 10^{38} cycles must be taking place every second within the Sun.

Though the Sun is losing a little over 4 billion kilograms per second, this is only equivalent to losing about five parts in 10^{20} of its mass per second or a little over one part in 10^{12} per year. In the 5 billion years since the Sun stabilised it has yet to lose 1% of its mass. The main change that has occurred is an increase in the helium concentration in and around the core. It is the proton–proton chain that provides the majority of the Sun's energy but other reactions, involving larger nuclei, also contribute.

Much research time and money has been devoted to trying to produce controllable conditions on Earth that might provide an energy source based on similar processes to those that take place in the Sun's core. While the science of hydrogen bombs was mastered some time ago, unfortunately, efficient fusion reactors remain elusive. The main difficulty is in creating the extreme conditions required to provide nucleons with enough energy so that mutual collisions allow them to overcome electrical repulsion and thus to fuse. Enormous lasers provide this energy in some systems but the input is so high that there is little more energy created than supplied. Nevertheless, steady progress towards better efficiency is being made. It should be noted that working nuclear power stations operate on fission rather than fusion reactions. In this case a large nucleus such as uranium (^{238}U) is split into two smaller nuclei. Mass is transformed and energy released. Fusion and fission reactions release energy when the nuclei involved are small or large respectively. The iron nucleus (^{56}Fe) represents the cross-over size which means that any nuclear reaction that it is involved in absorbs energy as mass is created. Such reactions are therefore unlikely to occur except under the most extreme conditions in the universe (for example, during supernovae, see Chapter 13).

Energy Transport Within the Sun

Most of the enormous quantity of energy that is created in the Sun's core is in the form of electromagnetic radiation, mainly gamma-rays. As discussed in Chapter 1, the Sun's electromagnetic radiation output peaks in the visible part of the spectrum and is considerably weaker at higher frequencies. The energy must be transported from the core to the surface and in so doing the average wavelength of light must be enormously red-shifted. Two main types of energy transport have been discussed in this book so far and these are also the processes of most importance in the Sun.

In Chapter 1 radiation was discussed. This involves the transmission of energy by coupled oscillating electric and magnetic fields. Energy in the core is produced in this form but the plasma density is such that the gamma-ray photons will not travel very far before encountering matter. In the core the plasma consists almost entirely of nuclei and electrons. The gamma-rays can be scattered by a moving charged particle. In so doing both the matter particle, most likely to be an electron, and the gamma-ray photon change directions and exchange energy. The rules that govern these processes are similar to those that determine the outcome of collisions between solid particles. For electron–photon collisions this usually causes a small red shift of the electromagnetic radiation. That is, the photons gradually lose energy to the plasma. The light thus bounces around in the core, being scattered by electrons in random directions at very frequent intervals. Rather than travelling at its unfettered speed of 300 million metres per second it becomes almost trapped in the core, effectively averaging only a few hundred metres per year! At the same time the gamma-rays gradually lose energy and are transformed into X-rays.

The region that surrounds the core is a little cooler and, though it is still unimaginably hot, some nuclei are able to attract a few electrons into orbits about them thus becoming ions. Larger, highly charged nuclei can best achieve this at high temperatures. The electrons trapped in ions may now interact with the X-ray photons in a different way. The electrons may wholly absorb a photon and with this extra energy once again escape the nucleus. Such is the density of the plasma that the electron will quickly be trapped by another nucleus or ion and the energy is again released as a photon of electromagnetic energy. Thus the photon continues its random walk from the core towards the surface through a chain of chance absorption and emission events. Going from the core outwards the temperature decreases and nuclei are able to trap more and more electrons, producing less highly charged ions. Electron recombination processes with weakly ionised atoms release less energetic photons so that the spectrum of ambient photons is further red-shifted to the ultraviolet and visible regions. The zone throughout which scattering and absorption processes dominate energy transfer is known as the radiative mantle and stretches from the core to a radius of about half a million kilometres.

Eventually, the temperature of the plasma drops sufficiently such that the vast majority of nuclei have trapped electrons, large nuclei having trapped many. The absorbence of such ions is considerably stronger than that of highly ionised species and the progress of energy transport via radiative processes is thus slowed further. However, the density of the plasma has decreased considerably, making it more mobile and allowing convection processes to operate. Convection has been described in Chapter 9 to explain the passage of heat through the giant planets. In the Sun, the magnitude of the temperature, pressure and volume of matter are all very much larger but the principle remains the same. Pockets of hot plasma rise to the surface and are replaced by cooler regions. The so-called convective shell delivers the energy created in the core (and transported throughout most of the Sun's volume by radiative transfer) to the surface. This last stage of the journey is much faster but, even so, the distance travelled from the core is at most 700 000 km and energy created there may take as long as a million years to arrive at the surface.

The Sun's surface region is known as the photosphere and is the place at which radiative transfer once again takes over. The difference is that there is very little matter to impede the radiation. The Sun's atmosphere absorbs a little energy but the vast majority of it now leaves as unhindered electromagnetic radiation. Eight minutes later it has reached the Earth's orbit and two months later it is passing through the Oort cloud at the extreme edges of the Solar System. In just a few years' time the light will probably be causing the Sun to appear in the night sky of a planet orbiting a nearby star.

The Sun's Surface

Convection currents, coupled with rapid rotation, cause the banding of the giant planets. As explained in Chapter 7, the Sun's rotation is very slow as it transferred most of its angular momentum to its planetary nursery early in its life. It rotates on its axis about once per month and this is too slow to cause hot and cool regions to be spread into bands. Instead, warmer and cooler regions of the Sun's surface remain in place and cause an effect known as granulation (see Figure 10.3). The photosphere has a slowly shifting, honeycomb-like network of subtly different cells. Each granule is irregular and angular in shape and about 1500 km across. The pattern on the surface gradually changes so that it is refreshed a few times per hour on average. Each cell is bright but bordered by darker, cooler gas returning to be heated deeper in the Sun.

Granulation is not the only feature of the Sun's photosphere. Much more prominent than this mild mottling effect are the dark blotches known as sunspots. These are much larger than granules at a few thousand or tens of thousands of kilometres across and

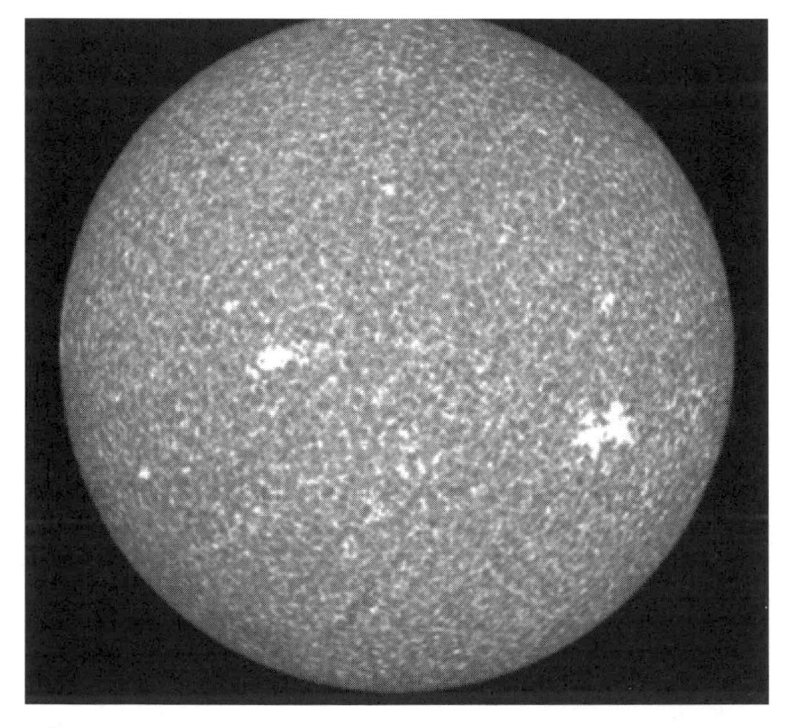

Figure 10.3 Photograph of the Sun showing granulation. The picture is taken by imaging a single wavelength of light (423 nm) that emphasises granulation and causes sunspots to appear as bright regions. (reproduced by permission of the Space Telescope Science Institute)

prevent the granulation effect from taking place in their presence. Their origin is not in heat transport but in a different plasma phenomenon. The motion of a charged particle creates a magnetic field. Plasma consists exclusively of charged particles, all of which are moving at high speeds due to the high ambient temperature. The Sun consequently has a strong magnetic field. On Earth it is easy to take a piece of wire, pass a current through it and measure the magnetic field associated with it. The result will comply with simple physical rules. In principle, the same is true of charged particles moving within the Sun. The difference is that the motion of these particles is highly complicated. Among other factors, the local magnetic field can be influenced by temperature, mobility of the local region of plasma and latitude. This latter variable is relevant because, like Jupiter again, the Sun experiences differential rotation whereby equatorial regions rotate more rapidly than polar zones.

To gain some understanding of the movement and nature of sunspots without a course in plasma physics it is necessary to accept that, through the various internal motions of the Sun, rope-like magnetic fields are set up. These ropes are twisted by the differential rotation of the Sun and occasionally break out of the surface, before arcing back in. When this happens the usual convection patterns are disrupted as the thermodynamic force of the rising plasma is not enough to disrupt the magnetic field lines. The part of the photosphere out of which this magnetic rope emerges and then returns is therefore not as strongly heated by hot plasma from below. Its temperature drops and, according to blackbody radiation laws (Chapter 1), it becomes much duller relative to its surroundings. Sunspots thus seem black next to the brilliance of the surrounding photosphere though their temperature has typically dropped to only about 4200 K.

WORKED EXAMPLE 10.3

Q. How much less power per unit area is radiated by a sunspot than by a typical region of the Sun's surface?
A. The average surface temperature of the Sun is 5800 K (see Worked Example 1.3) and the power per unit area radiated by an object is given by equation (1.3) so that the ratio between the radiated powers of two surfaces with temperatures of 5800 K and 4200 K is given by

$$\frac{P_{A_1}}{P_{A_2}} = \frac{\sigma T_1^4}{\sigma T_2^4} = \frac{T_1^4}{T_2^4} = \frac{5800^4}{4200^4} = 3.6$$

Thus, a sunspot radiates nearly four times less power per unit area as the rest of the Sun and consequently appears black. Observed without the bright background of the Sun, a sunspot would brightly glow red.

A close examination of sunspots reveals that they always appear in pairs, one of which is the equivalent of a magnetic north pole and the other a south pole, as in the opposite ends of a bar magnet, for instance (illustrated in Figure 10.4). During a period of about 11 years sunspots tend to be found at regularly varying latitudes, appearing in north and south hemispheres at equivalent latitudes that gradually decrease until they reach the equator. Magnetic north and magnetic south pole sunspots always follow each other around the Sun in the same order (though reversed between latitudes) during an 11-year cycle. At the start of the next cycle the sunspot polarities are reversed (in both hemispheres). Thus sunspots follow a 22-year cycle. This is known as the butterfly cycle after

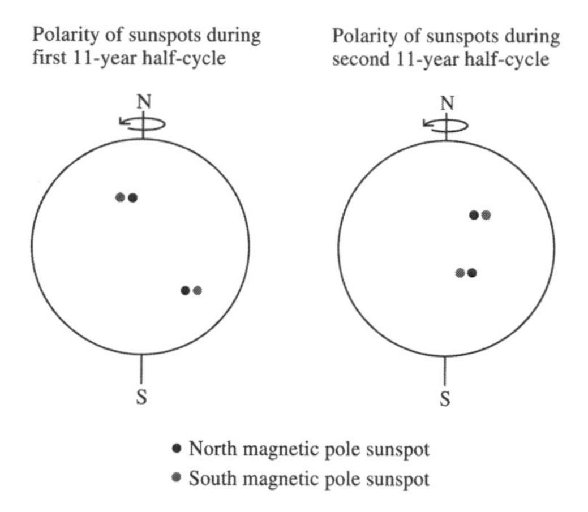

Figure 10.4 The magnetic polarity of sunspots

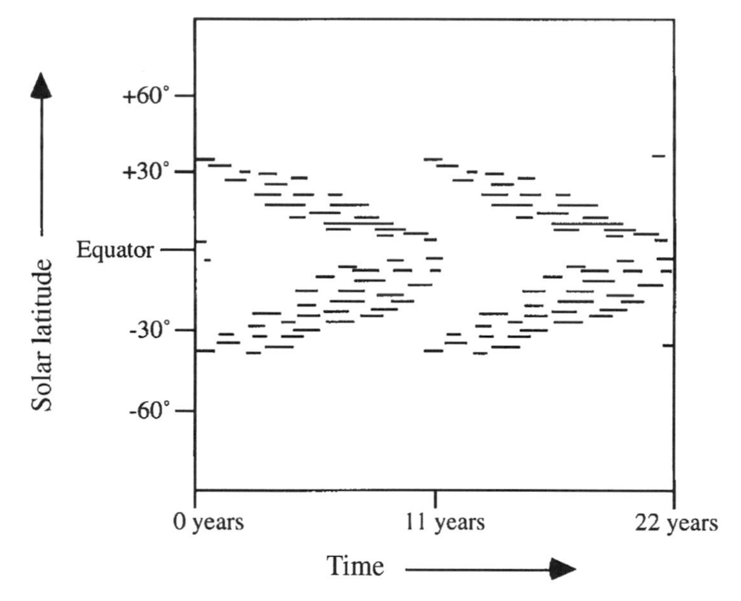

Figure 10.5 The butterfly diagram. Typical positions and (exaggerated) durations for sunspots are indicated as a function of time

the appearance of a plot of actual sunspot positions (Figure 10.5). This cycle is intimately associated with the Sun's differential rotation that causes the magnetic field lines to become twisted and thus break the surface in somewhat predictable locations. The fine details of many of these processes remain unclear.

The Solar Atmosphere

The Sun's magnetic field and its local twists and turns are associated with an almost bewildering array of phenomena on the surface of the Sun and in its atmosphere. Before describing the role of magnetism in their form and behaviour it is helpful to give a descriptive overview.

As for Jupiter, it is not a simple matter to define where the surface of the Sun ends and the atmosphere begins. The Sun's plasma becomes gradually thinner in moving outwards all the way from the core to the extreme outer part of the atmosphere. The photosphere is defined as being the region in which the density drops off quite rapidly by a factor of around a thousand in the space of a few hundred kilometres. It is also the region in which radiation once again becomes the dominant mechanism for energy transfer, now relatively unfettered by very high rates of absorption and re-emission. Through the photosphere the temperature drops from around 10 000 K to about 6000 K and continues to drop through the lower part of the atmosphere before beginning to slowly increase again (see Figure 10.6). The region of slow temperature variation in the Sun's lower atmosphere is known as the chromosphere. Through this region of 2000 km or so the temperature drops to about 4000 K and then creeps back up towards 10 000 K. The next sector of the solar atmosphere is delineated by a sudden temperature increase, by a factor of more than ten. After this step, the temperature continues to increase to above 1 million kelvin, the actual value varying significantly with time and position. This sparse but very hot region of plasma is known as the corona.

The chromosphere and corona together constitute the Sun's atmosphere. They are best observed during solar eclipses (see Worked Example 5.3). As the angular sizes of both the Sun and Moon vary slightly due to the non-circular orbits of the Earth and Moon, the Moon may block different portions of the Sun's atmosphere during solar eclipses. Sometimes the chromosphere can be seen, under which circumstances the corona is difficult to make out next to the much brighter inner layer. When the chromosphere is

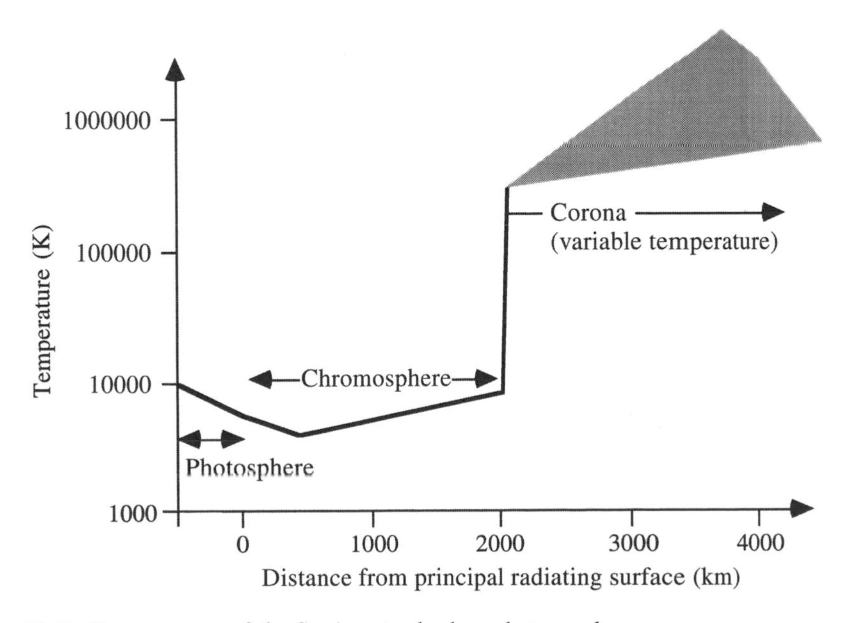

Figure 10.6 Temperature of the Sun's outer body and atmosphere

blocked by the Moon the corona's form is easy to identify. It is not a spherical shell like most atmospheres but is instead an irregular and time-varying collection of regions that stick out into space like tufts of unkempt hair. In some parts they may be visible more than a million kilometres from the Sun's surface while in other directions there may be almost no visible evidence of a corona at all. In fact visual evidence does not give the whole story as the corona stretches off into the Solar System becoming gradually more tenuous. Nevertheless, the corona is highly irregular.

There are three main additional features of the Sun's atmosphere that are easy to observe and important to describe. These are flares, spicules and prominences. Flares and prominences are usually observed in the active regions of the Sun near sunspots while a network of spicules fills the chromosphere and extend into the corona. Spicules are luminous tongues that extend upwards from the photosphere. They are arranged in grids, as if the Sun had been caught in a fine net. The net changes shape on the scale of about a day. Prominences are generally longer-lasting but much more sparsely scattered. They are essentially huge clouds of hydrogen gas that hang above the photosphere, extending into the corona. They are luminous but appear dark when viewed against the photosphere's brightness (sometimes being known as filaments under these circumstances) due to the absorption of light by the constituent hydrogen atoms. Light absorption is much stronger in filaments than for the corona as a whole because of the density of the filaments that have condensed from the much sparser surrounding corona. When viewed beyond the Sun's limb prominences are bright due to the re-emission of this same energy resulting in a more spectacular appearance. Flares also stretch from the photosphere out into the corona. These occasional events are eruptions of the Sun. They rise to a peak intensity in a few minutes and then fade over a period of an hour or so (Figure 10.7). Their intensity varies greatly but they may expel as much as 10^{30} J as well as many charged particles and reach temperatures of 10^7 K. As both flares and prominences occur in active regions of the Sun they may sometimes interact. A flare may cause a prominence to erupt itself with the latter being dispersed throughout the corona or further into space.

The behaviour of all the regions and features of the outer Sun are strongly influenced by local magnetic fields. As discussed above, sunspots occur in pairs that represent places in which the Sun's local field has looped out of and then back into the photosphere. Entry and exit spots have opposite polarities. In between, magnetic field lines loop through the

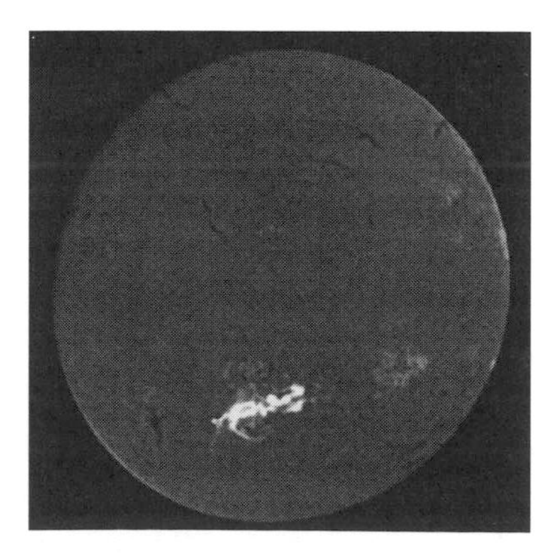

Figure 10.7 A massive flare observed on the surface of the Sun. The flare extended over 300 000 km across the surface and lasted for about an hour. Note how dim the Sun appears compared to the flare. This picture was taken using light at 656 nm (Hα) from Holloman Air Force Base (reproduced by permission of the Space Telescope Science Institute)

solar atmosphere. In a plasma, matter and magnetic fields are intimately connected. This is the reason that the differential rotation of the Sun causes the 22-year butterfly pattern. It also allows matter to be suspended by magnetic fields. Prominences are essentially just that. Gas is gravitationally attracted to the Sun but prevented from falling to the surface by the local magnetic field. The dynamics of prominences are actually somewhat more complex but this simple model does explain why prominences are associated with sunspots. Flares are also associated with such active regions. The formation of a flare is somewhat more complicated, however. A good analogy is with a dripping tap. If water had no surface tension then drips would not be stochastic; a slowly leaking tap would form a very weak stream of continuously flowing water. The Sun's magnetic field holds matter in place in the same way that surface tension holds water. In regions between sunspots north and south poles can get very close to each other. A sunspot pair does not consist of two symmetric, regular regions. They are irregular and their edges can overlap. Where this happens, the north and south polar fields suddenly cancel each other out. The annihilation of the field that is holding a ball of plasma in place suddenly frees it and it may shoot off into space. It is a very energetic tap that drips.

Spicules are associated with a global magnetic network. Above the photosphere a second granulation effect takes place and is known as supergranulation. The cells involved in this process are a few times larger than granules and magnetic field measurements show that fields are strongest at the cell boundaries, where the spicules are. This is again a case of a feedback interaction between the plasma and the magnetic field. The supergranules carry plasma from the centre of each cell outwards. Weak local magnetic fields are carried by this motion. Matter dives back down again at cell boundaries and this concentrates the magnetic fields in these positions. The stronger fields in these positions, in turn, concentrate matter to create rows of spicules.

The magnetic field of the Sun extends into the corona and controls its form. It should be recalled from Chapter 7 that extrasolar effects are strongly influenced by the Sun's magnetic field, for instance via the transfer of angular momentum from the nucleus to the disk of the Solar Nebula during the Solar System's formation. The corona is considerably sparser than the Solar Nebula was and much hotter. This implies that it will be strongly governed by the magnetic field due to its low mass and highly ionised state. In some regions of the Sun, for instance near sunspot pairs, magnetic field lines loop in and out of the Sun and thus hold the corona in place. In between, the magnetic fields are not tightly closed and the plasma of the corona is able to fly off into space. Plasma escapes from these coronal holes and creates what is known as the solar wind. Observing the corona during an eclipse, for instance, allows the Sun's magnetic field to be directly observed. In regions of high coronal intensity the field is tightly closed whereas dim regions imply more open fields where the plasma has escaped (see Figure 10.8). This effect is best observed by detection in the X-ray region (Figure 10.9). It is then not necessary to observe during an eclipse as the very hot plasma significantly outshines the cooler photosphere in this region of the electromagnetic spectrum according to blackbody theory (Chapter 1). It is not fully understood why the corona is so hot but it is thought also to be connected with the Sun's magnetic field. It should be noted that, despite its much higher temperature, the corona is significantly less energetic than the photosphere due to its much lower density. There is therefore no difficulty with energy conservation violation.

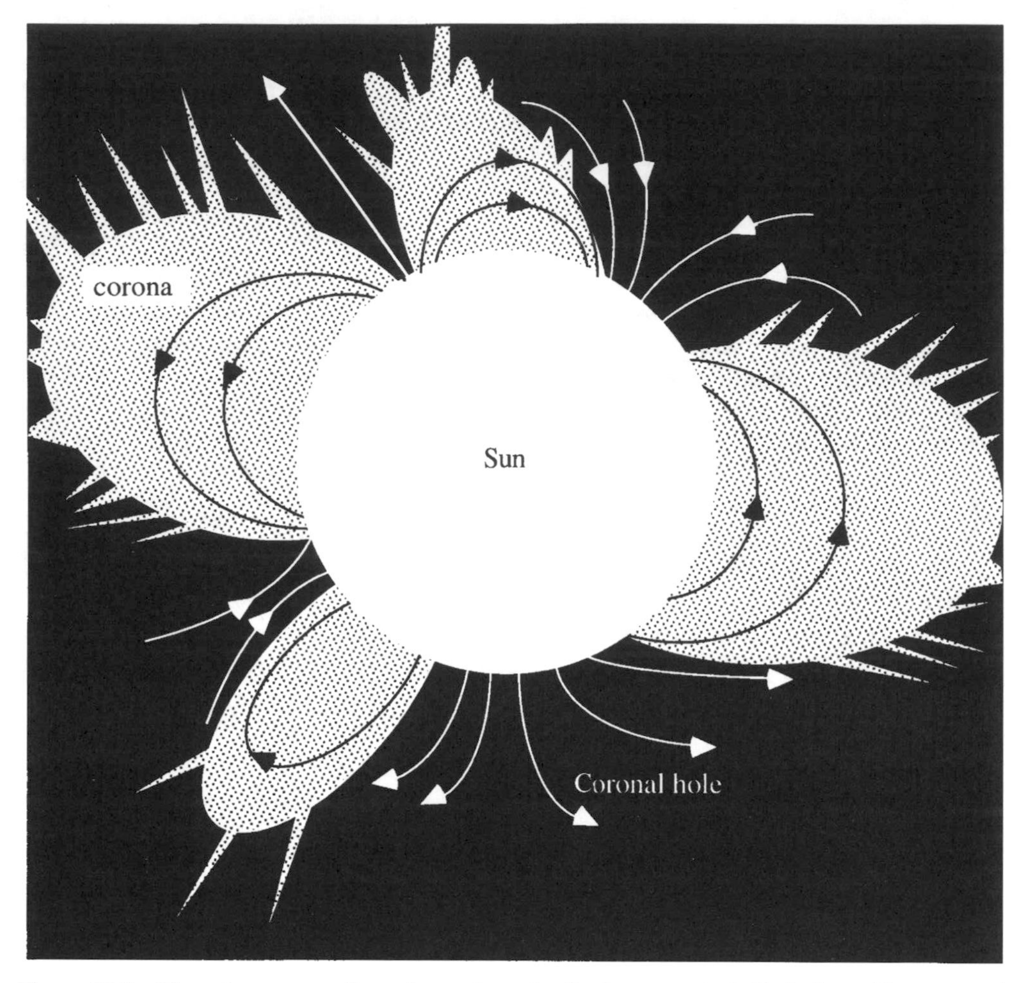

Figure 10.8 The solar corona. In regions where the Sun's magnetic field (indicated by arrowed lines) loops quickly back to the surface of the Sun the charged matter of the corona is held and causes the region to be bright. In regions where the field lines arc into (or from) space the matter is not held and a dark coronal hole appears

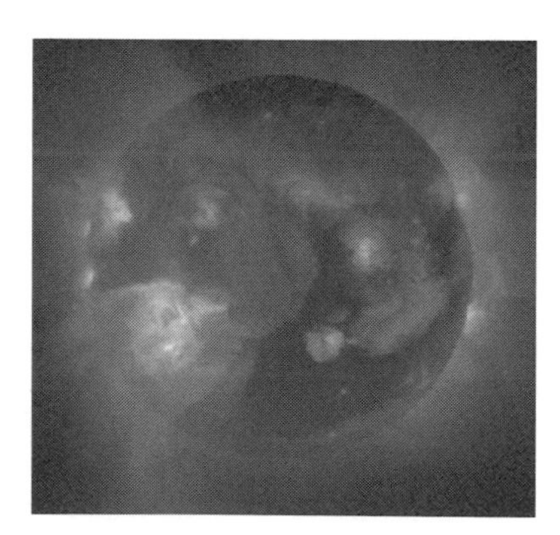

Figure 10.9 The Sun observed in X-ray wavelength radiation. The planetary surface is too cool to emit significantly in this region but the corona is much hotter. The corona's magnetically twisted shape can therefore be seen glowing (reproduced by permission of the Space Telescope Science Institute)

The Solar Wind

A large number of particles stream out of the coronal holes and the flux is markedly increased during solar flares. The majority of particles that leave the corona to constitute the solar wind are both charged and fast moving. In the same way that the light flux decreases with the square of the distance from the Sun then so does the particle flux. Cometary appearance reflects this. As mentioned in Chapter 5, a comet's tail always points away from the Sun and this is due to interaction with the solar wind. The charged particles cause gas and dust to break free from the cometary nucleus to stream away from the Sun. This explains why the tail precedes the nucleus when the comet is leaving the Solar System and why the tail is largest close to the Sun where the solar wind is strongest.

When the particles that make up the solar wind encounter a planet they may become trapped in one of two ways. The simplest way is to be gravitationally trapped. In the case of Mercury, which is close to the Sun, the solar wind contributes a significant fraction of the planet's (very weak) atmosphere. In the vicinity of a planet the charged particles are able to combine to make electrically neutral particles but for many planets the charged particles become trapped far above the atmosphere by the planetary magnetic field. These regions are known as the Van Allen radiation belts in the Earth's case and form the inner part of the magnetosphere. This is the region in which the Earth's magnetic field interacts with the magnetic field that the solar wind brings with it. The magnetosphere can be thought of as containing a very sparse plasma consisting mainly of protons (hydrogen nuclei) and electrons. The magnetosphere is stretched and distorted by the fact that the source of its particles, the solar wind, is directional, as is the Earth's magnetic field.

The Van Allen belts are continually being supplied with charged particles and can become overfilled. Here the dripping tap analogy becomes useful again. When the belts become overfilled a large pocket of plasma is released into space suddenly. At the same time the recoil spills particles into the Earth's atmosphere. Here they can collide with uncharged molecules or combine with oppositely charged particles to produce uncharged particles. These processes release light. The magnetic field of the Earth channels the particles toward the surface near the poles. This light can be seen, especially at high latitudes, when looking in the direction of the local celestial pole (which is close to being above the Earth's local magnetic pole). Such displays are known as aurorae; aurora borealis (or northern lights) in the northern hemisphere and aurora australis in the south. The position of coronal holes are irregular and not easy to predict so that the solar wind blows with greatly varying intensities. Aurorae are therefore doubly difficult to predict as the dripping tap is being fed by an erratic supply. One occasion on which aurorae can be predicted is after a large solar flare is observed. The light from a flare takes only a few minutes to arrive at the Earth but the particles follow a few days later. The large flux of particles has a good chance of overfilling the Van Allen belts and aurorae are likely to result at this time. Furthermore, flares are associated with sunspots, thus following an 11-year pattern.

What is known about stars largely derives from what is known about the Sun. What has been written about the Sun here barely scratches the surface of what is known. Some phenomena, such as solar vibrations to name but one, have not been mentioned. The overview that has been presented is sufficient, however, to allow this narrative to take its first steps outside the Solar System to investigate the universe of stars beyond.

Questions

Problems

1 The peak wavelength of electromagnetic radiation output from the Sun (radius, 7×10^5 km) is at 500 nm.
(a) What temperature is the Sun's photosphere?
(b) How much energy does the Sun radiate per second?
(c) If the average mass transformed during a nuclear reaction in the Sun's core is 8.8×10^{-30} kg, how many reactions take place per second?

2 (a) How much mass is converted to energy during each of the three stages of the proton–proton chain (reactions (10.4), (10.5) and (10.6))?
(b) One proton–proton chain involves reaction (10.4) taking place twice, reaction (10.5) taking place twice and reaction (10.6) taking place once. According to the answers in part (a), what is the total mass converted during one proton–proton chain?
(c) The masses of the reactants and products in the whole proton–proton chain, shown in reaction (10.7), have a mass defect of 4.8×10^{-29} kg. Why is this value different from the one calculated in part (b)?

Teasers

3 How many proton–proton chains would have to be completed to release the same amount of energy as an apple falling from a tree to the ground?

4 The nucleus with the highest binding energy per nucleon is ^{56}Fe. What happens if such a nucleus is involved in a nuclear fusion or fission reaction?

5 What is the ratio between the forces of electrostatic repulsion and gravitational attraction between two protons in free space? The mass and charge of a proton are 1.7×10^{-27} kg and 1.6×10^{-19} C, respectively. The universal gravitational constant, G, is 6.7×10^{-11} N m^2 kg^{-2} and the permittivity of free space, ε_0, is 8.8×10^{-12} C^2 m^{-2} N^{-1}.

Exercises

6 Explain how the energy created in the core of the Sun reaches the photosphere.

7 Explain what is happening when an aurora is visible in the Earth's sky.

8 What process generates the vast majority of the Solar System's energy and where does it take place? Give a brief explanation of the process.

9 What is the butterfly pattern?

10 Describe the Sun's atmosphere.

11 Studying Stars

The stage has been reached where the Solar System can be left behind and the universe of stars, galaxies and other exotic objects explored. A scaled-down version of the Solar System was discussed in Chapter 5 and it was mentioned that the inclusion of even the nearest star beyond the Sun rendered a scale model that includes both impractical. In other words, though the Solar System is enormous, it would appear as a mere blip on a model of even the most local part of the universe. Despite the Solar System's emptiness, it would provide a relatively significant contribution to the local mass density. The size of the Solar System is minute compared to the vastness of the universe but its density is appreciable relative to the emptiness of outer space.

Measuring Stellar Distances

A question that it is often worth asking is: how do you know? There are various methods of measuring distances of far-distant bodies and the technique employed mainly depends on the scale of the separation from the Earth. To make a measurement, perspective is required. To measure the length of this book a ruler is required so that the book's dimensions can be compared to a known scale. Looking out at the night sky's myriad point sources of light it is difficult to imagine an appropriate yardstick for measurement. The angular size of even the closest star is barely resolvable using the most sophisticated techniques. Even if angular size could be measured it would provide only a ratio of stellar radius to distance. Only the Sun's radius could be used as a definitive comparator but its size could certainly not be regarded as being a constant for all stars. The best yardstick is actually the Earth's orbital diameter. While this distance is still small on the scale of the universe, it is large enough to make a start on determining stellar separations. Its small size turns out to be convenient for the technique of trigonometrical parallax.

The celestial sphere was considered in Chapter 3 to be immutable with the exceptions of local stars that have measurable proper motions and the shift of coordinates caused by precession. Consider the following scenario, however. A nearby star is situated on the celestial sphere in the region of several faraway stars. When the Earth, the Sun and these stars are positioned perpendicular to each other (see Figure 11.1) the nearby star will appear in a specific position relative to the background of faraway stars. Six months later the perspective will have changed and the nearby star will appear to have shifted in the sky relative to the faraway stars. This works because the vast majority of stars are so far away

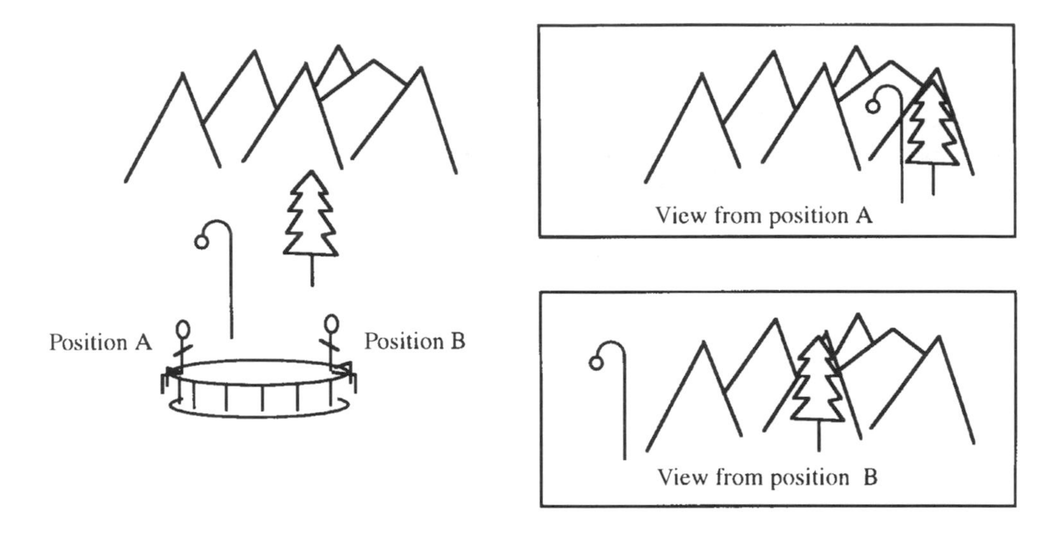

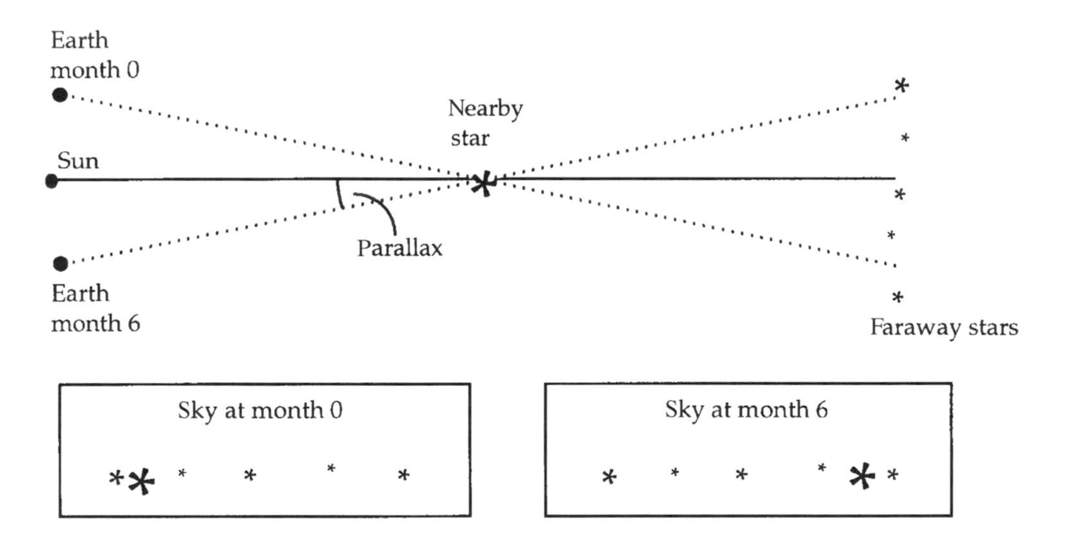

Figure 11.1 Perspective and trigonometrical parallax. Above, a person moves on a roundabout thus varying position considerably with respect to nearby objects but relatively little compared to faraway features. The nearby lamp-post therefore appears to move across the whole of the background mountains during one revolution. The tree in the middle distance changes its position slightly. The same effect is demonstrated for a star in the lower diagram. As the Earth changes its position in space due to its revolution about the Sun the apparent position of nearby stars appears to vary relative to faraway stars. In this diagram the relative distance of faraway stars has been underexaggerated but the parallax angle has been exaggerated (largest actual value for a real star is $0.77''$).

that the change in the Earth's viewing position effectively makes no difference to their position on the celestial sphere. There is a large backdrop of fixed stars. Only a few hundred can be seen to shift in position with an annual cycle using Earth-based telescopes, even with today's technology. This shift is less than two arcseconds even for the closest stars.

The magnitude of the shift of a star due to trigonometrical parallax is inversely proportional to the distance at which the star lies. To see this consider sitting on a children's roundabout looking at an extended view. Distant mountains might provide a backdrop against which nearer objects appear to move. A lamp post a few metres away will slide backwards and forwards across the mountains with a large oscillation as the roundabout rotates. A tree a few tens of metres away will oscillate at exactly the same rate, as determined by the motion of the roundabout, but the number of mountains across which the tree moves will be considerably less.

The angular shift in position of a star across the celestial sphere is used to define the unit of distance that is generally used for stellar studies. This unit is the parsec (from PARallax arcSECond) and is defined as being the distance of a star that has a parallax of one arcsecond (or, equivalently, the distance at which the mean radius of the Earth's orbit subtends $1''$). The parallax angle is actually half the total shift of a star's position in a full year as indicated in Figure 11.1.

Using the definition of a parsec it is easy to see that

$$d = 1/p \tag{11.1}$$

where the star's distance from the Earth (in parsecs) is d and p is its parallax. The closest star to the Earth is called Proxima Centauri and its parallax has been measured to be $0.77''$. It is simple to see therefore that Proxima Centauri is $(1/0.77) = 1.3$ parsecs from the Earth. The measurement of parallax is therefore not an easy matter as atmospheric resolving powers limit determination of stellar position to values close to the very scale on which real parallaxes are measured. This is compounded by proper motions that are typically somewhat larger than parallax values. For instance, Proxima Centauri has a proper motion of $3.85''$ per year and this must be subtracted from the star's overall motion to find the component due to trigonometrical parallax.

WORKED EXAMPLE 11.1

Q. How big is a parsec in astronomical units, metres and light years?
A. One parsec is the distance at which one astronomical unit subtends one arcsecond. Invoking the small angle approximation,

$$1'' \text{ (converted to radians)} = \text{separation/distance} = 1 \text{ AU}/1 \text{ parsec}$$
$$\therefore 2\pi/(360 \times 60 \times 60) = 1 \text{ AU}/1 \text{ parsec}$$
$$\therefore 1 \text{ parsec} = (360 \times 60 \times 60)/2\pi \text{ AU} = 2.06 \times 10^5 \text{ AU}$$

There are about 200 000 astronomical units in a parsec. Note that the closest star to the Earth beyond the Sun is 1.3 parsecs away, so Proxima Centauri is 260 000 times further away from the Earth than the Sun.

One astronomical unit is the Earth's mean distance from the Sun which is 150 million km. One parsec in metres is therefore $(1.5 \times 10^{11} \times 2.06 \times 10^5) = 3.1 \times 10^{16}$ m. One light year is the distance travelled by a light wave in one year:

$$\begin{aligned}\text{Distance travelled} &= \text{velocity} \times \text{time}\\ &= 3 \times 10^{8}\,\text{m s}^{-1} \times (365 \times 24 \times 60 \times 60)\ \text{s}\\ &= 9.5 \times 10^{15}\,\text{m}\end{aligned}$$

There are therefore ($3.1 \times 10^{16}/9.5 \times 10^{15}$ =) 3.3 light years in one parsec. In describing the enormous distances and sizes of the universe parsecs and light years can be used almost interchangeably as they represent units of the same order of magnitude. Obviously, correct units must be used in quantitative calculations.

Stellar Brightnesses

To determine the distance of the faraway stars other techniques must be used and these will be considered later in the book. For now, a large enough collection of stars can be measured using parallax such that it is possible to start studying variation in stellar properties.

The crucial stellar parameter that can be determined once the distance of a star is known is luminosity. Luminosity is a measure of how much light power a star radiates. In Chapter 1 calculations were performed to determine the power output of the Sun and the proportion of that power that is incident upon the Earth. As the energy is radiated isotropically it is spread evenly over any concentric spherical shell centred on the Sun. The same is true for any star. As the surface area of a sphere is proportional to the square of its radius then the energy flux at any point is inversely proportional to the square of the distance from the star. If a star's apparent intensity (how bright it appears from Earth) can be measured then, if its distance is known, its absolute (actual) intensity can be calculated.

WORKED EXAMPLE 11.2

Q. If a star identical to the Sun were at a distance of 10 parsecs from the Earth, what would the star's radiant intensity upon the Earth be, given that the solar constant is 1400 W m^{-2}?

A. The solar constant (Worked Example 1.4) describes the intensity of the Sun's power upon the Earth's surface. An identical star placed at a distance of 10 parsecs would radiate the same total power as the Sun but the energy would be spread over a much larger sphere by the time it reached the Earth. The size of the sphere is characterised by its radius, given by the distance between the Earth and the star. The surface area of the sphere is given by $4\pi r^2$ and the power is equally spread over this surface. Thus the power density decreases with the square of separation between the Earth and the star. The star is 10 parsecs from Earth, relatively close by universal standards, but this is ($10 \times 206\,000 \approx$) 2 million times further away than the Sun (see Worked Example 11.1). The intensity of light from the star incident upon the Earth is thus ($1400/(2 \times 10^6)^2 =$) 3.5×10^{-10} W m^{-2}. This apparently tiny quantity of light would still easily be detectable in the night sky, even with the naked eye.

As stellar separations and intensities can vary over many orders of magnitude it is more convenient to use logarithmic scales. Rather than using a scale based on SI units, astronomers utilise relative measures of intensity. The system that is used is to define the apparent intensity, i, of stars relative to the bright star Vega, so that $i_{\text{Vega}} = 1.0$. The conversion to a logarithmic scale is made in the following way:

$$m = -2.5 \log_{10} i \tag{11.2a}$$

Here, m is known as the apparent magnitude of the star and is an indicator of how bright the star appears in the night sky. As $i_{\text{Vega}} = 1.0$ then Vega has an apparent magnitude of 0.0. The few brighter stars (see Appendix 5) have negative apparent magnitudes, the brightest of all being Sirius at -1.5. Dimmer stars have larger apparent magnitudes. For instance, a star with $m = 2.5$ is ten times dimmer than Vega and one with $m = 5.0$ is one hundred times dimmer. The naked eye can detect stars up to around sixth or seventh magnitude ($m = 6$ or 7) enabling about 4000 stars to be visible without the use of a telescope.

It is clearly more useful to an astrophysicist to know absolute information about stars and so a conversion to absolute magnitude, denoted by M, is required. This proceeds simply from knowing that the apparent intensity is proportional to the absolute intensity, I, divided by the square of the separation between observer and light source:

$$i = \frac{c_{\text{p}} I}{d^2}$$

where c_{p} is an arbitrary constant of proportionality. Rearranging gives

$$id^2 = c_{\text{p}} I$$

and taking logarithms followed by multiplying by -2.5 gives

$$-2.5 \log_{10} id^2 = -2.5 \log_{10} c_{\text{p}} I$$

$$\therefore -2.5 \log_{10} i - 2.5 \log_{10} d^2 = -2.5 \log_{10} c_{\text{p}} - 2.5 \log_{10} I$$

By analogy with equation (11.2a), absolute magnitude and intensity can be related thus:

$$M = -2.5 \log_{10} I \tag{11.2b}$$

so that

$$m - 5 \log_{10} d = M - 2.5 \log_{10} c_{\text{p}}$$

To decide upon the connection between apparent and absolute magnitudes a second arbitrary definition must be made. The definition is made in such a way that absolute and apparent magnitudes are equal when the star is viewed from 10 parsecs. The (non-SI) unit for d must therefore be parsecs. Thus $2.5 \log_{10} c_{\text{p}}$ must be equal to $(5 \log_{10} 10 =)$ 5 to give

$$M = m - 5 \log_{10} d + 5 \tag{11.3}$$

When $d = 10$ parsecs, $m = M$, as required. A fully linked system has thus been developed to connect a star's apparent brightness to its actual intensity in which the only other information required is the star's distance.

WORKED EXAMPLE 11.3

Q. The Sun is 4.8×10^{10} times brighter than Vega as measured from Earth. (a) What is the Sun's absolute magnitude given that Vega is 8.1 parsecs from the Solar System? (b) How much more intrinsically bright is Vega than the Sun?

A. (a) The magnitude measurement system is based on Vega's relative intensity being unity. This means that the Sun's relative intensity, on the same scale, is 4.8×10^{10}. Therefore, invoking equation (11.2a):

$$m_S = -2.5 \log_{10} i_S = -2.5 \log_{10}(4.8 \times 10^{10}) = -26.7$$

There are 2.06×10^5 astronomical units in a parsec (see Worked Example 11.1) so that the Earth's separation from the Sun is $1/(2.06 \times 10^5)$ parsecs. This value for d, as well as the calculated value for m_S, can thus be inserted into equation (11.3):

$$\begin{aligned} M_S &= m_S - 5 \log_{10} d_S + 5 \\ &= -26.7 - 5 \log_{10}(1/(2.06 \times 10^5)) + 5 \\ &= -21.7 + 5 \log_{10}(2.06 \times 10^5) \\ &= 4.9 \end{aligned}$$

The Sun's absolute magnitude is thus +4.9.

(b) Vega's absolute magnitude can be calculated as for the Sun's using equation (11.3);

$$\begin{aligned} M_V &= m_V - 5 \log_{10} d_V + 5 \\ &= 0.0 - 5 \log_{10} 8.1 + 5 \\ &= 0.5 \end{aligned}$$

It is now possible to apply equation (11.2b) twice, to the cases of the Sun and Vega:

$$0.5 = -2.5 \log_{10} I_V \quad \text{and} \quad 4.9 = -2.5 \log_{10} I_S$$

$$\Rightarrow 4.9 - 0.5 = 2.5(\log_{10} I_V - \log_{10} I_S)$$

$$\Rightarrow I_V = 10^{\frac{4.9-0.5}{2.5}} I_S = 57 I_S$$

The Sun shines more than 50 times less brightly than Vega and is thus a fairly unremarkable star, except for its proximity to us.This loads to a general expression that gives the relative power output of two stars of absolute magnitudes M_1 and M_2:

$$I_1 = 10^{\frac{M_2 - M_1}{2.5}} I_2 \qquad (11.4)$$

WORKED EXAMPLE 11.4

Q. The stars Fomalhaut and Pollux appear to be equally bright when viewed from Earth. However, Pollux is 10.7 parsecs away and Fomalhaut is at a distance of 6.9 parsecs. What is the ratio of intensities of the two stars?

A. Though the absolute magnitudes of the two stars are not known the difference between them can be calculated and used to find the ratio of their output powers. It is known that $m_P = m_F$ so that, rearranging equation (11.3), and applying it to each star:

$$m_P = m_F = M_P + 5 \log_{10} d_P - 5 = M_F + 5 \log_{10} d_F - 5$$

$$\begin{aligned} \therefore M_P - M_F &= 5 \log_{10} d_F - 5 \log_{10} d_P \\ &= 5(\log_{10} 6.9 - \log_{10} 10.7) \\ &= -0.95 \end{aligned}$$

The final stage follows exactly as in Worked Example 11.3 above. The difference in absolute magnitudes can be simply substituted into equation (11.4):

$$I_1 = 10^{\frac{M_2 - M_1}{2.5}} I_2 \qquad \therefore I_F = 10^{\frac{M_P - M_F}{2.5}} I_P = 10^{\frac{-0.95}{2.5}} I_P = 0.42 I_P$$

Thus Pollux shines 2.4 times more powerfully than Fomalhaut. This is logical as both appear equally bright on the Earth even though Pollux is further away.

As a quick check it is worth confirming the inverse square law for intensities. If the stars provide equal powers per unit area at the Earth but Pollux is $(10.7/6.9 =)$ 1.55 times further away then the spherical surface over which the light of Pollux is spread is $(1.55^2 =)$ 2.4 times larger than that of Fomalhaut. Thus the light of Pollux is spread out by a factor of 2.4 more than that of Fomalhaut and so Pollux must have an output 2.4 times stronger so that both stars appear equally bright from the Earth. This exercise provides an alternative (and numerically easier) method of solving the original problem.

One difficulty in measuring stellar brightness originates from the fact that stars radiate energy in a manner similar to a blackbody. This means that stars radiate energy throughout the electromagnetic spectrum and to obtain a true intensity value a sum across all frequency ranges must be made. From Wien's displacement law (equation (1.4)) the peak wavelength will shift as a function of temperature so that broadband stellar spectra may vary quite dramatically. Under laboratory conditions this does not constitute a difficulty but when observing extraterrestrial bodies, particularly those at large distances, two related problems emerge. First, the Earth's atmosphere absorbs large amounts of energy especially in the infrared, ultraviolet and X-ray regions. This difficulty can be overcome by using a telescope mounted in space, for instance. For Earth-based observations interpolation can be used to predict the strength of the absorbed energy from the energy that arrives at unperturbed frequencies. A more serious problem is interstellar absorption and scattering. Though the density of matter in open space is very low, such is the distance through which light travels between stars that significant effects are observed. The effects are similar to those of the Earth's atmosphere in that light at particular wavelengths is absorbed and at all wavelengths is scattered. In the Earth's atmosphere, the non-linearity of this latter effect causes the sky to look blue. The same effect distorts the apparent colour of stars. Together, the effects of the Earth's atmosphere and the interstellar medium cause considerable difficulties. For the purposes of this book it is sufficient to say that they can be overcome, in particular for nearby stars that are the present focus. In more advanced books readers may see references to various systems that determine stellar magnitudes in terms of differences in intensities at different frequencies (colours). The simple model that culminates in equation (11.3) above is sufficient for this text.

Stellar Spectroscopy

A short diversion into technology is worthwhile at this point. In order to determine a star's surface temperature its emission spectrum must be recorded, corrected for absorption and scattering and compared to that of an ideal blackbody. In order to record such a spectrum it is necessary to direct the light from the telescope that has gathered it and split the radiation into its component colours. In Chapter 2, the difficulties caused by chromatic aberration were discussed. Light at different wavelengths is bent by differing amounts upon entering or leaving condensed matter. This can be used to advantage in a simple spectroscope. Light entering a prism (Figure 11.2(a)) is split into its component colours. By placing the prism in the beam and rotating it the dispersed beam will shift its position so that at any given position beyond the prism the frequency of light will gradually vary. By mounting a static optical detector behind narrow slits and rotating the prism a spectrum of the star thus can be taken. In fact modern spectroscopes use diffraction gratings rather than prisms to disperse the light. A diffraction grating is a mirror into which a large number of fine, parallel rulings have been drawn (or holographically produced). Through a combination of diffraction and interference the light is split, resulting in much higher instrumental resolution than is available from prisms. To enable high-resolution spectra to be taken it is advantageous to separate the dispersive element from the selective slits by as large a distance as possible as the separation of frequencies increases linearly with distance after dispersion. Spectrometers are therefore often long and bulky. It is inconvenient to place them on the end of a telescope and so they are often found at the Coudé focus (see Chapter 2). Light is usually detected by a photomultiplier but a more rapid technique is to use a linear array of charge coupled devices. In the latter case the slit is dispensed with and the dispersed beam is incident directly on the array, each pixel recording the signal due to a different wavelength range of the spectrum (Figure 11.2(b)).

Diffraction grating based spectrometers can be built that operate throughout the infrared, visible and ultraviolet parts of the electromagnetic spectrum (though internal components may often have to be changed to complete a full scan of such a large range). These three regions include the peak wavelength outputs of all stars. The very hottest stars have surface temperatures of around 50 000 K and peak in the ultraviolet while the coolest peak in the infrared. Due to absorption and scattering it is therefore the hottest and coolest stars that are hardest to characterise.

Difficulties in recording complete broadband spectra are considerable but spectroscopy gives a second chance to categorise stellar temperatures. To understand why this is so it is necessary to consider Kirchhoff's rules of spectroscopic analysis:

(1) Condensed matter, including gas or plasma at high pressure, emits a broadband, continuous spectrum of electromagnetic radiation. This has already been discussed for the Sun and stars and approximates to blackbody radiation for these objects.
(2) A gas (not at high pressure) emits a spectrum that is composed of individual contributions at discrete wavelengths. The intensity of such emission lines increases with temperature.
(3) When light that has a continuous spectrum passes through a low pressure gas it emerges with a spectrum that has a number of darkened lines at discrete wavelengths.

The rules are schematically illustrated in Figure 11.3.

Rules 2 and 3 both relate to the same phenomenon. Electrons bound to an atom or molecule are only able to have energies that are taken from a definite set of discrete values. These energies are specific to the atom or molecule. When an electron changes from one energy level to another its energy must therefore change by a specific amount, known as a quantum. The energy absorbed or emitted by an atom or molecule is often in the form of electromagnetic radiation, the quantum of which is known as a photon (see Chapter 1). In the case of rule 2, an atom is thermally (or otherwise) excited to a high energy state. Such

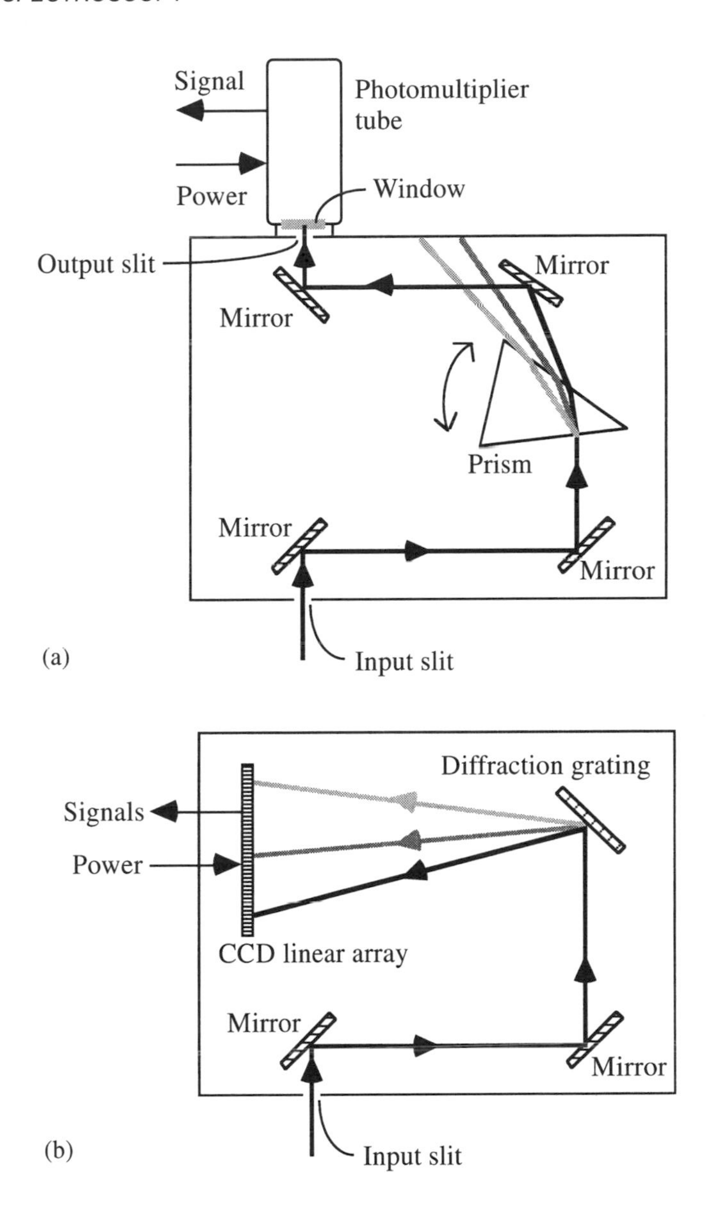

Figure 11.2 Spectrometers. (a) A simple prism spectrometer. The light is split into its component wavelengths by the prism. As the prism is rotated different wavelengths of light are selected by the mirrors and, most critically, by the output slit. Narrowing the output slit improves resolution by rejecting more of the dispersed light. The light intensity is measured by a photomultiplier tube. A spectrum is recorded by rotating the prism and measuring the variation in light intensity. (b) A more modern grating spectrometer. The light is split into its component colours by a diffraction grating. All wavelengths of light are simultaneously incident on a charged coupled device (CCD) linear array. Thus, each element of the array detects light at a different wavelength so that the whole spectrum can be recorded simultaneously

states are generally unstable, however. In relaxing to a lower energy state a photon is released according to Planck's equation:

$$E_P = hf = \frac{hc}{\lambda} \tag{1.5}$$

The photons released must therefore have specific wavelengths. Every atom or molecule has a characteristic series of lines in its emission spectrum according to its allowed energy levels. The same distinctive spectrum appears in rule 3. In this case the atoms or molecules extract photons of just the right energy from the continuous spectrum to allow electrons to be promoted to a higher energy level. Viewed in line, the continuous spectrum has dark lines where the photons have been removed. If the gas is viewed from sideways-on, the emission spectrum of rule 2 is once again observed. This represents the re-emission of the absorbed energy which is radiated isotropically. An example of this idea in action has already been seen in Chapter 10. Prominences appear darker when backlit by the Sun than when seen glowing beyond the Sun's limb.

Generally, the spectra of small atoms are simpler than those of larger atoms. Molecules have even more complex spectra that include occasional band features. In this case, a

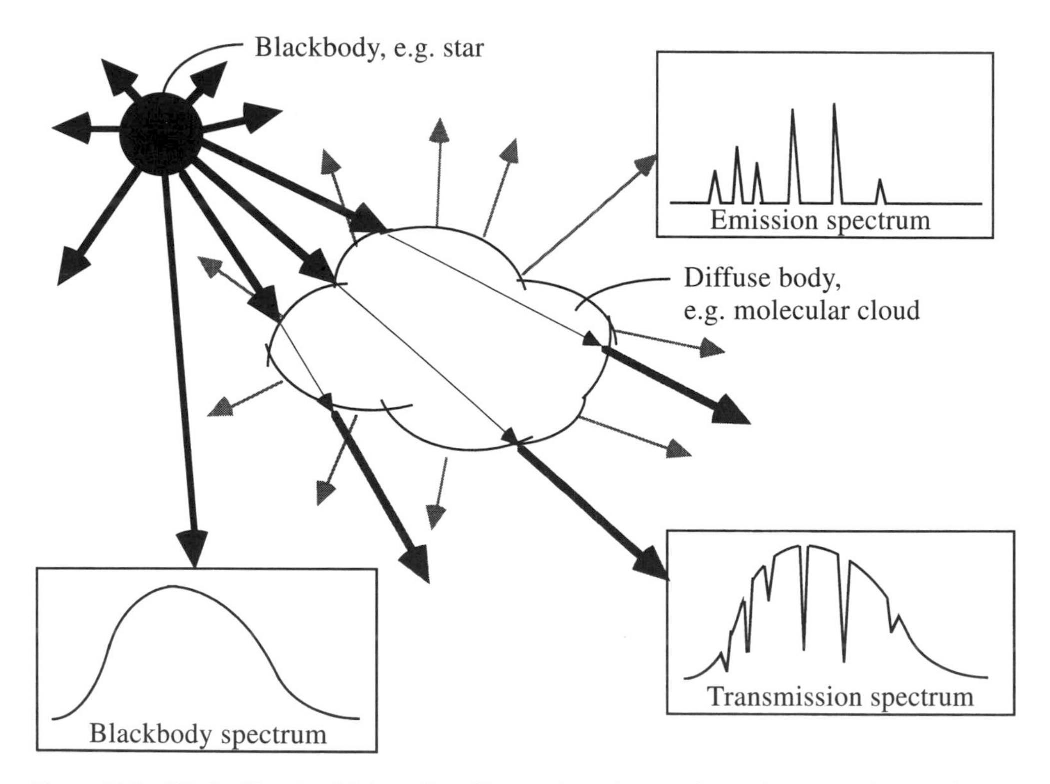

Figure 11.3 Kirchoff's rules. Light radiated by condensed matter is continuous and approximates to blackbody radiation. When such light passes through a low-pressure gas (such as a molecular cloud) discrete wavelengths are partially removed (see transmission spectrum). The light emitted by the gas is at discrete wavelengths, often corresponding to the wavelengths of light that show reduced transmission

whole portion of the electromagnetic spectrum may be absorbed or emitted. When a number of molecules join together to produce a dust particle the bands get increasingly broader until the particles begin to imitate condensed matter. The problem of absorption by interstellar dust is thus a serious one where the temperature is low and molecules are not thermally shaken apart. The Earth's atmosphere represents an intermediate case. The temperature is a few hundred kelvin so that triatomic molecules such as carbon dioxide, water and ozone may hold together. The effect of their absorption has already been discussed in detail but, in short, consists of a series of bites out of, in particular, the ultraviolet and infrared regions. Stellar atmospheres are much hotter than those of planets and are consequently populated by different, generally smaller species. For instance, the Sun's spectrum contains a series of dark lines (known as Fraunhofer lines) that correspond to the absorption spectra of principally hydrogen, calcium, iron and sodium atoms.

It is possible to determine the particles present in stellar atmospheres by comparing the positions of dark lines in stellar spectra with those obtained for species measured in the laboratory. By measuring the spectral width of features, information can be obtained on temperature and pressure. Line splittings sometimes occur in the presence of magnetic fields and this can provide further evidence of stellar behaviour. In the case of the Sun, where different regions of the surface can be imaged, this provides a handy way of studying local magnetic field variation, for instance in the vicinity of sunspots.

WORKED EXAMPLE 11.5

Q. An electron in an (imaginary) ion has three possible excited states having associated energies relative to the ground state of 4.81×10^{-19} J, 6.22×10^{-19} J and 7.13×10^{-19} J. If this ion is present in a stellar atmosphere calculate the six wavelengths at which absorption lines appear. Draw an emission spectrum for the same gas assuming that it is being excited by a strong white light source. Consider all line intensities to be equal.

A. The absorption transitions can best be understood by drawing an energy level diagram with arrows to indicate the 'quantum jumps'. Note that not all absorption transitions originate from the ground state as higher excited states will be occupied by electrons at the sort of high temperatures present in stellar atmospheres. Absorption transitions, that relate to the energy of the ion increasing through the absorption of a photon, are thus from all states to all possible states of higher energy as shown (with energies given in units of 10^{-19} J);

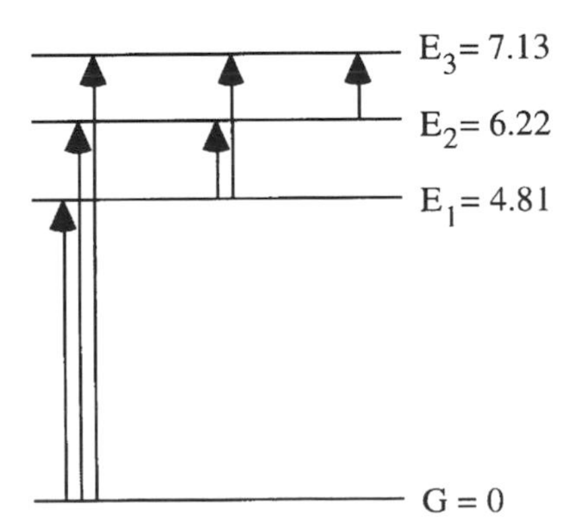

Equation (1.5) can be rearranged to give an expression for the wavelength of the absorbed photon in terms of the energy gap, ΔE, ($= E_p$, the photon energy), between the two levels involved in the transition:

$$E_P = hf = \frac{hc}{\lambda}, \ \therefore \lambda = \frac{hc}{\Delta E} = \frac{1.98 \times 10^{-25}}{\Delta E}$$

It is now an easy task to construct a table that gives the wavelengths corresponding to the (darkened) absorption lines in the stellar spectrum:

Transition	$\Delta E(\times 10^{-19}$ **J**)	λ (**nm**)
$G \rightarrow E_1$	$4.81 - 0 = 4.81$	411
$G \rightarrow E_2$	$6.22 - 0 = 6.22$	318
$G \rightarrow E_3$	$7.13 - 0 = 7.13$	277
$E_1 \rightarrow E_2$	$6.22 - 4.81 = 1.41$	1400
$E_1 \rightarrow E_3$	$7.13 - 8.81 = 2.32$	853
$E_2 \rightarrow E_3$	$7.13 - 6.22 = 0.91$	2180

The emission spectrum of the same gas in the lab will contain the same lines as the absorption spectrum of the gas in the stellar atmosphere (with the exception of line intensities, considered to be equivalent here for simplicity). A spectrum of the emission from the gas would therefore appear thus;

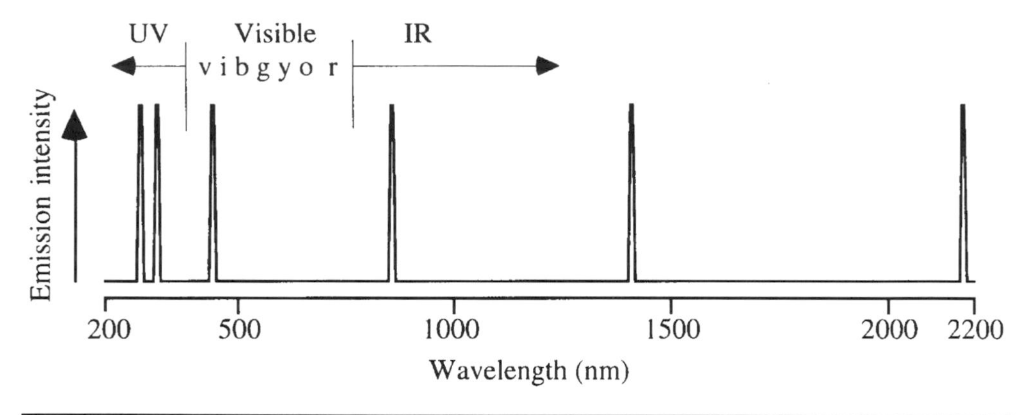

The very presence or absence of spectral features gives a great deal of information on the stellar temperature. A simple example of this is the appearance of the spectral fingerprint of any molecular species which is a certain sign of a cool star. At around 2500–3500 K molecular absorption lines and bands are very strong but fade away at higher temperatures. This is an indication that interatomic binding forces are overcome by thermally energised collisions and vibrations at higher temperatures and the molecule falls apart. As molecular lines disappear with increasing temperature neutral metal features take over. However, only small further increases in temperature sees neutral metal features quickly begin to be replaced by the spectra of singly ionised metallic atoms. Again thermal excitation is responsible for the ionisation. At still higher temperatures doubly ionised metal features begin to appear.

Such is the proportion of hydrogen present in stellar atmospheres that lines due to this atom are visible in almost all stars. However, the strength of particular lines can also give good insight into the star's temperature. An excellent line to study is the so-called Hα line in the middle of the red part of the visible spectrum. This corresponds to a transition from hydrogen's first excited state to its second. At low temperatures very few hydrogen atoms are thermally excited out of the ground state to populate the first excited state and so the Hα line is only faintly seen. As the temperature increases so too does the occupation of the first excited state. The Hα line thus becomes more intense (darker). It reaches a peak at about 10 000 K. Above 10 000 K thermal excitation is sufficient to cause some hydrogen atoms to lose their single electrons. That is, the hydrogen atom is ionised to produce independent electrons and protons no longer able to take place in simple absorption processes. As the temperature increases further, more and more hydrogen atoms are ionised and the Hα line decreases in intensity.

The most resistant element to ionisation is helium. Thus its lines dominate spectra in the high-temperature regime. Neutral helium can exist at temperatures up to 30 000 K but above this singly ionised helium lines can still be clearly seen.

The analysis of stellar spectra is highly complex. The examples given above show how the analysis can proceed. However, a large number of lines contribute to every stellar spectrum and each has its own idiosyncrasies. Complexity is often a good thing, however, as a large number of spectroscopic pieces allow the picture underlying the stellar jigsaw to be constructed. Stellar spectroscopy is of enormous importance in astronomy.

There are two main components in a stellar spectrum: a broadband blackbody curve, with contributions at all wavelengths, and absorption lines, mainly at discrete wavelengths. Both contributions to the spectrum indicate the temperature of the star and so either may be used. Stars are classified according to their spectral type (see Table 11.1) which must therefore reflect their temperature. Broadband spectra are the most obvious indication of this. As discussed in Chapter 1, according to blackbody theory very hot stars are blue, hot stars white, intermediate temperature stars yellow or orange and cool stars red. The details are not always quite so simple, however. Broadband spectra can be perturbed by surface conditions while line spectra are influenced by the luminosity of the star, the atmospheric density and variations in stellar chemical compositions among other factors. Nevertheless, the classification set out in Table 11.1 provides a very good guide to the correlations between stellar temperature, broadband spectra and sharp line features and is used by astronomers everywhere. Each class is subdivided into ten with subclass 0 having the hottest characteristics and subclass 9 the coolest. The Sun, for instance, is a G2 star.

Stellar Sizes

It is now worth reviewing the information that is available to the astronomer by making the three simple measurements of parallax, apparent magnitude and spectral type. Parallax allows the distance of the star to be calculated and thus the absolute magnitude can be deduced from the apparent magnitude. Absolute magnitude is a measure of the total power that is radiated by a body. A star's spectral type leads directly to its temperature. From the Stefan–Boltzmann law (equation (1.3)), the power radiated by the body per unit area can thus be found. It is therefore a simple matter to calculate the size of any star by dividing the total power radiated by the power per unit area radiated. It is reasonable to assume that all stars are spherical and so the stellar radius can be easily found from the surface area that results from the preceding division. The logic involved in such a calculation is diagramatically illustrated in Figure 11.4.

Table 11.1

Spectral type	Photospheric temperature (K)	Colour to the eye	Species responsible for absorption lines
O	30 000 +	Blue	Few lines. Dominated by ionised helium. Some highly ionised atoms. Hydrogen weak
B	11 000–30 000	Pale blue	More lines than O. Neutral helium dominant with hydrogen lines stronger than O
A	7500–11 000	White	Dominated by hydrogen. Some lines due to singly ionised metal atoms
F	6000–7500	Pale yellow	Hydrogen strong but weaker than A. Neutral and singly ionised metal atoms present
G	5200–6 000	Yellow	Many neutral and singly ionised metal atoms, especially ionised calcium. Hydrogen weaker than F.
K	3500–5200	Orange	As G but neutral atoms stronger than ions
M	~3500	Red	Molecules, especially titanium oxide, and neutral metal atoms
R,N,S	2500–3500	Red	Carbon and carbon compounds (R&N). Molecules, especially zirconium oxide and titanium oxide (S). Neutral metal atoms

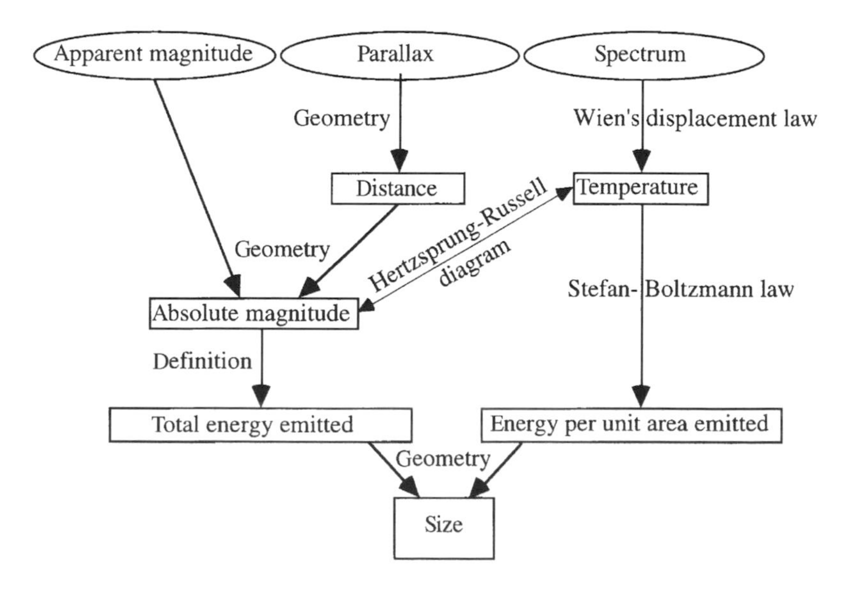

Figure 11.4 How physically measurable stellar parameters (in ovals) are used to calculate other stellar properties (in rectangles). There is no obvious connection between temperature and absolute magnitude and so the Hertzsprung–Russell diagram is used to investigate possible relationships

WORKED EXAMPLE 11.6

Q. Observations of Sirius show it to be an A1 star with an apparent magnitude of −1.46 and a parallax of 0.37″. How big is Sirius?
A. A parallax of 0.37″ implies that Sirius lies at a distance of (1/0.37 =) 2.7 parsecs. Application of equation (11.3) gives the absolute magnitude of Sirius:

$$M = m - 5\log_{10} d + 5 = -1.46 - 5\log_{10} 2.7 + 5 = 1.38$$

The absolute magnitude of Sirius is expressed using the standard relative scale but an absolute scale is required if the actual size of Sirius is to be calculated. The Sun's power output (4×10^{26} W, Worked Example 1.4) and absolute magnitude (4.9, Worked Example 11.3) have been calculated previously so that a comparison of these same parameters for Sirius can be made. The relative intensity of Sirius with respect to the Sun can be found using equation (11.4) (in Worked Example 11.4):

$$I_1 = 10^{\frac{M_2-M_1}{2.5}} I_2 \qquad \therefore\ I_{\mathrm{Si}} = 10^{\frac{M_{\mathrm{Sun}}-M_{\mathrm{Si}}}{2.5}} I_{\mathrm{Sun}} \qquad = 10^{\frac{4.9-1.38}{2.5}} I_{\mathrm{Sun}} \qquad = 26 I_{\mathrm{Sun}}$$

The power output of Sirius is therefore 26 times greater than the Sun at ($26 \times 4 \times 10^{26} =$) 1.0×10^{28} W. As Sirius is an A1 star its surface temperature must be about 11 000 K. Using the absolute temperature of the star's surface it is possible to calculate the power radiated per unit area using the Stefan–Boltzmann law (equation (1.3)),

$$\text{For } T = 11\,000 \text{ K}, \quad P_{\mathrm{A}} = \sigma T^4 = (5.7 \times 10^{-8}) \times 11\,000^4 = 8.3 \times 10^8 \ \mathrm{W\,m^{-2}}$$

Sirius radiates 8.3×10^8 W of power from every square metre of its surface so its total surface area must be ($1.0 \times 10^{28}/8.3 \times 10^8 =$) 1.2×10^{19} m^2. The surface area of a sphere is given by $4\pi r^2$ so that the stellar radius can be calculated thus;

$$1.2 \times 10^{19} = 4\pi r^2 \quad \therefore\ r = \sqrt{\frac{1.2 \times 10^{19}}{4\pi}} = 9.8 \times 10^8 \ \mathrm{m}$$

Sirius has a radius of just under 1 million km, about 40% greater than the Sun. Note that the calculation has been performed using only easily measurable data and presents a much simpler way of determining stellar radii than direct measurement.

The Hertzsprung–Russell Diagram

It would appear that all stellar parameters are now related. This may be true but do all combinations of temperature, luminosity and size that are self-consistent actually occur in nature? In other words, is there a connection between stellar temperature and luminosity other than the stellar radius that they imply? Without a knowledge of stellar dynamics it is impossible to answer this question. The quickest way to find the answer is to plot the data and see what emerges. Figure 11.5 does this. Such a plot is known as a Hertzsprung–Russell diagram. One observable, spectral type, is plotted along (the top of) the x-axis. It is known that a good correlation exists between spectral type and

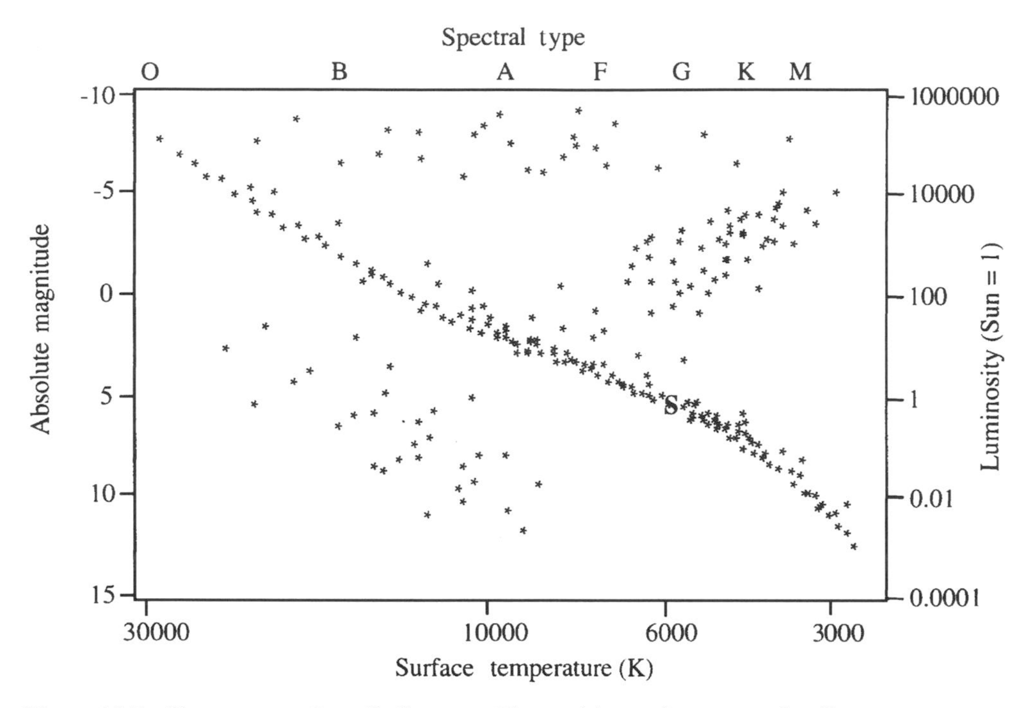

Figure 11.5 Hertzsprung–Russell diagram. The positions of stars on the diagram appear to concentrate in particular regions. This diagram is schematic though the data for the Sun (marked by an S) is real. Note the logarithmic scales for surface temperature and luminosity

temperature so that one implies the other. For convenience, temperature is plotted along the bottom of the diagram to show the correspondence with spectral type. In further Hertzsprung–Russell diagrams in this book spectral type will no longer be plotted and the physically quantifiable parameter, temperature, will be used alone. The historical convention of plotting temperature on a logarithmic scale increasing from right to left will be continued. The y-axis also has two sets of directly related scales, one shown on the left and the other on the right. Relative luminosity varies over several orders of magnitude and so must be represented on a logarithmic scale. As absolute magnitude is proportional to the logarithm of luminosity then this can be simultaneously plotted on the y-axis using a linear scale.

The Hertzsprung–Russell diagram shows clearly favoured regions for stars. Four groups can be identified. The principal diagonal belt is called the main sequence but to understand the naming of the other groups more thought is required. Any point on the graph, whether populated by stars or not, corresponds to a particular stellar temperature and luminosity from which a definite radius for any star that occupies this position can be calculated. Figure 11.6 shows the results of a set of calculations in which positions of equal radius are shown (in terms of the Sun's radius). Superimposed on top of this plot are the four groupings. They are clearly distinguished by their sizes. Relative to the main sequence one group of stars are very small but generally hot. They are therefore known as white dwarfs. The group of stars directly above the main sequence are cool and therefore

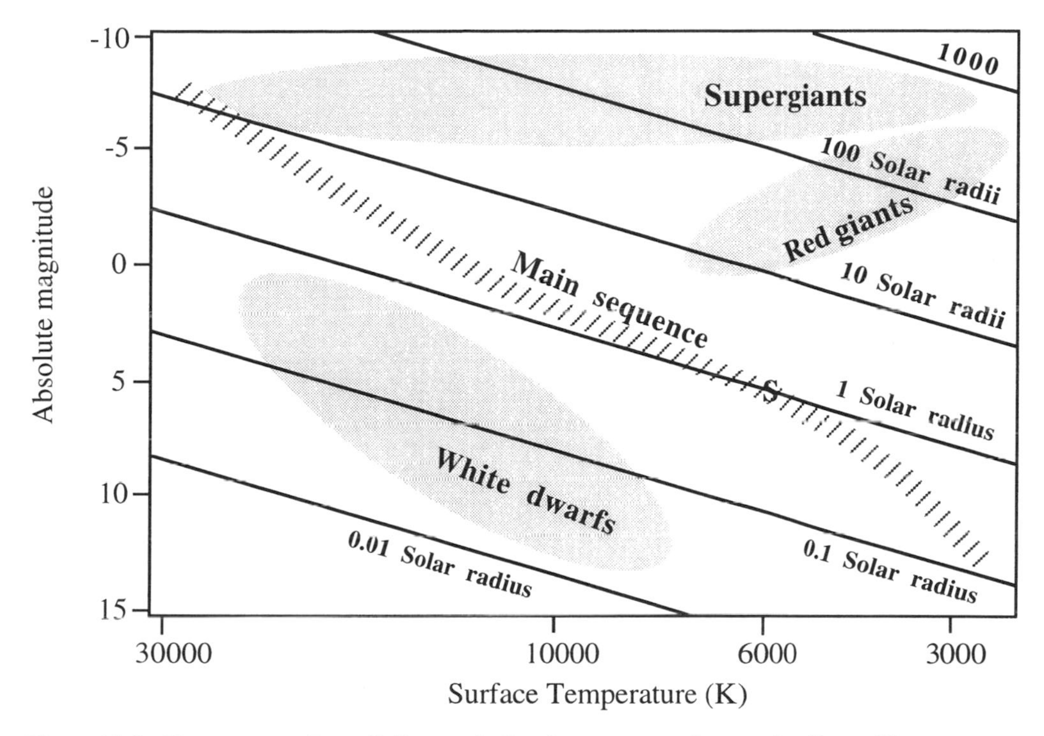

Figure 11.6 Hertzsprung–Russell diagram indicating star groupings and stellar radii

reddish but large. They are therefore known as red giants. The group at the top are very large with varying colour and are therefore known simply as supergiants.

WORKED EXAMPLE 11.7

Q. A star has a photospheric temperature of 12 500 K and an absolute magnitude of 0.0. It has a parallax of 0.1 arcsecond as viewed from the Earth. (a) Using a Hertzsprung–Russell diagram, estimate the radial size of the star. (b) Calculate the power radiated by the star into space. (c) What is the star's apparent magnitude?

A. (a) In order to estimate the star's radius without long calculations such as in Worked Example 11.6, a Hertzsprung–Russell diagram showing lines of equal radius can be used. The star's position on the graph can be plotted and thus its radius estimated. In the diagram the Sun's position is marked by an S and the other star by a cross.

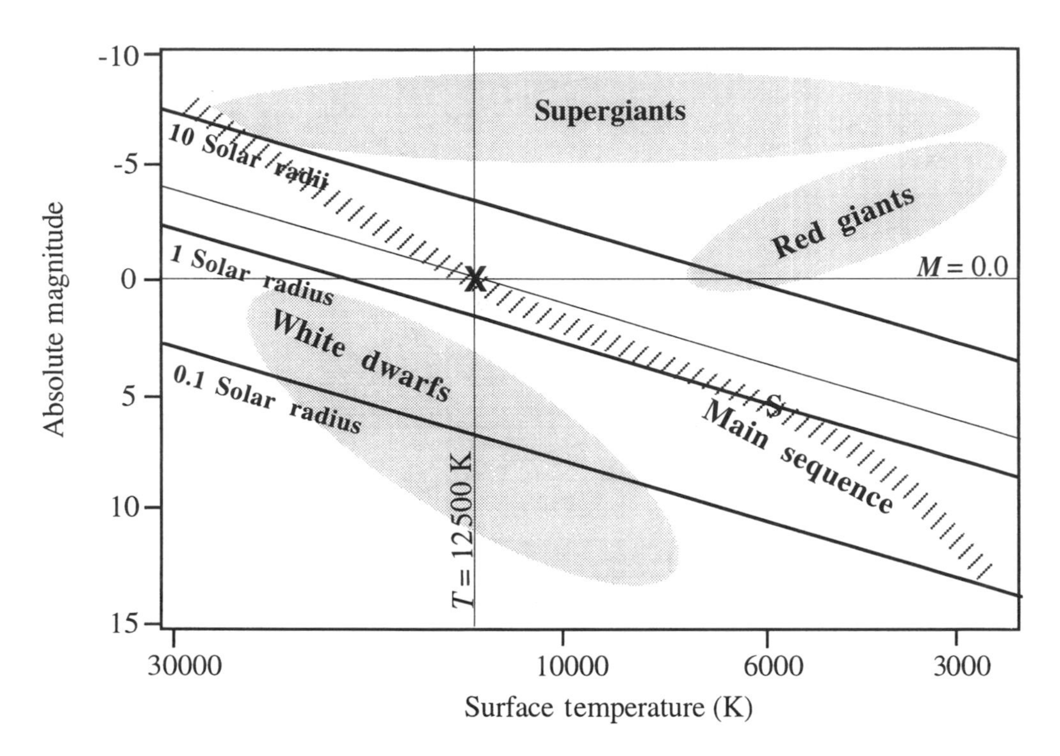

A parallel line to the standard lines of equal radii drawn through the star allows its radius to be estimated. Remembering that the lines are logarithmically separated, the star has a radius about twice that of the Sun. From its position on the diagram, it is a main sequence star.

(b) Using the Stefan–Boltzmann law (equation (1.3)), the power radiated per unit area by a surface at a temperature of 12 500 K is given by;

$$P_A = \sigma T^4 = 5.7 \times 10^{-8} \times 12\,500^4 = 1.4 \times 10^9 \text{ W m}^{-2}$$

The star's radius is twice that of the Sun at 1.4×10^9 m so that the total power radiated by the star is

$$P = P_A A = 1.4 \times 10^9 \times 4\pi(1.4 \times 10^9)^2 \approx 3.4 \times 10^{28} \text{ W}$$

(c) The star has a parallax of 0.1″ so that its distance is (1/0.1 =) 10 parsecs. Substituting into equation (11.3) and rearranging gives the star's apparent magnitude as

$$\begin{aligned} m &= M + 5 \log_{10} d - 5 \\ &= 0.0 + 5 \log_{10} 10 - 5 \\ &= 0.0 \end{aligned}$$

So the star has numerically equivalent absolute and apparent magnitudes (as expected for a star at 10 parsecs) at 0.0.

WORKED EXAMPLE 11.8

Q. A star has a photospheric temperature of 5000 K and an absolute magnitude of −2.5. It has a parallax of 0.03 arcsecond as viewed from the Earth. (a) Using a Hertzsprung–Russell diagram, estimate the radial size of the star. (b) Calculate the power radiated by the star into space. (c) What is the star's apparent magnitude?

A. (a) The process is identical to that of Worked Example 11.7. The new star's position is marked by a cross. Again remembering that the lines are logarithmically separated, the star has a radius of about 40 times that of the Sun. From its position on the diagram, it is a red giant.

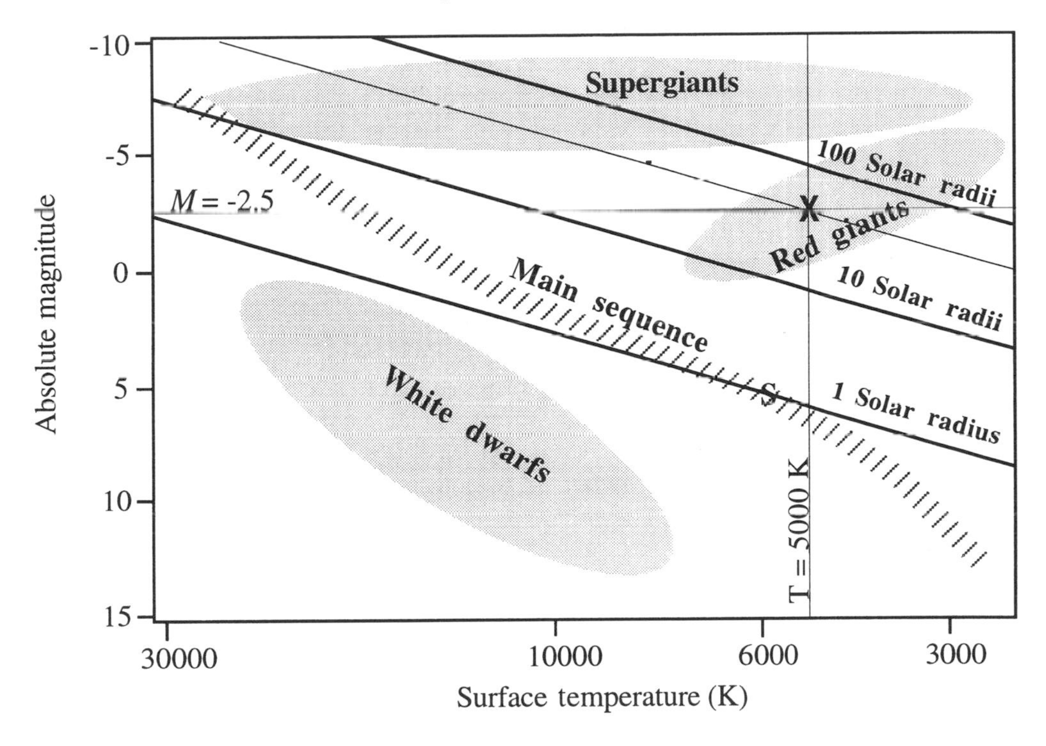

(b) The power radiated per unit area by a surface at a temperature of 5000 K is given by

$$P_A = \sigma T^4 = 5.7 \times 10^{-8} \times 5000^4 = 3.6 \times 10^7 \text{ W m}^{-2}$$

The star's radius is 40 times that of the Sun at 2.8×10^{10} m so that the total power radiated by the star is

$$P = P_A A = 3.6 \times 10^7 \times 4\pi(2.8 \times 10^{10})^2 \approx 3.5 \times 10^{29} \text{ W}$$

(c) The star has a parallax of 0.03″ so that its distance is (1/0.03 =) 33 parsecs. The star's apparent magnitude is therefore given by

$$\begin{aligned} m &= M + 5\log_{10} d - 5 \\ &= -2.5 + 5\log_{10} 33 - 5 \\ &= 0.1 \end{aligned}$$

It is interesting to compare the stars studied in Worked Examples 11.7 and 11.8. The absolute magnitudes of the two stars differ by 2.5 which implies that the red dwarf, having the smaller value of M, should be more intense by a factor of ($10^{\frac{2.5}{2.5}}$ =)10. Comparing the answers to part (b) shows that the rough calculation performed on the basis of estimations from the Hertzsprung–Russell diagram confirm the intensity relationship well. The power output of the red giant is indeed ten times that of the main sequence star. The two stars have almost identical apparent magnitudes despite the greater power output of the red giant. The red giant appears to be equally bright because it is further away than the main sequence star. Identification by eye would be easy, however, because the red giant, at 5000 K, is a K star and would be orange whereas the main sequence star, at 12 500 K, is a B star with light-blue coloration.

Projecting into Deep Space

By examining subtleties in the spectra of known distance stars it becomes possible to determine the localities of those at much greater distances. The best example of this is the width of absorption lines[1]. Larger stars have narrower lines because they have more rarefied atmospheres. Line broadening is determined by, among other factors, the particle collision rate and this decreases with particle density (and increases with temperature). For any given temperature (obtained from the broadband spectrum), the width of spectral lines can be correlated with the size of the star. This pinpoints the star's absolute magnitude. By comparing its absolute and apparent magnitudes the star's distance can be determined. The range of useful study can in this way be enormously increased.

To learn more about stars it now becomes important to think about the forces that hold them together and eventually rip them apart. So far little mention has been made of stellar mass, for instance. Mass is not necessarily proportional to size as density can and does vary between stars of different types. Observational techniques can begin to give information on mass and on stellar evolution but the easiest step to take next is to make a few predictions based on simple physical ideas.

Questions

Problems

1 Rigel has an absolute magnitude of -8.1 and an apparent magnitude of $+0.1$. How far from the Sun is Rigel?

2 Sirius is the brightest star in the night sky with an apparent magnitude of -1.46 and a parallax of $0.377''$. What is the distance between Sirius and the Earth?

3 The stars Vega and Pollux have the same absolute magnitudes though Vega has an apparent magnitude that is 1.0 less than that of Pollux. What is the ratio of the distances of Pollux and Vega?

4 A star has a radius twice that of the Sun. Its surface temperature is twice that of the Sun.
(a) How many times more power does the star radiate than the Sun?
(b) If the Sun's absolute magnitude is $+4.9$, what is the absolute magnitude of the other star?

5 A star has a photospheric temperature of 12 000 K and an apparent magnitude of 7.8. It has a parallax of 0.07 arcsecond as viewed from the Earth.
(a) Is the star visible to the naked eye?
(b) What is the star's absolute magnitude?
(c) What type of star is it?
(d) Using a Hertzsprung–Russell diagram, estimate the radial size of the star (one solar radius = 700 000 km).
(e) Calculate the power radiated by the star into space.

[1] Linewidth refers to the range of wavelengths contained within a spectral 'line'. A large linewidth means that light is detected across a broad, continuous range of wavelengths. In astronomical spectroscopy absorption and emission linewidths are always small relative to the wavelength of the light but may vary a great deal.

Teasers

6 If the Sun's apparent magnitude is -26.7 and its absolute magnitude is $+4.85$, how many astronomical units are there in a parsec?

7 Pluto orbits the Sun at a mean distance of 39.5 astronomical units. The apparent magnitude of the Sun as viewed from Earth is -26.7. Using no further information, what is the apparent magnitude of the Sun as viewed from Pluto?

Exercises

8 (a) Explain why there is a good correlation between the sharp line and broad band spectra of most stars.
(b) Explain why the line spectra of red stars often contain features due to the presence of oxide molecules whereas blue stars do not.

9 Draw a rudimentary Hertzsprung–Russell diagram, explaining what is plotted and labelling the main groupings of stars. Explain what measurements can be made of a local star in order to plot its position on the diagram and what calculations are required along the way. Using this same information, the star's size can be calculated. Explain how. Draw on the diagram a few lines of equal radius.

10 How can the distances of stars beyond the reach of the technique of trigonometrical parallax be determined?

12 Stellar Birth and Early Life

During the next two chapters the lives of stars will be explored. It will be shown that the pockets of the Hertzsprung–Russell diagram into which stars fall are representations of different stages of their lives. Not all stars visit all populated regions of the diagram. A star's mass determines where it first settles and how it evolves thereafter. Once a star appears on the Hertzsprung–Russell diagram there is little to prevent it from living a preordained life. Fate may be paramount but this does not stop the journey from being interesting or complicated.

A simple description of the birth of the Solar System has already been given in Chapter 7 and extended a little in Chapter 10 for the evolution of the early Sun. This is a good start in an attempt to understand stellar formation as almost all stars form in similar ways. Later in this chapter the simple ideas already expounded will be developed and broadened for stars of different mass.

Conditions for Stellar Stability

Before explaining the basis of the complex and still imperfectly understood models for the processes that lead to the establishment of a stable star it is worth starting with a bigger idea that can provide a foundation for the less concrete and more complicated concepts that follow. To do this, the formation process will be left to one side for a while and the moment at which stability has been reached will be considered. It is a general rule of physics that a system is stable when all forces acting on it or within it are balanced. To properly model a star it is necessary to consider the balance of forces throughout the whole structure. For now a simpler picture is sufficient. To maintain a constant size compressive forces must equal expansive forces. Stars consist of plasmas, very hot soups of ions, electrons and nuclei. The plasma created by the detonation of a hydrogen bomb expands very rapidly. It explodes. The continuous explosion of metaphorical hydrogen bombs in stellar cores exerts an enormous outward force. At the outside of a star this manifests itself as a plasma pressure[1]. Inward gravitational forces hold the star together.

[1] The outward forces (generally referred to here as 'thermodynamic forces') include classical particle pressure and the quantum mechanical concept of radiation pressure. The latter results from the fact that, though photons have no mass, they do possess momentum and therefore exert a force on a body with which they collide. In a star the overall photon flux is outwards and so radiation pressure acts in the same direction as classical particle pressure. The proportional contribution of radiation pressure is greater in more massive stars.

Stars are spherical because gravity is a central force. Any non-trivial deviations from spherical symmetry would cause forces to fall out of equilibrium. An overall net force would then act to restore shape.

Stellar masses can vary over about three orders of magnitude. A mass of greater than 100 solar masses cannot equilibrate as a star and is likely to either lose mass through violent eruptions (see Figure 12.1) or split into two or more close-by stars during gravitational contraction. The most massive star so far observed is thought to have formed from a cloud with a mass about 200 times greater than the Sun. However, it has subsequently shed much of this mass into a surrounding nebula which is more than a parsec across. Nevertheless, the star still shines with a luminosity equivalent to about 10 million Suns. A mass of less than 0.1 solar mass is not sufficient to stoke the stellar core to high enough temperatures and pressures to initiate nuclear reactions. Stars of about 0.1 solar mass have been observed and are often known as red dwarfs. Bodies with a mass less than 0.1 solar mass may resemble Jupiter. Failed stars, known as brown dwarfs (Figure 12.2) when they form discretely, probably litter space but are very difficult to detect due to their very low temperatures and consequent dimness.

In between about 0.1 and 100 solar masses, stars may sustain nuclear reactions in their cores while maintaining an equilibrium between inward and outward forces. Clearly the gravitational force attempting to squeeze a 100 solar mass star is greater than that in a

Figure 12.1 Perhaps the most brilliant star ever photographed (centre). This infrared image of a 100 solar mass star was taken by the Hubble Space Telescope. The star is thought to have once had a mass equivalent to 200 Suns but that the instability of such an enormous body resulted in much mass being released in outbursts to create the clouds that surround the star (reproduced by permission of the Space Telescope Science Institute)

Figure 12.2 A brown dwarf or 'failed star' orbiting a red dwarf star. This photograph of the environs of Gliese 229B attempts to remove the glare of the star (left) to reveal the light reflected by the much fainter brown dwarf. The brown dwarf is separated from the star by about the Sun–Pluto separation and has a mass about 50 times that of Jupiter (reproduced by permission of the Space Telescope Science Institute)

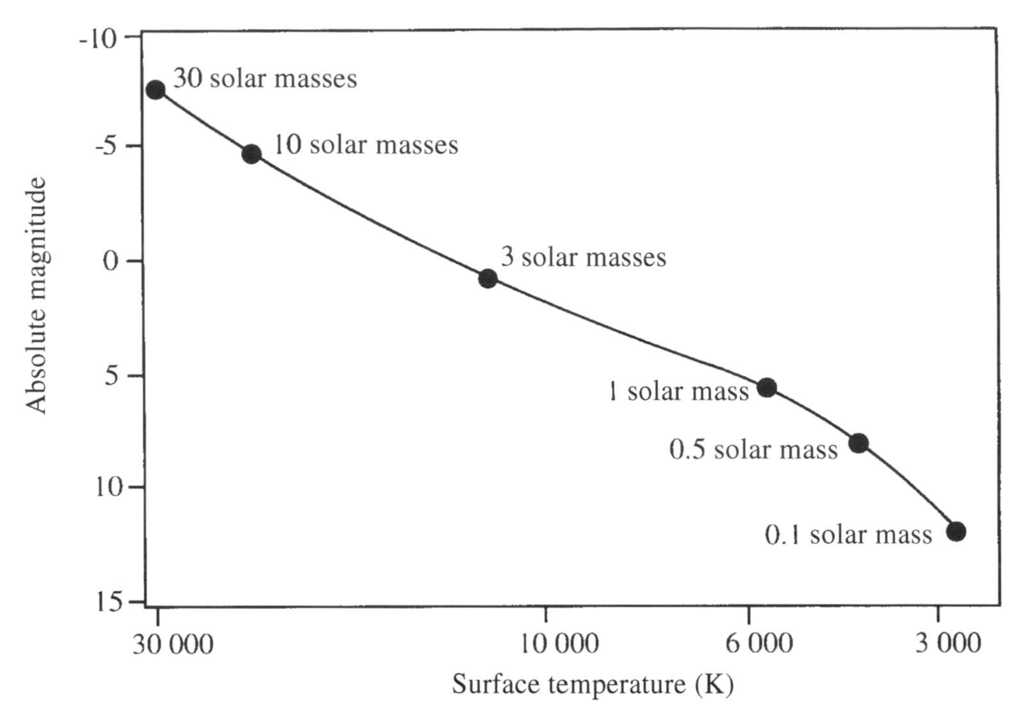

Figure 12.3 Hertzsprung–Russell diagram showing the zero-age main sequence. The diagram indicates the temperatures and magnitudes at which stars of different mass first reach equilibrium

smaller star. It follows that to maintain equilibrium the outward thermodynamic forces must also be greater in large stars. This implies that they must be hotter and therefore brighter and must be converting mass to energy at a faster rate. The perfect balance between forces implies that any star of a given mass and composition will have a definite plasma pressure, nuclear reaction rate and temperature. A small increase in mass will cause a small shift in the position of the equilibrium, making the star a little brighter and a little hotter. Plotting the equilibrium position of all possible star masses produces a line across the Hertzsprung–Russell diagram. This is not a straight line as relationships between stellar parameters are rarely linear, but it is smooth and continuous. Such a plot is shown in Figure 12.3. It is immediately obvious that the line in Figure 12.3 strongly resembles the diagonal grouping of stars in Figures 11.5 and 11.6, known as the main sequence. The curve for newly equilibrated stars is known as the zero age main sequence. The main sequence must therefore be the starting point for all stars. As stellar composition changes with time, due to nuclear reactions in the core, equilibrium positions on the Hertzsprung–Russell diagram change and the star drifts off the main sequence.

Measuring Stellar Masses

So far the concept of the zero-age main sequence and the positions of stars of various masses upon it has been based on theory. The paradigm is simple and compelling but experimental evidence is always a good way of removing any doubt. The missing measurable is mass. Pretty much the only way of definitively determining the mass of a large body is through its gravitational interaction with other bodies. Imagine observing the Solar System from a distance of a few hundred astronomical units. The periods of the orbits of the planets about the Sun are given by Kepler's third law (equation (6.7)). By observing the motion of any planet about the Sun, the Sun's mass, M, can be simply determined. The problem is that the nearest other star is not just a few hundred astronomical units away. It is a few hundred thousand astronomical units away. Stars do have planetary systems but almost all are too far away to be clearly seen (for now). Something else is required.

It is fortunate that a large proportion of stars form in pairs and remain locked in orbit about each other. Some stars, including the Sun's nearest neighbour, Alpha Centauri, are bound into orbital systems with two or more partners. When stars are involved in such associations it is possible to study orbital motion in a similar way to a planetary system to work out the stellar masses involved. Individual members of stellar pairs do not dominate the mass distribution in the same way that the Sun dominates the Solar System. Binary star systems behave more like large versions of the Pluto–Charon system. Under these circumstances the full version of Kepler's third law as given by equations (6.9) is required. In order to individually calculate the masses of both stars accurate observation is required. If only the separation of the partners and not their individual orbital characteristics can be determined then only the sum of the masses can be found (see Chapter 6).

The best chance to determine a binary system's full orbital characteristics is if both partners can be observed separately. Such a system is known as a visual binary. Only nearby binaries are close enough so that both stars can be resolved by telescope. Perhaps the most famous visual binary is Sirius, which is only 2.6 parsecs away. The eye sees only the brightest partner, Sirius A, but it has a much dimmer companion, Sirius B, that can be resolved easily using a telescope.

Distant binary systems that cannot be resolved telescopically can still be studied. Even when the stars are so far away that they appear as a single point source of light to a telescope, the spectrometer can come to the rescue. Each star has its own distinctive spectrum. At any given time the stars are likely to have different velocities relative to an observer on Earth. The Doppler effect shifts wavelengths of spectra by amounts related to the radial velocity of the stars. By observing the compound spectrum of the system over a period of time all the spectral lines should be observed to shift according to one of two patterns. Each set of lines must therefore correspond to one of the two stars. By examining the behaviour of each set the radial motions of the stars can be determined and this leads directly to their full orbital characteristics if one extra piece of information is known. That is the orientation of the plane of the stellar orbits. If the observer is in the same plane as the stars then the radial velocities calculated correspond directly to actual velocity components of the orbital motions. If the plane of the binary star is skewed from the line connecting the observer and the stars then the measured radial velocities will be less than the actual velocities (see Figure 12.4). To see this most clearly consider a system that orbits in a perpendicular plane (as if on a far-distant wall). The radial velocities are always zero and the Doppler shift technique is rendered useless. The best chance of knowing a star system's orientation is if it is in the plane of observation. Twice per orbit the stars will pass in front of each other and the total luminosity will dim. This is known as

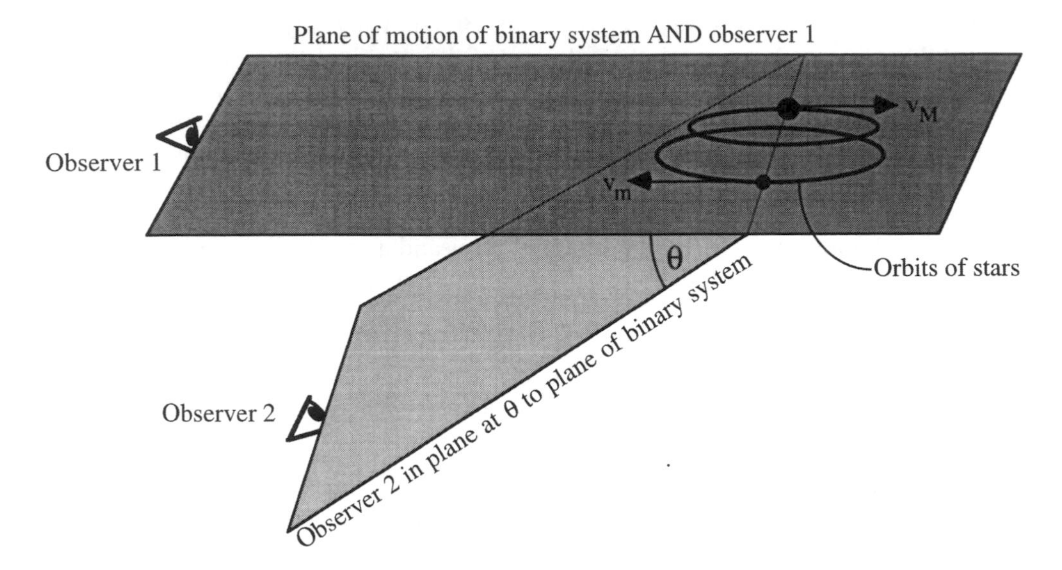

Figure 12.4 Two observers watching the motion of a binary star system. The lateral velocity of the stars can be measured using the Doppler shift. For observer 1 these velocities (at the moment shown) are v_M and v_m while for observer 2 they are $v_M \cos\theta$ and $v_m \cos\theta$. For distant spectroscopic binaries it is often not possible to ascertain the value of θ and so the full dynamics of the system cannot be determined

an eclipsing binary[2]. For angles between 0° and 90° little can be done to determine the orientation. Nevertheless, there are sufficient numbers of eclipsing and visual binaries to check the mass–luminosity relationship predicted from theory above. As is often the case, once such a relationship is established it is possible to work backwards, in this case to find binary orientation angles from measured luminosities to determine the stellar masses. Observations of binary systems have been successful enough to confirm the simple ideas that led to the explanation of the zero-age main sequence expounded above.

WORKED EXAMPLE 12.1

Q. Observations of Sirius show it to be a binary star. Only Sirius A is visible with the naked eye but the motions of both stars can be tracked by telescope. The separation of the stars is (on average) 20 AU and the orbital period is 50 years. Observations of Sirius A show it to have an orbital radius of 7 AU. What are the masses of Sirius A and B?

A. With the data provided, the solution to this problem merely requires substitution into equations (6.9). The total mass of Sirius A and B can be derived from equation (6.9a);

$$T^2 = \frac{4\pi^2}{G(M+m)} r^3 \Rightarrow M + m = \frac{4\pi^2}{GT^2} r^3$$

---

[2] The light curves of eclipsing binaries can also be used to determine stellar radii from the rate at which the total intensity decreases as one partner shifts in front of the other and gradually blocks its light from reaching the observer.

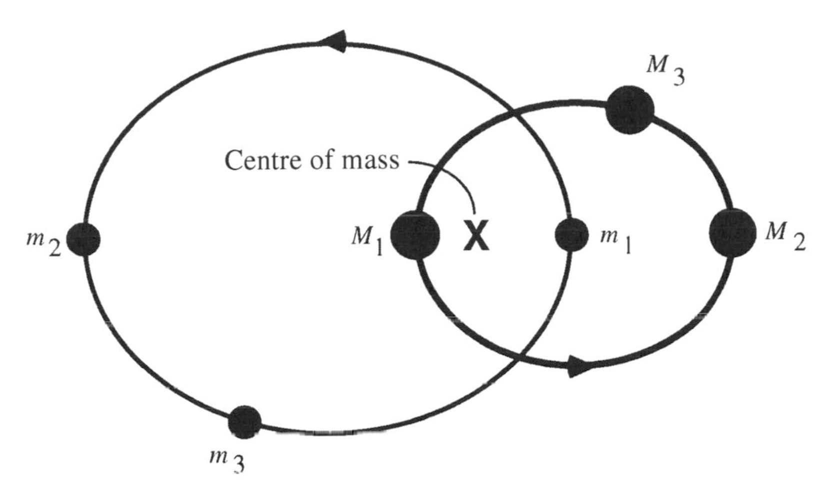

Figure 12.5 The motion of a binary star system. The stars both move in elliptical orbits about the centre of mass (barycentre) which is stationary (relative to the system but may be moving through space). The positions of the stars are indicated by like subscripts for three different positions during the orbits

Converting 20 AU to metres and 50 years to seconds gives

$$M + m = \frac{4\pi^2}{(6.67 \times 10^{-11})(50 \times 365 \times 24 \times 60 \times 60)^2}(20 \times 1.5 \times 10^{11})^3 = 6.4 \times 10^{30}\,\text{kg}$$

To obtain the masses of Sirius A and B either equation (6.8), that gives the position of the centre of mass of the two bodies, or equations (6.9b) or (6.9c), that describe the motion of the individual bodies, can be used. Here equation (6.9b) is used:

$$T^2 = \frac{4\pi^2(M+m)^2}{Gm^3}d_M^3 \Rightarrow m^3 = \frac{4\pi^2(M+m)^2}{GT^2}d_M^3$$

$$\therefore m^3 = \frac{4\pi^2(6.4 \times 10^{30})^2}{(6.67 \times 10^{-11})(50 \times 365 \times 24 \times 60 \times 60)^2}(7 \times 1.5 \times 10^{11})^3 \Rightarrow m = 2.2 \times 10^{30}\,\text{kg}$$

Thus the mass of Sirius B (the smaller mass) is 2.2×10^{30} kg and, from the total mass above, the mass of Sirius A is 4.2×10^{30} kg. This corresponds to 1.1 and 2.1 Solar masses respectively. Remember that the mathematics used in this book is only derived for circular motion though actual motions are elliptical. The final results are the same for both but the derivation is more complicated for ellipses (see Figure 12.5).

The Mass–Luminosity Relationship

A simplified plot of the mass–luminosity relationship for main sequence stars is shown in Figure 12.6. For all but the smallest stars a definite relationship seems to exist. Both axes of the plot are logarithmic and so the slope of the line gives the power relationship

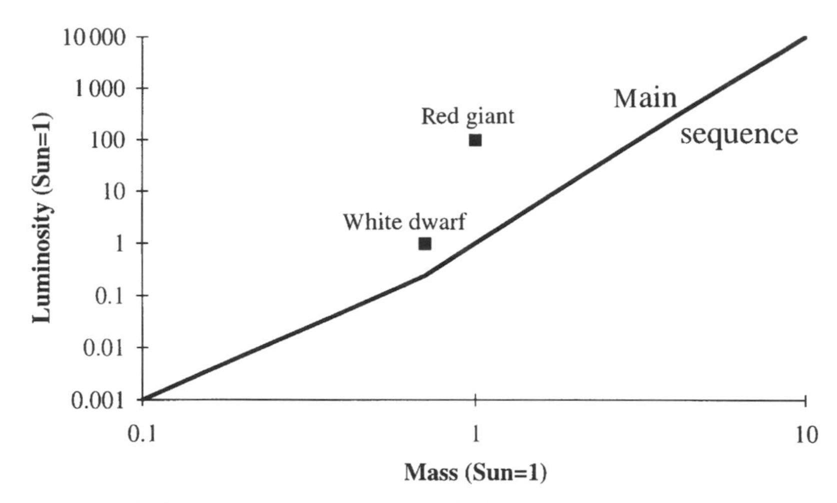

Figure 12.6 Idealised mass–luminosity relationship for stars. Main sequence stars are indicated by the line. The line corresponds to equation (12.1) for more massive stars. The actual spread of main sequence stars is somewhat greater than the thickness of the line here. Red giants and white dwarfs are shown not to behave in the same way by using single examples though both categories vary greatly in their luminosity (and mass for red giants). Note the logarithmic scales for both axes

between the two. The slope for the main part of the graph gives

$$L \propto m^4 \tag{12.1}$$

As expected, luminosity, effectively representing the rate at which nuclear reactions are taking place within the star, rapidly increases as a function of mass as gravitational forces heat and compress the star to greater extremes. Two points are important to remember. The approximate relationship represented by equation (12.1) is non-linear so that it does not translate simply onto the Hertzsprung–Russell diagram (as the non-linearity of the mass on the zero-age main sequence in Figure 12.3 shows). Second, stars from other groupings on the Hertzsprung–Russell diagram do not fit within this simple paradigm. For instance, Sirius B has a mass similar to the Sun but a luminosity more than 300 times smaller. Sirius B is a white dwarf. Sirius A, however, sits just off the centre of the curve in Figure 12.3, having a mass of 2.1 times that of the Sun and a luminosity about 25 times greater (*Note*: $2.1^4 \sim 20$).

WORKED EXAMPLE 12.2

Q. If stars are stable between masses of 0.1 and 100 solar masses, what is the range in luminosities of stars?

A. Applying equation (12.1) and using the solar intensity and mass as a unit, the range of stellar intensities is between ($0.1^4 =$) 10^{-4} solar intensities and ($100^4 =$) 10^8 solar intensities. This implies that intensities range over a factor of 10^{12} (a trillion). The extremes of these estimates (in particular the upper extreme) are unusual and on the edge of stability so that the vast majority of young stars have luminosities between 10^{-3} and 10^5 solar intensities, ranging over a factor of 100 million.

Interstellar Space and Stellar Birth

The ideas that underlie the physical nature of a newly stabilised star are simple and there is a wealth of data to back up the theories. Stellar wombs are much harder to observe. The first reason for this is that bodies warm up as they gravitationally collapse and so it takes some time before a collapsing molecular cloud is hot enough to radiate sufficient energy to make it observable. Once a nucleus has been formed and the protostar is significantly gravitationally warmed the period required to reach stability is measured in millions of years, a short time by astronomical standards. Observation of the night sky is like taking a snapshot of time. There is a constant cycle of stars from birth to their eventual death. Short-lived stages are therefore less commonly observed. A good analogy is to consider how many baby humans you know compared to adults. The baby stage is short-lived and so you don't know very many at any given moment though they are constantly being born. A more important reason that pre-main sequence stars are rarely seen is concerned with the nature of the interstellar medium. Newly forming stars are often hidden among the absorption and scatter of molecular clouds.

The interstellar medium is highly inhomogeneous, containing regions with densities varying from 100 particles per cubic metre to as many as 10^{11} particles per cubic metre at the centre of molecular clouds. Temperatures may vary from 10 K to more than 10^7 K. The interstellar medium is the hardest region of space to study because it is so tenuous. A typical cubic metre of air close to the Earth's surface would contain more than 10^{24} molecules and even though its temperature is low it is easy to detect optically due to its absorption characteristics (much to the irritation of Earth-based astronomers).

Much of space is filled with a very low density plasma. The low density means that absorption is very weak, even when observing strong light sources at large distances through the plasma. If the light source is far enough away then weak absorption lines due to this tenuous matter can be seen though the light may have passed through many different regions. Analysis of interstellar plasma absorption lines show that these extensive regions are very hot, typically around 10^7 K. It may seem strange that such isolated regions of space are so hot but it should be remembered that the total energy contained in such a region is very small due to the sparsity of matter (cf. the solar corona). The interstellar medium is heated by occasional events, such as passing shock fronts from supernovae (discussed in the next chapter). A shock wave occurs when a sudden disturbance enters a region moving at a speed greater than that of sound in the medium. Sonic booms caused by supersonic aeroplanes are phenomena caused by shock waves. The resulting collisions between constituent particles causes rapid heating. Hot regions of space may be expected to emit radiation and they do but, again due to the very low density of the matter, the quantity of energy emitted is very small. This matter is known as coronal interstellar gas.

There are many types of gas regions between stars. A patchwork of zones having different temperatures and densities exists. How much of each type of region exists is still not known. It is known that all regions are dominated by hydrogen, either in molecular form in cold places, atomic form at intermediate temperatures or separated into a plasma of protons and electrons in the coronal interstellar gas. Generally, cooler regions have larger particle densities, the larger absorption being due to greater amounts of matter that prevent energy from reaching the centre of such clouds thus keeping them cool. In other words, such clouds insulate themselves. Lower temperatures allow a greater variety of molecules to stabilise. Molecular variety also aids cooling as different molecules have absorption bands at different wavelengths thus giving greater spectral shielding from nearby electromagnetic energy sources that might cause heating. Many molecules have been detected in space that are unstable on Earth. In space they are capable of surviving in low-temperature regions where the density may be high by interstellar standards but is still billions of times smaller than in the Earth's atmosphere. The average rate and energy of intermolecular collisions (that might overcome interatomic forces and split the molecule apart) is therefore significantly reduced so that the molecule can survive. It is mainly from these cold, dark clouds that stars eventually condense.

Watching the early stages of the collapse of a molecular cloud is an extremely difficult task. Temperatures are very low and so there is little emission of electromagnetic radiation. Study through analysis of the light absorbed by molecular clouds is also difficult because, though these regions are still relatively devoid of matter, their moderate density is combined with an enormous size. The total number of particles in a molecular cloud is therefore immense and the total absorption very strong. Training a telescope on such a region yields only blackness until stars start to form within it. This is consistent with the observation noted above that these regions are well insulated from electromagnetic radiation and therefore cold. It should be noted, however, that these regions come in many shapes and sizes so that smaller clouds are less well insulated and therefore warmer and less opaque. Processes in such regions are easier to observe and study.

It is not always clear what triggers a molecular cloud to begin a period of stellar condensation. It is easy to state blithely that the overall gravitational force on the cloud is inwards so that it will inevitably collapse but there are many other factors that can promote expansion or a steady-state situation. For instance, the overall force on the planets of the Solar System is towards the centre but no cataclysmic implosion has ever taken place. Generally, an external nudge is required to change the status quo and allow gravity to take over. Shock waves are again thought to be of importance as sparks that ignite the fuse of stellar birth.

A good example of a giant molecular cloud that has been studied closely is in the Orion constellation. Immediately in front of this cloud is the Orion nebula. The Orion nebula was until recently (astronomically speaking) part of the giant molecular cloud behind it. It now contains a number of large, bright O stars. Such stars emit a large quantity of ultraviolet radiation that is sufficient to heat up a region of the surrounding molecular cloud. Most of the molecules are broken apart by the flux of ultraviolet photons and a plasma forms. The parent molecular cloud is, as always, dominated by hydrogen and so the plasma is dominated by hydrogen ions (protons) and electrons. Unlike in the coronal interstellar medium, there is a sufficiently high density for reasonably frequent electron–proton recombinations to take place, each of which releases a photon characteristic of such a process. The nebula therefore glows with the emission spectrum characteristic of a hydrogen discharge. Other species also make contributions to the spectrum enabling the chemical composition to be determined quite accurately.

The bright O stars at the centre of the Orion nebula (visible with a small telescope as the Trapezium cluster) are able to ionise a region about a parsec across (see Figure 12.7). At the edge of this emission nebula, where the photon flux from the newly formed stars is sufficiently reduced, a fairly sudden change of environment occurs. One effect of a hot plasma containing fast-moving particles coming into contact with a cool molecular cloud

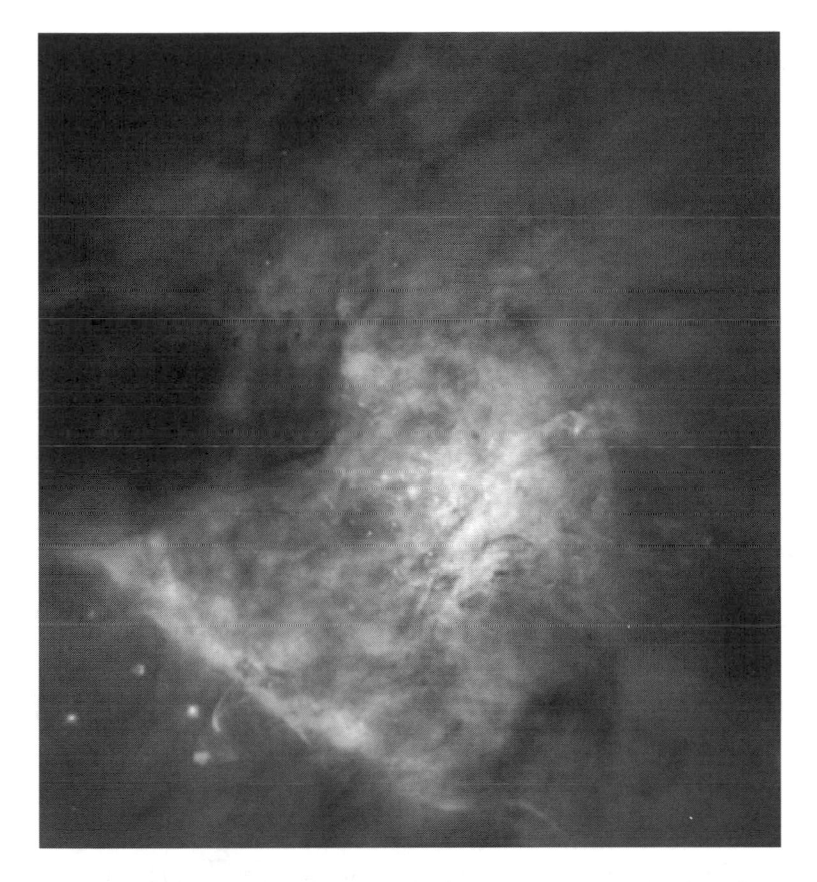

Figure 12.7 The Orion nebula. This photomosaic shows a one-parsec portion of a star-forming molecular cloud. The cloud is illuminated by young stars such as the four most massive in the region, shown in the centre of this image and known as the Trapezium. Much of the structure of the nebula is caused by bipolar outflows from forming stars. Such disruption triggers cloudlet collapse elsewhere in the molecular cloud thus creating further star birth (reproduced by permission of the Space Telescope Science Institute)

is the production of shock waves. These shock waves are able to destabilise other regions of the molecular cloud and this stimulates further star formation. Observing in the infrared indeed shows that new stars are beginning to form just behind the emission nebula. Star formation thus proceeds via a domino effect in this case.

There are a number of regions where stellar nurseries can be observed but Orion is a particularly good one as the plasma region is on the edge of the giant molecular cloud that faces the Earth. Had it been on the other side it would have been obscured by the cloud. The recent placement of telescopes in space will benefit the study of these dark and difficult to observe regions perhaps more than any other field and a much better understanding of stellar birth-triggering processes should quickly emerge. The single mechanism detailed above, though now well established, cannot explain all star birth-triggering processes.

Some of the ideas associated with the collapse of molecular cloudlets have already been discussed in describing the formation of the Sun and its planetary system. Solar simulations are especially useful as an enormous amount of evidence is available close at hand. The problem is that there is a very large range of stellar sizes beyond that of the Sun so that different processes may occur at various stages of formation when different stars are considered. Time scales also vary greatly. The role of this book is to explain the main ideas and so a generalised, simplified paradigm will be given that is a rough approximation to present ideas for all protostars.

Pre-equilibrium Stars

As a molecular cloudlet contracts, simple Newtonian mechanics shows that the central density increases faster than the outer reaches. This in turn hastens the collapse of the core. The release of gravitational potential energy must also be concentrated at the centre and the core heats up. Once the core reaches a few hundred kelvin it becomes possible to observe such objects by detecting radiation in the infrared. They can be seen in molecular clouds but only faintly as the protostars are still surrounded by large volumes of gas and dust from which they contracted.

Eventually the protostar becomes hot enough to start to create a region around itself that has sufficient thermal energy to break apart dust particles. This region is gradually cleared of matter so that there is some separation between the nucleus and the rest of the cloudlet. The protostar still does not resemble a real star. It is not hot enough to be a plasma and is therefore, despite its low density, still quite opaque. Energy transfer proceeds throughout the body through convection to produce a central, swirling ball of gas. Matter continues to fall onto the core from the surrounding cloudlet that has been contracting at a slower rate. As the star's temperature rises its luminosity increases rapidly (even if this cannot be seen very easily outside the molecular cloud). It may seem surprising that a solar mass protostar with a surface temperature ten times smaller than the Sun's could be ten times brighter than it. This is because the protostar is still very large and therefore has a much larger radiating surface area. The convective nature of the protostar allows rapid energy transport and this feeds the energy output.

The protostar and its surrounding cloudlet are also rotating at increasing rates in order to conserve angular momentum. This causes the combination to become lenticular and eventually to develop an outer disk of material. About the same time as the disk appears a number of dynamical instabilities may occur that begin to remove much of the rest of the surrounding dust. Strong outward winds, directed along the spin axis in both directions remove much of the remaining dregs of the cloud (see Figure 12.8). These bipolar outflows are more directed and violent for larger stars. Members of one class of smaller protostar known to be losing mass through stellar winds are known as T Tauri-type stars. They have also been observed to be surrounded by disk-like structures that could be planetary nurseries.

As stellar winds and bipolar outflows blow off the remnants of the surrounding cloud the new bodies that are about to become stars emerge from their womb. All that remains before the main sequence is reached is further contraction, the core always becoming hotter and denser. Now the stellar luminosity decreases with decreasing surface area. The surface temperature stops increasing as energy transfer becomes more and more inefficient in the bulk of the protostar which is now a true plasma. When the core reaches

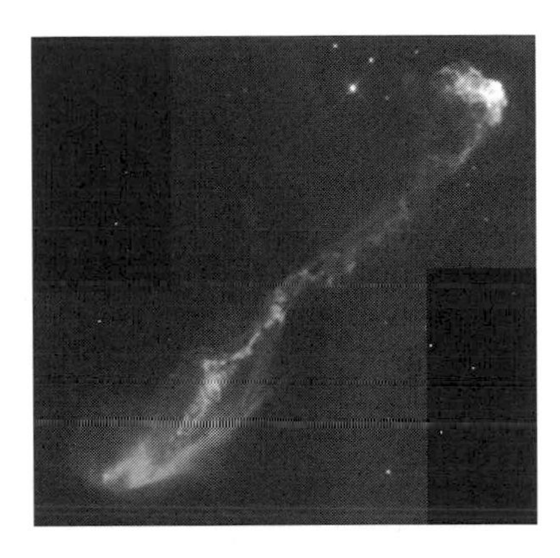

Figure 12.8 Strong bipolar outflows (jets) from a forming star. The jets emerge along the spin axes of the system and are perpendicular to the plane of the accretion disk. The scale of this photograph is of the order of ten times larger than the Solar System (reproduced by permission of the Space Telescope Science Institute)

about 8 million Kelvin, nuclear reactions begin and, after some instability, an equilibrium is reached on the zero age main sequence.

Life on the Main Sequence

The processes occurring in the Sun detailed in Chapter 10 are typical for newly equilibrated smaller type stars. In fact all stars on the main sequence produce their energy through fusion reactions that convert hydrogen nuclei (protons) to helium nuclei. All main sequence stars must also maintain an equilibrium throughout their body so that energy flows and temperature gradients vary in such a way that prevent a collapse in one part of the star that might lead to a general collapse of the whole structure. Balances between forces throughout the star must be maintained. The scale and style of the balancing act varies with stellar mass. The description of the Sun's interior given in Chapter 10 provides a reasonable model for medium and small mass stars and presents the principles relevant to all. The word 'model' is very appropriate as interiors of stars of non-solar size can really only be investigated inside computers. Only simple parameters such as mass, size and surface temperature can be measured due to the distance at which stars other than the Sun lie. Physics that takes place under conditions that can only be created on Earth for a matter of moments relies heavily on theory. Even models of the Sun remain problematic. Though neutrino detection is extremely difficult, experiments that have achieved this have consistently produced smaller yields than would be expected from models of fusion activity in the Sun's core. Models of smaller and larger stars are also sure to be imperfect but where such deviations lie will not be known until better data can be obtained.

One important difference between small and large stars is the way in which they convert hydrogen to helium. The mechanism that mediates this process in the Sun, the proton–proton chain, has already been fully described. In larger stars that have more compressed and hotter cores, a process known as the carbon–nitrogen–oxygen cycle begins to take over. The fusion of four hydrogen nuclei to form a helium nucleus remains the net result but this process begins with a ^{12}C nucleus that successively fuses with protons to produce ^{13}N (which spontaneously decays to ^{13}C), ^{14}N and ^{15}O (which spontaneously decays to ^{15}N). Fusion with a fourth proton regenerates the original ^{12}C and a ^{4}He nucleus.

WORKED EXAMPLE 12.3

Q. Show that the proton–proton chain and carbon–nitrogen–oxygen cycle are equivalent in terms of starting and finishing products.

A. The sequence of reactions are written below with reactants and products that do not take place in the next step of the reaction chain in bold (note that product in the last step becomes reactant in the first step). The sum of bold particles, those that are permanently destroyed or created, are shown below for each chain;

Proton–proton chain

$$^{1}H + {}^{1}H \rightarrow {}^{2}H + \mathbf{e}^{+} + \nu \qquad (10.4)$$
$$\mathbf{^{1}H} + \mathbf{^{1}H} \rightarrow {}^{2}H + \mathbf{e}^{+} + \nu \qquad (10.4)$$
$$\mathbf{^{1}H} + {}^{2}H \rightarrow {}^{3}He + \gamma \qquad (10.5)$$
$$\mathbf{^{1}H} + {}^{2}H \rightarrow {}^{3}He + \gamma \qquad (10.5)$$
$$^{3}He + {}^{3}He \rightarrow \mathbf{^{4}He} + {}^{1}H + {}^{1}H + \gamma \qquad (10.6)$$

$$\mathbf{^{1}H + {}^{1}H + {}^{1}H + {}^{1}H \rightarrow {}^{4}He + 2e^{+} + 2\nu + 3\gamma}$$

Carbon–nitrogen–oxygen cycle

$$\mathbf{^{1}H} + {}^{12}C \rightarrow {}^{13}N + \gamma \qquad (12.2)$$
$$^{13}N \rightarrow {}^{13}C + \mathbf{e}^{+} + \nu \qquad (12.3)$$
$$\mathbf{^{1}H} + {}^{13}C \rightarrow {}^{14}N + \gamma \qquad (12.4)$$
$$\mathbf{^{1}H} + {}^{14}N \rightarrow {}^{15}O + \gamma \qquad (12.5)$$
$$^{15}O \rightarrow {}^{15}N + \mathbf{e}^{+} + \nu \qquad (12.6)$$
$$\mathbf{^{1}H} + {}^{15}N \rightarrow {}^{12}C + \mathbf{^{4}He} \qquad (12.7)$$

$$\mathbf{^{1}H + {}^{1}H + {}^{1}H + {}^{1}H \rightarrow {}^{4}He + 2e^{+} + 2\nu + 3\gamma}$$

The particles destroyed and created are therefore the same for each chain. Note that each positron created will quickly be annihilated after collision with an electron to produce two further gamma photons.

Whichever route is taken by hydrogen nuclei to become helium nuclei it is clear that, despite the enormous sizes of stars, they must run out of hydrogen eventually. The core must slowly fill up with helium and the critical balance between outward thermodynamic and inward gravitational forces must eventually be shifted in one direction or another. The details of what happens at this stage are discussed in the following chapter. For now, a simpler consideration is the period of time before this shift in equilibrium takes place. The luminosity of a star on the main sequence is approximately proportional to the fourth power of its mass (equation (12.1)). The vast majority of a star's power output is provided by the conversion of mass to energy in the core. A star with a mass ten times greater than the Sun must therefore be converting mass ten thousand times more rapidly. As hydrogen's conversion to helium is the principal transformation in all stars then hydrogen nuclei must run out ten thousand times faster. However, the 10-solar mass star arrives on the main sequence with ten times more hydrogen than the solar-mass star and so it would be expected to exhaust its supply one thousand times more quickly. Such a calculation is highly simplified as it ignores the onset of other nuclear reactions and several other factors but ends up working quite nicely in this case. The Sun's lifetime on the main sequence should be about 10 billion years (so it is about half-way through this period) whereas a ten solar mass star will continue hydrogen burning in a fairly static position on the main sequence for around 10 million years. Massive O and B stars survive in their initial state only for a few million years whereas very small stars may survive well beyond 10 billion years.

An interesting consequence of the large variation in main sequence lifetimes is the false impression that results from examining stellar populations in different parts of the main sequence. There are fewer massive stars. This does not mean that fewer are formed. A look at the Hertzsprung–Russell diagram is a snapshot in time. High-mass stars have short lifetimes and so soon disappear off the diagram giving the impression that few are formed. The presence of a massive star in a molecular cloud is thus a positive sign for future star birth as the short lives of such stars provide rapid stirring of the local environment.

Open Clusters

It is frequently the case that a large region of a molecular cloud will be disturbed in such a way that a number of stars will be formed at about the same time. A good observational indication that such a cluster has a recent date of origin is wispy nebulosity in the region. As the stars collapse from the same molecular cloud there will inevitably be a little cloud material left over. Stellar winds will tend to blow the residue away but this process may take some time, especially in a region where a large number of stars form together so that outflow currents may combine in some places and cancel in others to form eddy regions where molecular matter can accumulate. The presence of this nebulosity allows open clusters to be distinguished from much older and more compact globular clusters (that will be discussed later). A typical open cluster is a few parsecs across and contains a few tens or hundreds of stars. They are of great importance as they allow theoretical models to be checked. The presence of leftover gas is a good start in establishing models for the

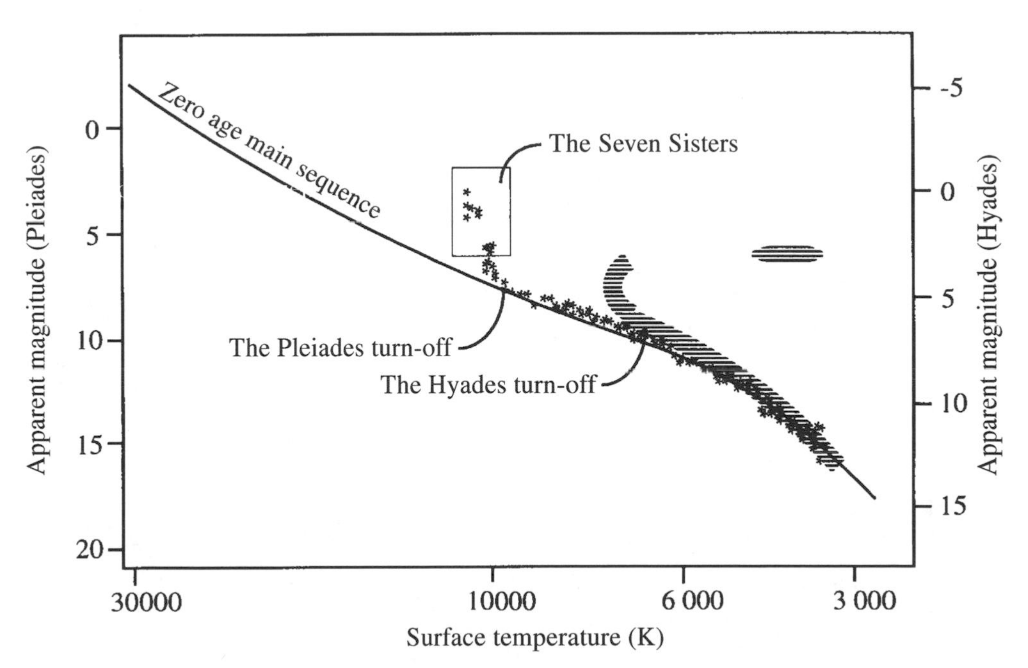

Figure 12.9 The Hertzsprung–Russell diagram for two open clusters, the Pleiades and the Hyades. The apparent magnitudes of stars within a cluster can be plotted rather than absolute magnitude as all members of a cluster are approximately the same distance from Earth. Different axes are required for each cluster, however. The position of the turn-off from the main sequence indicates the age of the cluster, the Hyades (indicated by shading) being older than the Pleiades (indicated by individual stars). The stars of the Pleiades that are visible with the naked eye, known as the Seven Sisters, are shown in a box

collapse of molecular clouds but, more importantly, Hertzsprung–Russell diagrams can be drawn that contain data only for members of individual open clusters. The approximate age of different clusters can be estimated initially from the quantity of remaining gas. Plots for very young open clusters nicely reproduce the theoretical zero-age main sequence.

Perhaps the most famous open cluster is the Pleiades, the brightest components of which are visible to the naked eye in the constellation Taurus. As the stars of the Pleiades are spread over only about 4 parsecs, a definite determination of the distance of each is not essential as a conversion from apparent magnitude to absolute magnitude according to equation (11.3) simply requires the addition or subtraction of the same number, related to the cluster distance, for each star. Open clusters at initially unknown distances can therefore be studied (and from their properties their distance determined if required). The Hertzsprung–Russell diagram for the Pleiades is therefore shown in Figure 12.9 with apparent magnitude rather than absolute magnitude plotted. A fit to a zero-age main sequence shows a good fit for stars with surface temperatures less than 10 000 K. Hotter stars do not fit. This corresponds well with the ideas expounded above. Main sequence

Figure 12.10 The Pleiades open cluster. Note the nebulosity that remains in the region of the cluster. This indicates that the cluster is fairly young (photograph David Malin, © Anglo-Australian Observatory)

stars with surface temperatures of 10 000 K are expected to have a supply of hydrogen that will last for about 100 million years. Hotter (larger) stars have already begun to run out and drifted off the main sequence. Cooler (smaller) stars remain in place much longer. The Pleiades cluster is thus about 100 million years old, relatively young, and this is corroborated by the small amounts of nebulosity present and the fact that the cluster is still relatively compact (see Figure 12.10).

A much older cluster is the Hyades. The stars that make up this cluster have drifted further apart than in the Pleiades and no nebulosity remains, immediately indicating a greater age. The region of the Hertzsprung–Russell diagram occupied by the stars of the Hyades therefore covers a smaller region of the main sequence, down to stars a little more massive than the Sun. The cluster can thus be aged at about 4 billion years. Studies of open clusters go a long way to proving the ideas discussed in this chapter. As a cluster ages its turnoff from the main sequence progresses to smaller stars as the larger members sequentially begin to run out of hydrogen. Thousands of open clusters can be observed and all of the data obtained can be accounted for using the ideas explained in this chapter.

The Hertzsprung–Russell diagram for the Pleiades and the Hyades shows the presence of stars away from the main sequence. Life goes on without hydrogen fusion reactions in stellar cores. In fact this is when stars become really interesting. Read on!

Questions

Problems

1 (a) A star with a photospheric temperature of 6300 K has a radius of 1 million km. Calculate the power radiated by this star.
(b) This star is similar to the Sun. Briefly explain how energy created through mass conversion reaches outer space. Ignore the effects of the stellar atmosphere.

2 The separation of the stars in a binary system averages 35 AU and their orbital period is 100 years. Close observation shows that one of the stars has an orbital radius of 7 AU. What are the masses of the two stars?

3 Two stars in a binary system are found to be orbiting about their barycentre with orbital radii of 1.0×10^8 km and 2.0×10^8 km. Their orbital period is 5 years. What are their masses?

4 (a) Proportionally, how much more power does a two solar mass main sequence star radiate than the Sun (surface temperature 5800 K)?
(b) If the star's surface temperature is 8000 K, how many times larger is its radius than the Sun's?

Teasers

5 (a) According to simple theory, why is the zero-age main sequence on the Hertzsprung–Russell diagram a single line rather than, for instance, an area, a series of areas or a series of lines?
(b) What might (and does) vary a little from star to star to give some 'thickness' to the zero-age main sequence?

6 Observations of the spectra of an eclipsing binary system reveal that the total light output of the pair dips once every ten years. During the dips, the Hβ spectral line is observed at 486.13 nm. At other times this spectral line splits into two, one component varying by ± 0.02 nm and the other by ± 0.01 nm. What are the masses of the two stars, assuming both have circular orbits?

Exercises

7 What is the relationship between mass and temperature for young stars? Qualitatively, why is this the case?

8 Draw a rudimentary Hertzsprung–Russell diagram, marking on it the following;
(a) On the appropriate axes, the direction of increasing temperature and increasing stellar intensity.
(b) A line representing the zero-age main sequence.
(c) Where stars would first appear upon formation and stabilisation for the following masses; 4×10^{29} kg, 4×10^{30} kg, 4×10^{31} kg.

9 What is unique about the Sun?

10 (a) Describe how interstellar space varies and in what regions stars form.
(b) Describe the early stages in a star's formation until it stabilises as a main sequence star.

13 Stellar Evolution and Death

The course of stellar evolution up until leaving the main sequence is qualitatively similar for all stars. They collapse from molecular clouds, ignite nuclear fusion reactions in their core to convert hydrogen nuclei into helium nuclei, settle into equilibrium on the zero-age main sequence and then drift off as their supply of hydrogen declines. The main differences between stars of different sizes are quantitative. Large stars do everything more quickly than small stars. Large stars have higher temperatures and pressures and do things in a more spectacular way, from the strength of their bipolar outflows during formation to their power output after stabilisation. This trend continues until the end but the routes that stars take to their deaths vary greatly depending on their mass, as does the nature of the end itself.

After the Main Sequence

After the stellar core has become too heavily depleted of hydrogen to support either the proton–proton chain or the carbon–nitrogen–oxygen cycle a sequence of other processes follow. Each subsequent process needs more extreme conditions than its predecessor. In the early stages of a star's life, higher temperatures and pressures are generated in larger stars. The same remains true for the later phases of a star's history. As the demands for higher temperatures become increasingly difficult to meet, smaller stars are unable to generate the necessary conditions and the start of the end of their lives ensues. Larger stars continue through more stages but, such is the pace at which they consume ingredient nuclei, most have scattered their matter across space and stopped producing energy before small stars have even begun to shift from their equilibrium position on the main sequence. In what follows a number of examples of stars of different masses are described to show how the step-wise evolution takes place and how this is limited in smaller stars.

In describing stellar properties, this book has taken the Sun as its principal example. In projecting the future there is less to be gained from the Sun's proximity. Observation of similar stars elsewhere in the universe is required to obtain the necessary information. Nevertheless, solar-mass stars provide a good first example of stellar evolution, demonstrating how different eras come and go without the complexity of large stars or the less interesting case of very small stars.

A Solar Mass Star Becomes a Red Giant

So far, only the dynamics of the star as a whole have been considered but, in describing stellar evolution after the main sequence, a discussion of the balance of forces within internal regions of the star becomes necessary. For nuclear fusion reactions to take place, colliding nuclei have to have high speeds to overcome their electrical repulsion. Nuclei move fastest at the centre of the star where temperatures peak and so, in the early stages of stellar evolution, hydrogen is most rapidly depleted in the core. There is little convective mixing between the core and the radiative mantle and so the proportion of helium in the core increases. Each proton–proton cycle reduces the total number of nuclear particles by three as four protons are required to produce one helium nucleus. At the same time, two electrons are removed through positron annihilation. The core shrinks and releases gravitational potential energy thus becoming even hotter. The heat generated in the core is sufficient to provoke nuclear reactions in a spherical shell surrounding the helium-rich core. As the core continues to contract its temperature increases further, eventually fusing the last of its hydrogen nuclei to produce an inert helium core. The scenario is now one of a contracting core providing gravitational potential energy to maintain reactions in a surrounding shell of hydrogen-rich nuclei.

The balance between gravity and plasma pressure favours gravity in the core but the outer part of the star begins to expand as expansive thermodynamic forces created in the active shell tip the balance further out. In fact the shell has become so hot that reactions in this region proceed mainly through the carbon–nitrogen–oxygen cycle as in much larger main sequence cores. The overall energy output increases and the star becomes more luminous. Increased emission is achieved through the star's increase in surface area despite the fact that the surface temperature slowly drops. This drop in surface temperature occurs because the energy source becomes further and further removed from the photosphere. The star moves off the main sequence, slowly at first while fusion reactions continue in the core but ever more rapidly as the core becomes inert. Eventually, the star's surface temperature drops to as low as 3000 K at which point its luminosity is more than a thousand times greater than it was as a main sequence star and its radius more than a hundred times greater. The star has become a red giant. In the Sun's case, Mercury will be engulfed by plasma. The path of the star across the Hertzsprung–Russell diagram is shown in Figure 13.1.

WORKED EXAMPLE 13.1

Q. By how much does a Sun-like star increase its luminosity in becoming a red giant?
A. If the star's surface temperature decreases by a factor of two (from about 6000 K to about 3000 K) then, according to the Stefan–Boltzmann law (equation (1.3)), the power radiated by the star per unit area will decrease by a factor of (2^4 =) 16. If the stellar radius increases by a factor of (say) 150 then the surface area increases by a factor of (150^2 =) 2×10^4. The surface area is 20 thousand times greater but radiates 16 times less power per unit area so that the overall increase is by a factor of a little over 1000. Note that these numbers are indications for a typical star but are subject to variation in individual cases.

At some stage during the process of a solar mass star becoming a red giant its helium core adopts a matter state not encountered so far in this book. It becomes a degenerate

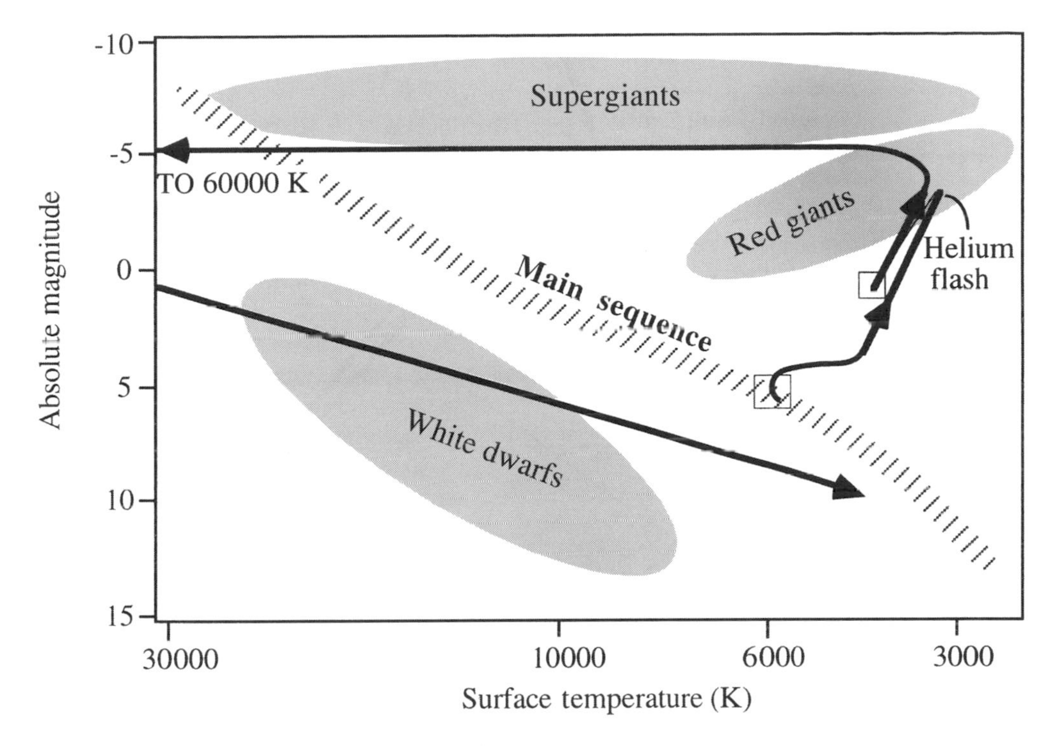

Figure 13.1 Hertzsprung–Russell diagram showing the evolution of a solar mass star. The positions of greatest stability are shown in boxes; on the main sequence (for about 10 billion years) where hydrogen fusion reactions dominate energy production and immediately after the helium flash where helium fusion reactions dominate energy production

electron gas. This is essentially an unusual quantum mechanical state, the details of which are beyond the scope of this book but the properties of which influence the behaviour of the star at several stages. Electrons in their usual state on Earth are bound to nuclei to form atoms. Only certain energy levels are allowed within any given atom (see Chapter 11) and the number of electrons that take on any particular energy is strictly limited by a rule called Pauli's exclusion principle. In a dense plasma ($\sim 10^8$ kg m^{-3} or more) the electrons are no longer associated with particular nuclei but rather with a much larger region through which they can move. Pauli's exclusion principle still applies to limit the number of electrons that can possess different energies. Though there are considerably more energy states available for occupation in a dense plasma than in an atom, as density increases the very large number of electrons that move throughout the star eventually run out of energy levels that are within reach of their usual thermal energies (see Figure 13.2). The electrons must therefore take an energy greater than would be the case for the usual thermal distribution at that temperature. As the energy of an electron determines its speed of motion and this in turn determines the pressure exerted, the result is that the electron pressure no longer depends on the temperature but only on the density. This is because, under these extreme conditions, the plasma density determines the energy states available. Electron degeneracy pressure thus prevents the core of the star from collapsing completely.

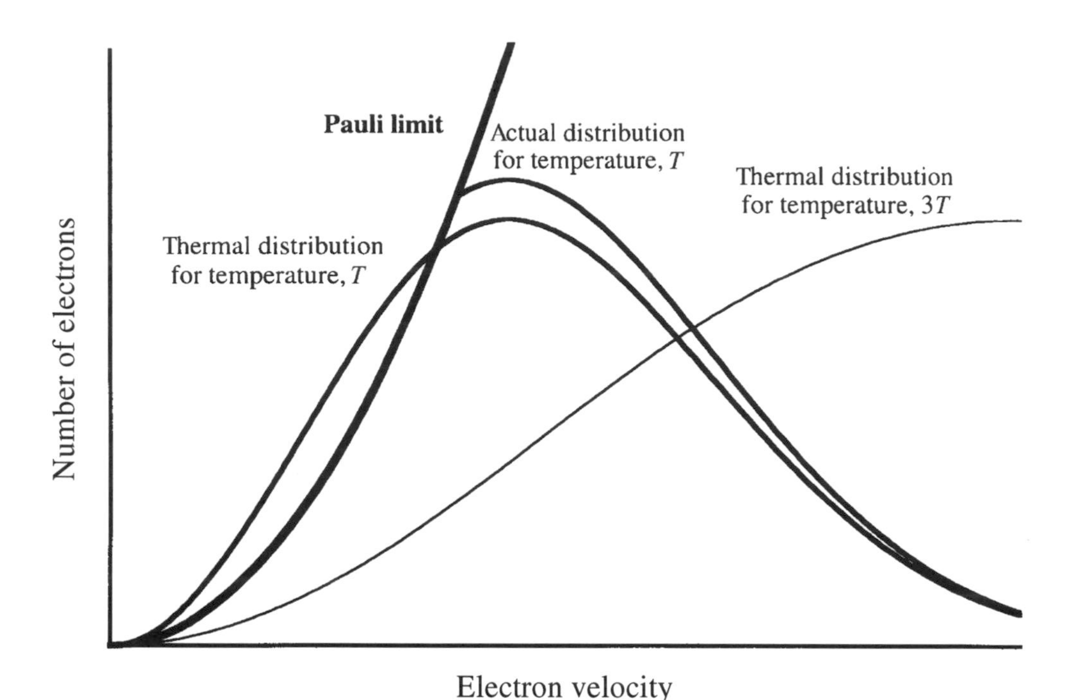

Figure 13.2 The thermal distributions of electrons in a very high density plasma at different temperatures and the effect of the Pauli limit. The Pauli limit (thickest line) determines the maximum number of electrons that have any given velocity. For a given temperature, T, the normal thermal distribution (medium-thickness line) indicates that this limit should be transgressed for low velocities. The normal thermal distribution cannot therefore be maintained, the distribution is pushed to higher velocities (as indicated by the other medium-thickness line) and the gas is said to be electron degenerate. If the temperature is increased to $3T$ (for instance) there are no electron velocities for which the thermally distributed population (thin line) is greater than the Pauli limit and the gas is no longer electron degenerate. This diagram is drawn for a particular electron density. A different electron density would result in a different Pauli limit so that, for electron degenerate stars, the gas pressure (determined by the electron velocity distribution) is determined by the density but not by the temperature. For instance, increased density would cause the slope of the Pauli limit to decrease resulting in electron degeneracy at higher temperatures

At the core of a solar mass red giant the temperature eventually reaches 200 million kelvin. This marks the approximate temperature at which a new nuclear fusion process may occur. It is known as the triple alpha process (^{4}He nuclei are known as a α-particles). Three helium nuclei fuse to produce a carbon (^{12}C) nucleus with an attendant loss of mass and production of energy. When this process begins, the energy released immediately heats the core, causing the kinetic energy of nuclei to increase. This causes an increased rate of nuclear reactions leading to more heating, leading to a further increase in reaction rates and so on. This is another example of runaway dynamics caused by positive feedback, previously discussed for the greenhouse effect (see Chapter 8). In earlier stages of stellar evolution, increases in temperature result in thermodynamic forces taking the upper hand and causing expansion. In this case, the electron gas, which bears most of the strain of holding up the star's core, does not expand because it is degenerate so that its pressure depends only on density and not on temperature. Because the core does not expand it is unable to cool and so the onset of helium fusion reactions is a rapid, runaway process. The so-called helium flash lasts for only a matter of seconds but in that time a large part of the core begins to take part in carbon-producing reactions. To the observer, the star does not change its appearance in seconds as the action takes place in the core and so takes much longer to manifest itself externally.

WORKED EXAMPLE 13.2

Q. Compare the energy released through a single triple-alpha reaction to that released during a proton–proton chain.

A. From Chapter 10, the energy released during a proton–proton chain is 4.3×10^{-12} J. To calculate the energy released when three helium nuclei fuse to form a single carbon nucleus their masses are compared. From tables of atomic masses, the mass of a ^{4}He atom is 4.0026 atomic mass units and the mass of a ^{12}C atom is, by definition, exactly 12 atomic mass units. Note that these masses actually include the mass of attendant electrons in both cases. Outer electrons play no part in nuclear reactions (they are not even attached in stellar cores) but the mass of 12 electrons is included on both sides of the reaction and so will not influence the calculation. The mass destroyed is therefore

$$(4.0026 \times 3) - 12.0000 = 7.8 \times 10^{-3}\ \text{amu}$$

The conversion of atomic mass units to SI units is that $1\ \text{amu} = 1.66 \times 10^{-27}$ kg so that the mass discrepancy is $(7.8 \times 10^{-3} \times 1.66 \times 10^{-27}) = 1.29 \times 10^{-29}$ kg. Applying equation (10.2),

$$E = mc^2 \quad = 1.29 \times 10^{-29}(3 \times 10^8)^2 \quad = 1.2 \times 10^{-12}\ \text{J}$$

The energy released in fusing three helium nuclei into one carbon nucleus is thus about four times less than that released in fusing four protons to form a helium nucleus.

As the core's temperature increases after the helium flash electrons gain kinetic energy so that they can eventually occupy energy levels according to a thermal distribution again. Now the core does expand as it is no longer electron degenerate. A new equilibrium position on the Hertzsprung–Russell diagram is established. As the core has expanded then so too has the hydrogen fusion shell around the core. Its temperature and density consequently decrease and the rate at which hydrogen fusion reactions proceeds decreases substantially. Though the star now has two regions in which nuclear reactions are taking place its overall luminosity decreases due to the reduction in the much more energetic hydrogen-based reactions. The scenario is reversed from that immediately after main sequence life. The core expands due to increased activity but the outer regions of the star contract as gravity takes the upper hand while shell reactions are subdued. This contraction is quite fast and is almost never seen but leaves the star with a surface temperature of around 4000 K and a reduction in luminosity of around a factor of one hundred. The new helium core star is quite stable and is known as a clump star as such stars form a small group just above the main sequence. In this region the stellar core gradually increases its carbon concentration until an inert carbon core forms. It is surrounded by concentric shells in which helium (closest) and then hydrogen (further out) nuclear fusion reactions take place.

The anatomy of a solar mass star with an inert carbon core is comparable to that when the inert helium core appeared. After the helium in the core is depleted sufficiently that the triple-alpha process ceases, the core will begin to contract, thus becoming hotter. The star responds in exactly the same way as it did to having a contracting helium core and returns to the red giant region of the Hertzsprung–Russell diagram. This time there is no new onset of nuclear reactions to save the star. One solar mass is not enough to be able to crush the core sufficiently to cause heating that will allow carbon to take place in fusion reactions. Very large nuclear velocities are required to overcome the electrical repulsion between carbon nuclei, a repulsion that is much greater than that between two hydrogen nuclei (36 times, see equation (10.1)) or helium nuclei (nine times) due to carbon's larger nuclear electrical charge.

The Death of a Solar Mass Red Giant

As a star expands through the red giant region of the Hertzsprung–Russell diagram it must eventually become unstable. The balance between gravitational and thermodynamic forces can no longer be maintained throughout the star. The outer envelopes are expelled leaving behind a star that is no longer able to maintain shell nuclear reactions. Much of the story of stellar evolution relies on observations of stars whose behaviour is static on the scale of an astronomer's lifetime. The almost invariant Hertzsprung–Russell diagram has to be interpreted using computer models without direct evidence of the processes that are occurring deep within stars. The death of a red giant is an exception to this. The star produces around itself a sphere of expelled gas known as a planetary nebula (see Figure 13.3). The central remnant star grows hotter and

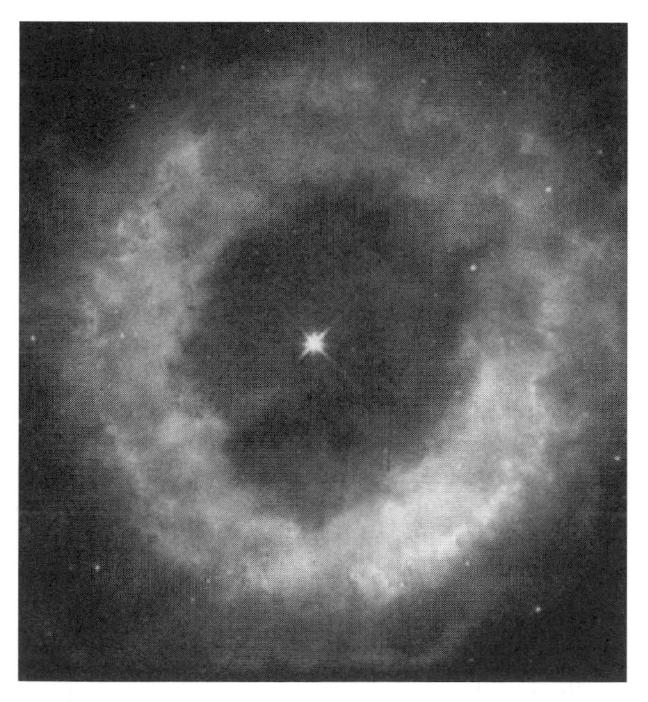

Figure 13.3 The planetary nebula NGC6369 as photographed by the Hubble Space Telescope. The nebula is about 0.1 parsec across. Not all planetary nebulae are quite so spherical! (reproduced by permission of the Space Telescope Science Institute)

hotter as it gravitationally collapses and so the planetary nebula is heated more and more. As the matter of the planetary nebula is bathed in radiation from the central star it is able to absorb light and reradiate it. By observing the Doppler shift of the spectra from different parts of the planetary nebula (receding, moving perpendicular to the observer's direction, moving towards the observer, etc.) the speed at which the planetary nebula is expanding can be measured. It then becomes easy to piece together the history of such stars without complex computer models. A small planetary nebula usually contains a star with a temperature of around 10 thousand K and from the speed of expansion it can be shown that it formed just a few thousand years ago. Larger planetary nebulae must have formed at an earlier date and are found to contain a much hotter central star, perhaps as hot as 100 thousand K. The oldest observable planetary nebulae are aged around 20 thousand years. The process of planetary nebula production is not explosive though the early stages have never been observed because the central star is initially too cool to illuminate the surrounding gas sufficiently. As the central star collapses it converts gravitational potential energy to thermal energy and becomes hotter and hotter so that the planetary nebula glows more and more. Eventually the nebula dissipates and the star enters a new stage of its life.

WORKED EXAMPLE 13.3

Q. The speed of expansion of particles in a planetary nebula is measured spectroscopically (using the Doppler effect) to be 20 km s^{-1}. If the planetary nebula has a diameter of 0.1 parsec, when did it start to form?

A. To solve this problem it is necessary to assume that the particles in the nebula have travelled at the same speed throughout their journey. It is then a simple matter to calculate how long ago they set off. If the diameter of the nebula is 0.1 parsecs then the particles have travelled 0.05 parsecs (the distance from the centre). Their speed is 20 km s^{-1} and so, converting to metres, the time elapsed is given by

$$\text{Time} = \frac{\text{Distance}}{\text{Speed}} = \frac{0.05 \times 9.5 \times 10^{15}}{20 \times 10^3} = 2.4 \times 10^{10}\ \text{s} \approx 800\ \text{years}$$

Evolution to a White Dwarf

The remnant of a planetary nebula expulsion event is a carbon (and oxygen formed by simple nuclear reaction not discussed here) core that is surrounded by a helium envelope. The helium envelope cannot support nuclear reactions for very long as it is quickly cooled by finding itself close to the outside of the star. The star then has no source of internal energy and all that can prevent its continued gravitational collapse is the effect of electron degeneracy. The star becomes completely electronically degenerate and stops contracting. The electrons are unable to reduce their energy because there are no available lower energy states and so energy transport in such a star is via the much heavier nuclei. Such processes must therefore be quite slow so that the star cools at a low rate.

In planetary nebula remnant stars the usual gas thermodynamics that involve pressure and volume changes do not apply because electron degeneracy is the most important opponent of gravity. An interesting illustration of this fact is that the size of an electronically degenerate star decreases with increasing mass. Increasing mass requires increasing resistive forces. For main sequence stars this results in a high nuclear fusion rate that keeps the stellar radius relatively large. For electronically degenerate stars outward pressure forces are increased by greater density and so more massive stars must become smaller. In terms of Figure 13.2, increasing density decreases the slope of the Pauli limit line so that a greater number of electrons are forced to take on non-thermal energies and the pressure increases until equilibrium with inward gravitational forces is reached.

The initial equilibrium size that is reached by a planetary remnant star is maintained indefinitely. That energy is radiated from the surface of such a star is unimportant because temperature is no longer relevant to the star's hydrostatics. The star has no remaining energy source however. As it radiates energy away it fades, following a line of equal radius along the Hertzsprung–Russell diagram through the white dwarf region until it disappears. Such white dwarfs never stabilise at any one position on the Hertzsprung–Russell diagram but simply slide through it. After a long period of time the star will have radiated away almost all of its energy and will come into thermal equilibrium with its surroundings. Such bodies, still maintaining their size through electron degeneracy pressure, are sometimes known as black dwarfs. The Sun's final fate is to be such a cold, dark object.

Low-Mass Stars

The Sun's main sequence lifetime will be about 10 billion years. The lifetime of smaller stars is longer but the age of the universe is thought to be only about 15 billion years. Opinions on the age of the universe vary but even by taking higher estimates the main sequence lifetime of a half solar mass star is longer. There is therefore no possibility of observing evolved very small stars, even for those that formed early in the universe's history. Ideas of what will occur are based entirely on models. Theory suggests that the temperatures and pressures in the core of the star never become sufficient to allow the triple-alpha process to occur. That is, there is no helium flash. The star's first loop up towards the red giant region is its last. When it runs out of hydrogen and develops an inert helium core a similar process to the one that creates carbon white dwarfs occurs. Matter is ejected to form a planetary nebula and the helium-based remnant becomes fully electron degenerate before fading through the white dwarf region to finally become a black dwarf. Unfortunately, theorists will have to wait a few billion years to have this hypothesis experimentally tested.

High-Mass Stars

In contrast to stars of somewhat less than one solar mass that are unable to generate conditions to continue along the evolutionary chain as far as the Sun then more massive stars progress further. More massive stars generate higher core temperatures that enable fusion reactions to take place between more strongly charged (and therefore more massive) nuclei. Higher core temperatures also imply that electrons are able to take energies within a thermal distribution without being limited by the Pauli exclusion principle so that the plasma does not become electron degenerate (see Figure 13.2). The progress of more massive stars is consequently increasingly complex as more stages are included but the result of the onset of each stage becomes less dramatic. Stages also follow on from each other much more quickly for stars of increasing stellar mass as the dominant nuclei at each step are transformed more rapidly due to the more extreme conditions generated by greater gravitational contractional forces.

A star with a mass of around three solar masses will leave the main sequence almost a hundred times more quickly than a solar mass star. Its surface temperature will then steadily decrease but its luminosity increase due to stellar swelling will be moderate. It will not develop an electron degenerate core but the triple-alpha reaction will slowly turn on as the star becomes a red giant. Consequently, no helium flash will take place but a more gentle shift in the star's surface temperature will shift its position on the Hertzsprung–Russell diagram back towards the main sequence. The processes that control this change are as for smaller stars. Core expansion causes the shell of material that supports hydrogen fusion reactions to expand. The shell becomes cooler so that the nuclear reaction rate decreases a little and the outer envelopes of the star contract as a consequence. The closer proximity of the photosphere to the energy source causes the effective temperature to increase. As for lower mass stars a second movement towards the red giant region ensues once helium fusion reactions become dominant. Such a temporary diversion of the star's track across the Hertzsprung–Russell diagram is known as a blue loop excursion and several such loops may take place as new nuclear processes begin to occur in large stars. For a star of around three solar masses the onset of carbon fusion reactions occurs after the core's density has once again increased due to gravitational contraction and this time has become electron degenerate. A carbon flash occurs. In larger stars core temperatures are higher and no carbon flash occurs, the onset of carbon fusion reactions being marked by a more gentle blue loop excursion.

The first fusion reaction involving carbon as a starting material involves the capture of a ^{4}He nucleus to produce ^{16}O. The star now begins to develop an onion-like structure that becomes ever more complex as new core reactions begin to take place. The star can be thought of as being gravitationally differentiated, like a planet, with large nuclei at the core, hydrogen close to the surface and other nuclei distributed according to mass in between. When the first carbon fusion reactions begin to take place the core consists mainly of carbon and oxygen. Immediately surrounding this is a region dominated by helium, the innermost shell of which supports the triple-alpha process. Around the helium envelope is the rest of the star which remains principally composed of hydrogen. The innermost shell of this region supports hydrogen fusion reactions. Though three solar mass stars develop through more stages than solar mass stars the same fate eventually befalls them, developing planetary nebulae and cooling as white dwarfs until eventually disappearing from sight. The path of such a star across the Hertzsprung–Russell diagram is shown in Figure 13.4.

The most massive stars slide quickly across the Hertzsprung–Russell diagram from the main sequence to the supergiant region producing more and more massive nuclei at an ever-increasing rate. For instance, a 25 solar mass star derives its energy from hydrogen fusion reactions for around 7 million years during which time it moves off the main sequence (Figure 13.5). A helium core lasts for a period ten times shorter and carbon core reactions dominate for a thousand times shorter still. An oxygen core may last for less than a year and a silicon core for perhaps only one day. Eventually an iron core forms. The ^{56}Fe nucleus is the most stable of all nuclei. Fusing other nuclei to ^{56}Fe requires energy input as does splitting it apart in fission reactions. The iron core should be truly

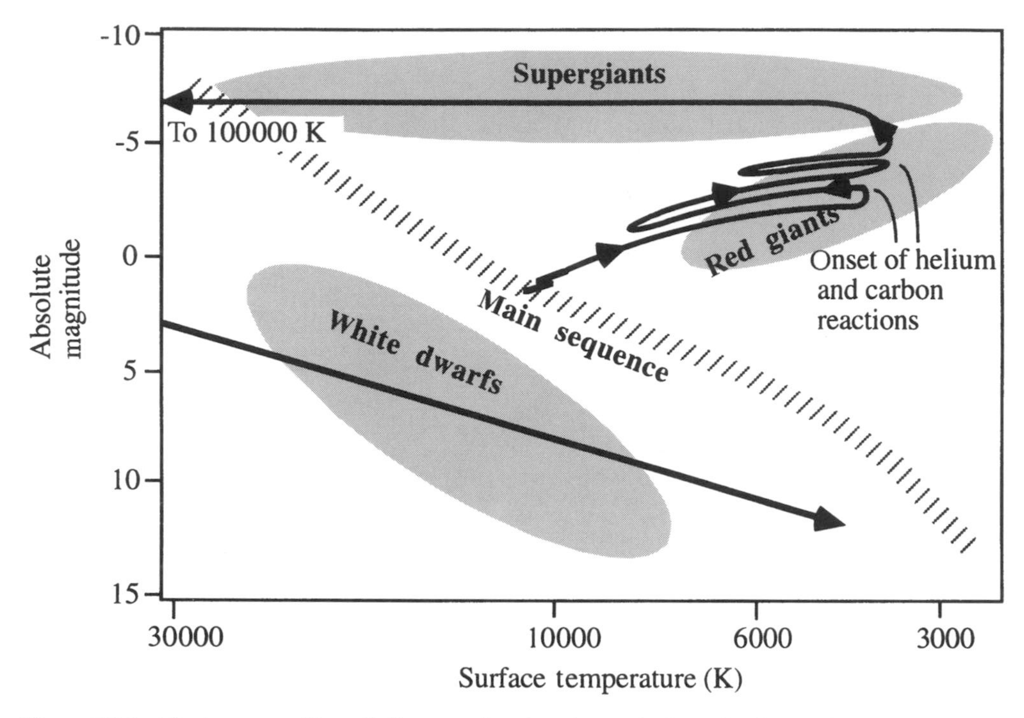

Figure 13.4 Hertzsprung–Russell diagram showing the evolution of a three solar mass star

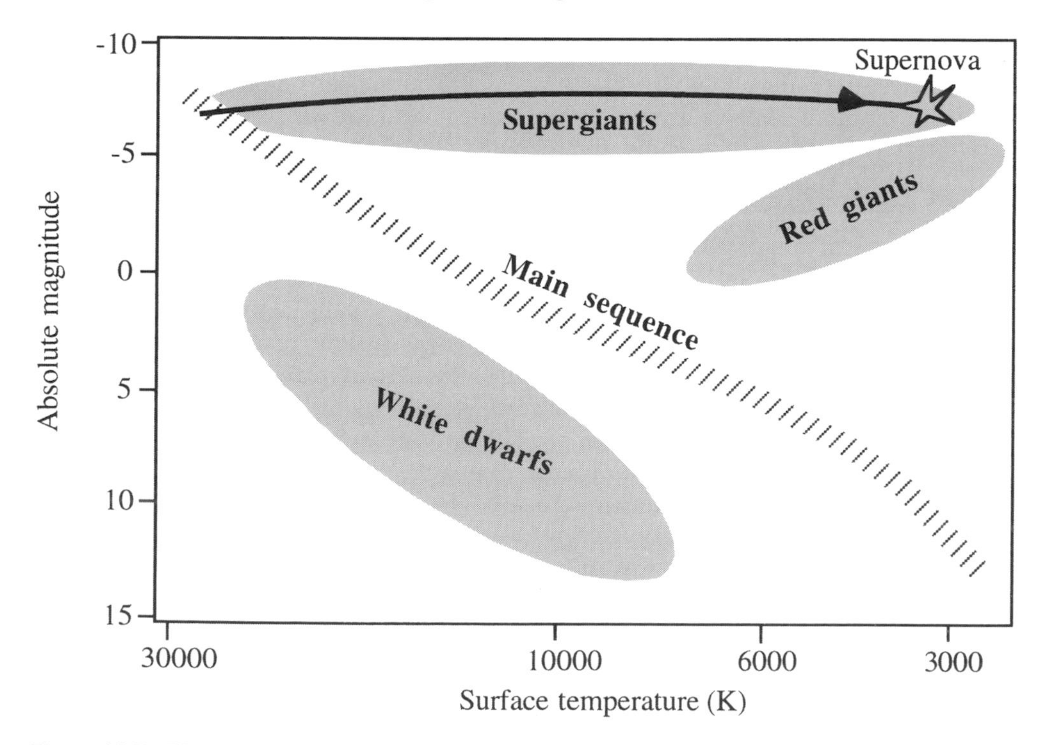

Figure 13.5 Hertzsprung–Russell diagram showing the evolution of a 25 solar mass star

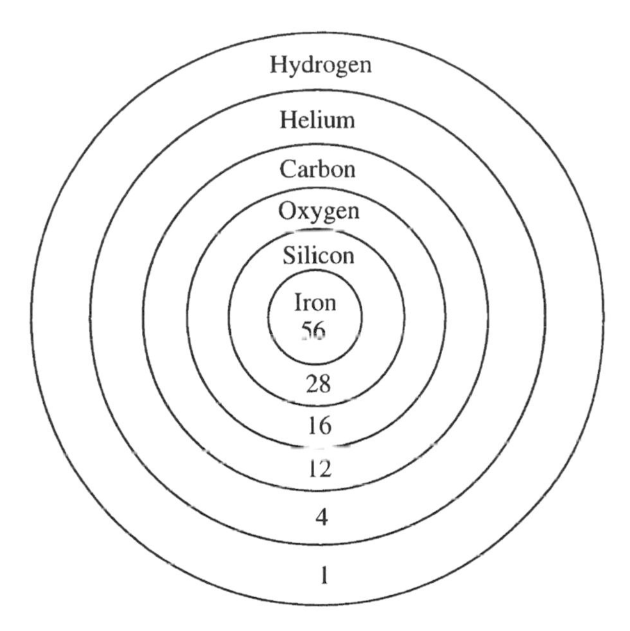

Figure 13.6 Cross-section through a high-mass star immediately before supernova. Dominant nuclei are indicated by name and atomic number. Other nuclei are also present and the differentiation is not quite as pronounced as indicated here

inert and the differentiated star, as schematically represented in Figure 13.6, might be expected to quietly form a massive white dwarf with an accordingly tiny radius. Nothing could be further from the truth.

Neutron Stars and Black Holes

There is an upper limit to the mass of a white dwarf. The so-called Chandrasekhar limit is about 1.4 solar masses. In other words, after a star has jettisoned its outer envelopes into a planetary nebula, if the central remnant has a mass of greater than 1.4 solar masses it will overcome electron degeneracy pressure. The pressures become so great that protons fuse with electrons to produce neutrons. The inert star collapses to a tiny size at which point the neutron energies become degenerate. The radius of the newly formed neutron star is around 10–20 km. The vast majority of the matter in the neutron star is in the form of neutrons, in either a fluid or solid lattice (see Figure 13.7). The density of such a star is obviously incredibly high even by the standards of the already unimaginably high values discussed for other situations earlier in the book.

WORKED EXAMPLE 13.4

Q. Compare the densities of a white dwarf, a neutron star and a main sequence star where all three bodies are of about 1.4 solar masses.

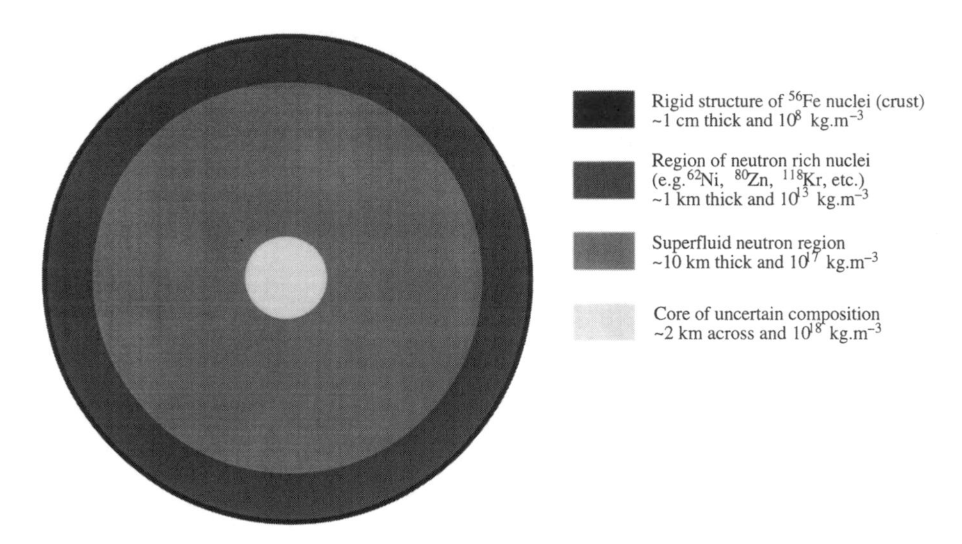

Figure 13.7 Model for the internal differentiation of a neutron star. Note that the largest stable nuclei of Ni, Zn and Kr on Earth have masses of 64, 70 and 86 respectively. Densities of greater than about 10^{10} kg m^{-3} promote the combination of protons and electrons to produce neutrons

A. A white dwarf of (just under) 1.4 solar masses corresponds to the most massive white dwarf that can withstand electron degeneracy pressure and therefore has the smallest possible radius for a white dwarf. This is similar to the Earth's size (radius $\sim 10^7$ m). Density is given by

$$\text{Density} = \frac{\text{Mass}}{\text{Volume}} = \frac{M}{\frac{4}{3}\pi r^3} \approx \frac{3 \times 10^{30}}{4(10^7)^3} \approx 10^9 \text{ kg m}^{-3}$$

The largest neutron stars (those of minimum mass at (just over) 1.4 solar masses) have a radius of around 20 km. Density is therefore given by

$$\text{Density} \approx \frac{3 \times 10^{30}}{4(2 \times 10^4)^3} \approx 10^{17} \text{ kg m}^{-3}$$

A main sequence star of 1.4 solar masses will have a similar density to the Sun (see Figure 11.5 which shows that slightly more massive main sequence stars than the Sun have slightly larger radii). The Sun's density is given by

$$\text{Density} \approx \frac{2 \times 10^{30}}{4(7 \times 10^8)^3} \approx 10^3 \text{ kg m}^{-3}$$

The densities of the three bodies of the same mass are therefore radically different. The Sun's density is about that of water on the Earth's surface whereas a white dwarf has a density one million times greater. A teaspoon of white dwarf material has a mass similar to a car. The density of a neutron star is a further 100 million times greater so that if the Earth were shrunk to the same density it would fit into a sports stadium. It should be noted that the density of white dwarfs and neutron stars do not vary greatly throughout their structure whereas the Sun is much denser in its core than at its photosphere.

Larger mass neutron stars have smaller radii for reasons entirely analogous to white dwarfs. Like electron degeneracy pressure, neutron degeneracy pressure also has limits.

Once gravitational forces overcome neutron degeneracy pressure there remains nothing to prevent the collapse of the star to a singularity. That is, the dimensions of the star become zero. The behaviour of such an entity can only properly be described using general relativity but, as a simple approximation, consider the Newtonian gravity field around a singularity. In Chapter 6 the escape velocity from the surface of a planet was derived:

$$v_e = \sqrt{\frac{2GM}{r_p}} \tag{6.14}$$

The planetary radius, r_p, can be replaced by a general distance from the centre of mass and the equation becomes applicable to any point in space relative to any centre of mass. It becomes possible to calculate the Schwarzschild radius, r_S, of a singularity. This represents the distance from the singularity at which the escape velocity is equal to the speed of light:

$$c = \sqrt{\frac{2GM}{r_S}}$$

$$\therefore \quad r_S = \frac{2GM}{c^2} \tag{13.1}$$

Calculations for the smallest mass possible to form such an entity via stellar death vary a little. The calculation depends on the mixture of nuclei present in the star immediately before its collapse and comes out at between 3 and 5 solar masses. For a 5-solar mass star the Schwarzschild radius is about 15 km. This implies that a photon created anywhere within a sphere of radius 15 km about a 5-solar mass singularity will be unable to escape. Similarly, photons encroaching within this limit will be trapped. The speed of light represents the maximum speed of motion for any object and so nothing can escape from within the Schwarzschild radius. The entity is known as a black hole.

WORKED EXAMPLE 13.5

Q. Estimate the smallest possible Schwarzschild radius for a black hole that forms as the result of a supernova. The mass of the Sun is 2×10^{30} kg.

A. Equation (13.1) shows that the Schwarzschild radius of a black hole is proportional to the mass of the collapsing core. The smallest possible Schwarzschild radius is therefore given by the smallest possible mass that is capable of overcoming neutron degeneracy pressure, thought to be about 3 solar masses. Inserting numbers into the equation:

$$r_S = \frac{2GM}{c^2} = \frac{2(6.67 \times 10^{-11})(6 \times 10^{30})}{(3 \times 10^8)^2} \approx 9000 \text{ m}$$

A Schwarzschild radius of 9 km is the smallest possible but even enormous black holes formed in this way will have very small sizes on an astronomical scale.

Supernovae

The evolution of a star to a neutron star or a black hole is rarely gentle. A supernova refers to the sudden and explosive collapse of a large star. Though nuclear fusion reactions occur in shells around newly inert iron cores there is nothing to counterbalance the inward gravitational force on the core and so it simply collapses. The energy of gravitational contraction can no longer be used to promote new reactions and the ^{56}Fe nuclei break apart with an attendant increase in mass:

$$^{56}\text{Fe} \rightarrow 13\,^{4}\text{He} + 4\,\text{n} \tag{13.2}$$

This increase in mass requires an input of energy that is provided by the gravitational collapse. For stars of around 10–20 Solar masses most of the energy of collapse is required to provide the energy input first to allow the break-up of iron nuclei and then to fuel the proton/electron fusion reaction that takes place in the helium nuclei created and produces the neutrons that eventually make up a neutron star:

$$\text{p} + \text{e}^- \rightarrow \text{n} + \nu \tag{13.3}$$

The gravitational energy that is left over creates an enormous explosion. While the inside of the star collapses to create a neutron star or perhaps a black hole, the outer envelopes are flung into space, creating shock waves for many parsecs around. The collapse of the core takes less than a second. Other nuclear processes take place in the outer envelopes of large stars during this second to create nuclei more massive than ^{56}Fe. Space is thus enriched with a mixture of nuclei from hydrogen to uranium. The heavy elements found on Earth, for instance the iodine in a medical kit or the mercury in a thermometer, must have been fabricated in a supernova explosion many billions of years ago, scattered across space and then gravitationally regathered to form a new generation star and planetary system.

According to some calculations, the collapse of iron cores in stars with masses greater than about 20 solar masses leaves no energy for explosion and the whole star collapses to produce a massive black hole. The star simply switches off and vanishes.

WORKED EXAMPLE 13.6

Q. Estimate the amount of gravitational potential energy released during a supernova.

A. The mass of the supernova *remnant* will be assumed to be about 2 solar masses thus forming a neutron star with a radius of about 15 km. Equation (6.13) expresses the amount of potential energy released in moving an object of mass m to the surface of a sphere of mass M and radius r from outside the effective gravitational field of the sphere:

$$W = \frac{GMm}{r} \tag{6.13}$$

The potential energy released during a supernova is equal to the difference in the amount of gravitational energy released in forming the neutron star from diffuse gravitationally independent particles and the gravitational energy released in forming the core of the star just before collapse in the same way[1]. To calculate the potential energy of formation of a star it is necessary to consider the energy released in building the body by adding very thin spherical shells of increasing radius to an ever-increasing central ball. The density of the stars are taken to be uniform so that, from Worked Example 13.4, it may be expressed in terms of its total mass, M, and radius, R, as:

$$\text{Density} = \frac{M}{\frac{4}{3}\pi R^3}$$

The mass of a spherical shell of very small thickness, dr, with a radius of r is given by

$$m_{\text{shell}} = \text{volume} \times \text{density} = 4\pi r^2\, \mathrm{d}r \frac{M}{\frac{4}{3}\pi R^3} = \frac{3M}{R^3} r^2\, \mathrm{d}r$$

In adding such a layer to a central nucleus the energy released, W_{shell}, is given by equation (6.13) where m is replaced by m_{shell} and M by the mass of the central nucleus, M_{nucleus}:

$$M_{\text{nucleus}} = \text{volume} \times \text{density} = \frac{4}{3}\pi r^3 \frac{M}{\frac{4}{3}\pi R^3} = \frac{M}{R^3} r^3$$

Substituting into equation (6.13) gives;

$$W_{\text{shell}} = \frac{GM_{\text{nucleus}} m_{\text{shell}}}{r} = \frac{G\left(\frac{M}{R^3} r^3\right)\left(\frac{3M}{R^3} r^2\, \mathrm{d}r\right)}{r} = \left(\frac{3GM^2}{R^6}\right) r^4\, \mathrm{d}r$$

To find the total energy, W_{total}, the sum of the energy released by the addition of each shell must be found by integrating over the full radius of the star:

$$W_{\text{total}} = \frac{3GM^2}{R^6} \int_0^R r^4 \mathrm{d}r = \frac{3GM^2}{R^6} \left[\frac{r^5}{5}\right]_0^R = \frac{3GM^2}{5R}$$

The energy release during a supernova, W_{super}, is therefore given by the difference in the energy of formation of the neutron star, W_{neutron}, and the energy of formation of the initial core, W_{initial}:

$$W_{\text{super}} = W_{\text{neutron}} - W_{\text{initial}} = \frac{3GM^2}{5R_{\text{neutron}}} - \frac{3GM^2}{5R_{\text{initial}}} \approx \frac{3GM^2}{5R_{\text{neutron}}}$$

The final approximation above is made because the radius of the core before collapse is much larger than the radius of the neutron star that it forms so that its energy of formation is very small compared to that of the neutron star. The calculation has assumed uniform density within both the initial and final bodies. This is acceptable for the neutron star but may not be for the parent star. Calculations which assume graded densities do not radically change the expression for W_{initial} (e.g. if the density is assumed to decrease with the square of separation from the centre then $W_{\text{initial}} =$

[1] Such an approach is allowed in a so-called conservative field like a gravitational or electric field. In such a case, the quantity of energy released or consumed is determined only by the initial and final conditions and the order of events is irrelevant. Ripping apart the parent star into gravitationally independent particles and then reassembling them to form the remnant is therefore energetically equivalent to going directly from the parent to the remnant.

GM^2/R, only 70% greater, so that $W_{\text{initial}} \ll W_{\text{neutron}}$ remains firmly true). The final expression given above therefore gives a good approximation to the energy released during a supernova. Using the values given above:

$$W_{\text{super}} = \frac{3GM^2}{5R_{\text{neutron}}} = \frac{(3 \cdot 6.67 \times 10^{-11})(4 \times 10^{30})^2}{5(1.5 \times 10^4)} \approx 4 \times 10^{46}\ \text{J}$$

As explained above not all of this tremendous amount of energy will be emitted as radiation. Most (or possibly sometimes all) is taken up in the formation of neutrons and expulsion of neutrinos. Nevertheless, an interesting comparison is the amount of energy that the Sun will radiate in a 10 billion-year period (its predicted lifetime on the main sequence) which is (4×10^{26} W $\times$ 3600 s $\times$ 24 hours $\times$ 365 days $\times$ 10^{10} years $=$) 10^{44} J. If a supernova radiates approximately 1% of the potential energy released as electromagnetic radiation (a reasonable estimate) then it will emit as much radiation in the space of a few days as the Sun does in its entire main sequence lifetime.

The stories of stellar evolution described above require experimental evidence as corroboration. In what follows, the most important observations and interpretations are described and explained.

Clusters as Evidence for Stellar Evolution

Young open clusters were discussed in the previous chapter as proof of the theories of early stellar stabilisation. The Pleiades were given as an example and compared to a much older open cluster, the Hyades in Figure 12.9. The different turnoff position from the main sequence indicates the age of the cluster. This also provides excellent information for the study of later evolution. Though the passage of individual stars cannot be tracked across the Hertzsprung–Russell diagram in a reasonable period of time, clusters of differing age can be studied. The behaviour of stars of different masses can thus be inferred. Figure 13.8 shows a Hertzsprung–Russell diagram for a number of clusters of different ages that demonstrates how stars of different ages occupy the regions predicted by the theoretical curves given in Figures 13.1, 13.4 and 13.5.

Some of the data in Figure 13.8 is for a different type of cluster known as a globular cluster. These groups of stars are very old and quite different in appearance from open clusters. In globular clusters, hundreds of thousands of stars are crammed into a small spherical region with concentrations as high as a thousand stars per cubic parsec at the centre. The age of these clusters is such that the only stars that remain are of solar mass or less, most others having evolved into white and then black dwarfs (Figure 13.9). Most globular clusters are over 10 billion years old, close to the age of the universe. This is important for several reasons. First, advanced stellar evolution can be studied. Second, these stars, largely free of material that has been recycled by former stars, can be contrasted with those that contain heavier elements produced by defunct stars that have enriched space by ejecting matter as they died. Stars such as those in globular clusters are known as population II stars and contain less than one part in a thousand of nuclei other than hydrogen and helium. Population I stars that have formed from recycled material and are more commonly associated with open clusters may contain as much as 2% heavier elements. The proportion of heavy elements in a star affects reabsorption processes in stellar interiors. This influences energy transport which in turn influences

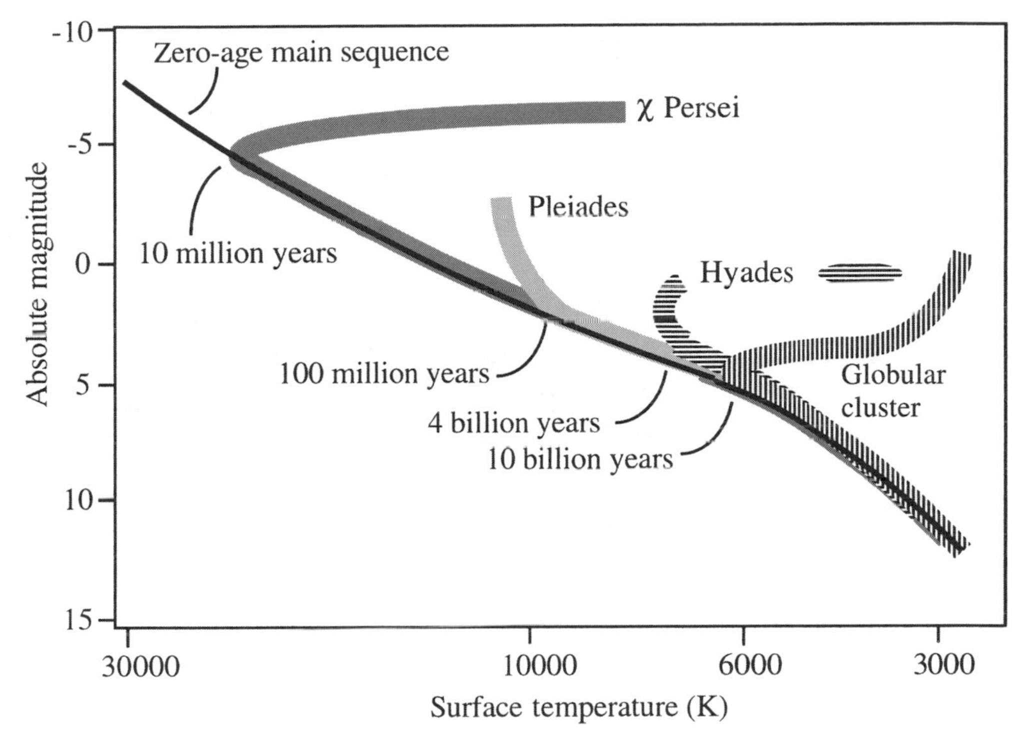

Figure 13.8 The Hertzsprung–Russell diagram indicating the positions of the stars of three open clusters and a globular cluster. The time from main sequence stabilisation to turn-off is indicated for each cluster

the rate at which stars evolve. The picture created above for a generic star following a preordained path through life dependent exclusively on mass is therefore not strictly true. The proportion of heavy elements varies from star to star and this blurs evolutionary tracks. However, routes across the Hertzsprung–Russell diagram are merely perturbed by element variation and not radically changed.

Convective processes within stars cause a little mixing of the differentiated zones. Convection is particularly strong in giants and supergiants as outer envelopes become well removed from the energy source. Heavier nuclei are dredged up from close to the core at such times and when they reach the surface will contribute to the sharp line spectra of the star. Data obtained in this way fits well to the paradigms for nucleosynthesis given above.

Observations of White Dwarfs

A useful check on the ideas presented on white dwarfs, beyond the direct observation of very young such objects that are forming planetary nebulae, is to measure their masses. As discussed in Chapter 12, the best way to do this is by considering the orbital motion of visual binaries. The masses of white dwarfs are always found to be less than 1.4 solar masses, the Chandrasekhar limit, substantiating the electron degeneracy theory. The

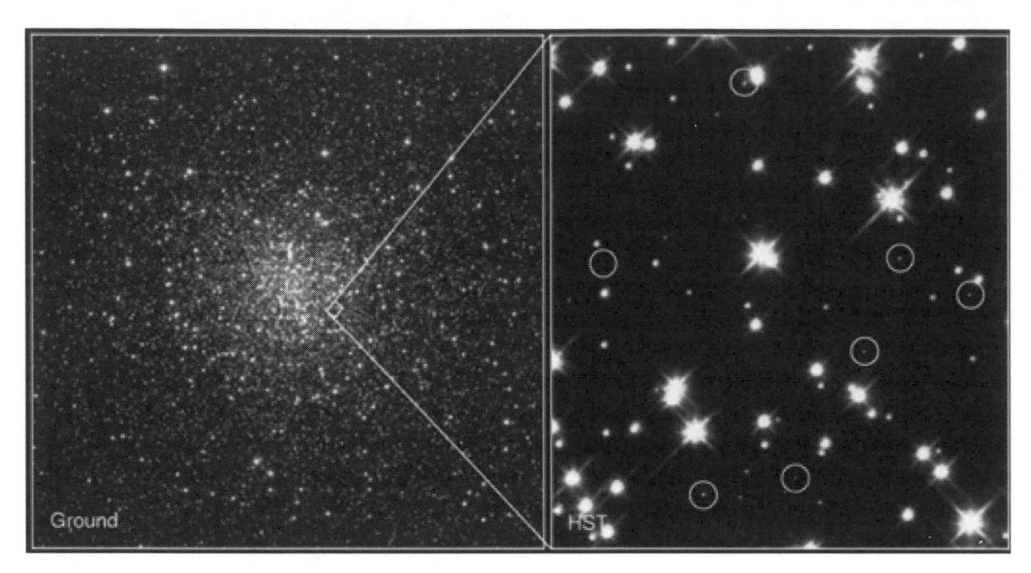

Figure 13.9(a) The globular cluster M4 (left) with a close-up (right) of part of the cluster taken by the Hubble Space Telescope. Circled are white dwarf stars.

radius of a white dwarf can also be determined from its position on the Hertzsprung–Russell diagram. This provides a second method for mass determination because, as discussed above, mass and radius are directly related in white dwarfs. There is a strong clustering of white dwarf masses at just over half a solar mass. This corresponds well with the calculated mass of a carbon-based core of a medium-mass star and also to observed masses of remnant stars in planetary nebulae that have evolved from lower-mass stars. Observation and a number of interrelated theories on white dwarfs are thus mutually consistent.

Observations of Supernovae and their Remnants

The end product of a massive star is a neutron star or a black hole. Both present considerable observational difficulties. By definition, even light cannot leave the region around a black hole. The supernova explosion that signals the end of a massive star's life is much easier to see but suffers from the opposite problem to most stages of stellar evolution. It doesn't last long enough. For every massive star that forms there is a 10 million-year wait before the sky blinks for a few days and then returns to normal. Only one such event has occurred in the latter half of the twentieth century within a radius of about 100 thousand parsecs of us and only six such events have been observed within 10 thousand parsecs throughout recorded history. Perhaps the most famous occurred in 1054 and was so bright that it was visible during the day for three weeks. The material that was scattered through surrounding space formed the Crab nebula in Taurus and is still visible today.

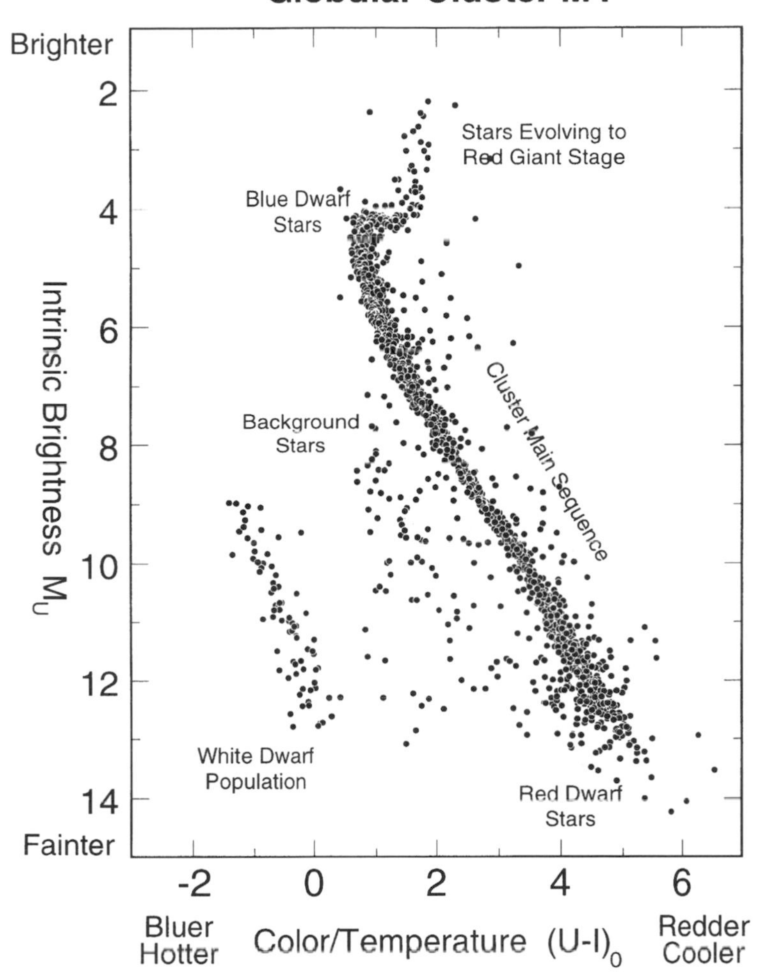

Figure 13.9(b) A part of the Hertzsprung–Russell diagram for this globular cluster (facing) shows main sequence stars and the white dwarfs. Such systems are important for our understanding of stellar evolution (reproduced by permission of the Space Telescope Science Institute)

Perhaps the single most important chance event in astronomical history was observed on Earth on 24 February 1987. The observed event had taken place 170 thousand years earlier but the light arrived just as astronomers had assembled a welcoming committee composed of detectors and imaging equipment that operated across the electromagnetic spectrum as well as special equipment capable of counting neutrinos. As soon as the super-new (super-'nova') star was spotted the world's observatories could focus their attention upon it. The progenitor star was named Sanduleak and was a B3 supergiant with a surface temperature of 17 000 K, a luminosity 100 thousand times greater than the Sun, a mass around 20 times that of the Sun (based on standard models of stellar evolution) and a radius about 50 times larger than that of the Sun. It occupied an active region of star formation and the shock waves from its death have probably destabilised surrounding areas into further cloud collapse. A supernova created through the mechanism described above is labelled type II and models suggest that neutrinos are able to escape from the intense conditions during stellar core collapse some time before photons. The first light to be observed from a supernova should be characteristic of hydrogen from the expanding outer envelopes of the exploding star whereas the spectrum should develop features associated with larger nuclei later. All of this was observed with the Sanduleak supernova and the remnant nebula can now be observed spreading into the surrounding region (see Figure 13.10). Not everything fit perfectly. The evolution of the light intensity followed a typical pattern consisting of a sharp peak, representing the explosion itself, and a slow decrease, partially powered by the radioactive decay of nuclei produced during the supernova, but the peak intensity was significantly less than would be expected. Furthermore, a star so close to the main sequence would not have been expected to have evolved sufficiently to take place in a supernova explosion. It should be remembered that the whole of stellar evolution theory is just that—theory. Amendments must always be made to models so that they give full and comprehensive explanations as nature gradually reveals itself. Theory may have advanced sufficiently in time for the next local type II supernova so that it will behave exactly as predicted.

While astronomers are waiting for the next bright supernova further evidence can be gathered elsewhere. Some information can be gained from more distant supernovae though their study is practically more difficult. Supernova remnants are also useful. They represent yet another type of nebula and shouldn't be confused with those discussed previously. A supernova remnant has a spherical shape similar to a planetary nebula though considerably larger and without a bright star at its centre. A supernova remnant is also considerably more ragged, an indication of the much greater forces that created it. The best indicator that a nebula is a supernova remnant is that part of its emission has a different characteristic from other sources. Synchrotron radiation refers to the electromagnetic energy emitted by charged particles (in this case electrons) as they spiral in magnetic fields. The emitted electromagnetic radiation is polarised in this case. This means that the electric field of the waves (see Figure 1.1) is always in the same direction for any given electron direction. As electrons tend to move in particular directions (mainly outwards) then supernova remnants appear differently when viewed through perpendicularly oriented, polarisation-sensitive materials (such as Polaroid as used in sunglasses). Synchrotron radiation also has an entirely different spectrum from blackbody radiation and is determined by the speed of the electrons and the strength of the magnetic field rather than by temperature alone. As the electrons emit photons they lose energy and therefore speed. The electrons eventually lose all of their energy, stop moving and stop emitting radiation.

The Crab nebula was formed by the supernova observed in 1054. The Doppler shift of the spectral lines in the outer parts of the nebula indicates that the speed of expansion is around 1500 km s^{-1}. The particles that give rise to these spectral lines are atoms that have been excited by the synchrotron emission of the fast-moving electrons. How can the electrons still be moving at high speeds nearly a thousand years after the supernova? They cannot. There must be a continuous source of electrons to power the synchrotron radiation.

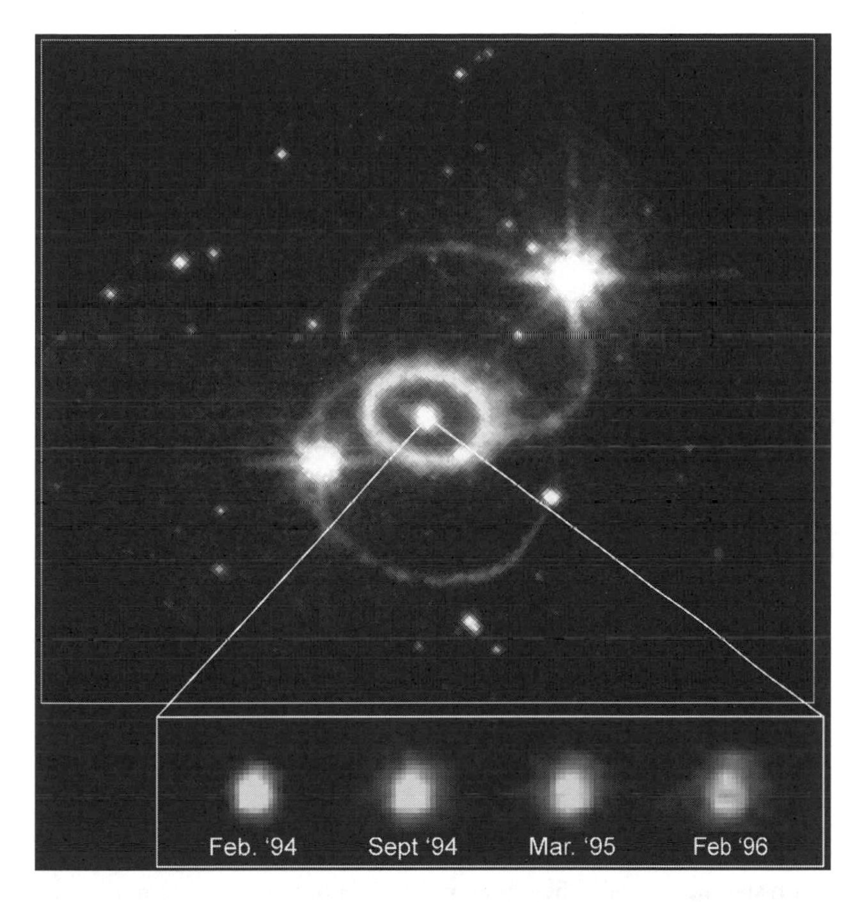

Figure 13.10 The 1987a (Sanduleak) supernova photographed ten years after explosion, now 0.03 parsec across. The matter expanding from the remnant core is moving at about 10 million kilometres per hour (reproduced by permission of the Space Telescope Science Institute)

Pulsars

The source of the Crab nebula's electrons is a pulsar (PULSAting Radio star) at its centre. Pulsars produce radio pulses with a very precise gap between the bursts of energy on the order of seconds or less. Located at the centre of a supernova remnant and still able to emit electromagnetic radiation, as well as electrons, there is only one obvious candidate to be the pulsar. That is a neutron star. Despite the enormity of the explosion that creates a neutron star simple physical laws must still be obeyed. Angular momentum must be conserved. In the same way that the rotation rate of a collapsing molecular cloud increases then so must the collapsing core of a supernova remnant. Such is the scale of the decrease in physical size of the core that its rotation period decreases to a second or less. The rotation period of the Crab nebula neutron star is only one thirtieth of a second, assuming it to be equal to the pulsar period. The model required to explain the

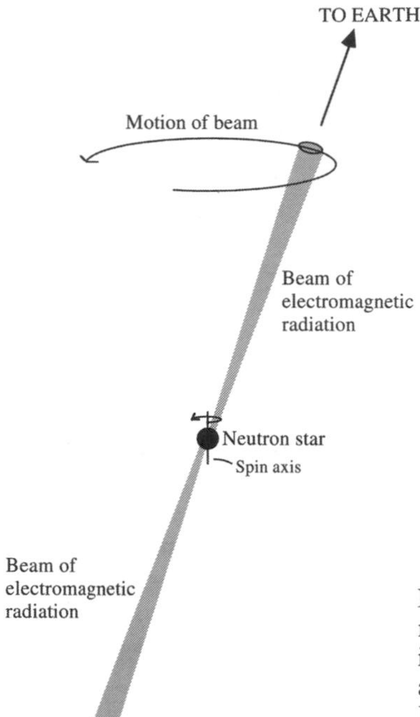

Figure 13.11 A pulsar. As the neutron star spins rapidly on its axis a beam of electromagnetic radiation is emitted in a particular direction (determined by the axis of the magnetic field of the neutron star). The Earth is in line with the beam once per rotation and so periodically receives a pulse of energy

coincidence in these two time periods is analogous to the projection of the magnetic fields of planets such as Jupiter where the spin and field axes are not aligned (see Figure 9.1). For a neutron star the magnetic field is enormously more powerful than in a planet. This accelerates electrons (and other charged particles) away from the neutron star along the magnetic field axes. The field axes are being whirled around by the rapid rotation of the planet and so the electrons are sprayed into space like a speeded-up lawn sprinkler. The usual analogy is to a lighthouse. The electrons emit synchrotron radiation in the direction of their motion and this produces two beams of electromagnetic radiation travelling from the neutron star in opposite directions along the magnetic field axes. Some of this radiation is scattered throughout the nebula whereas the rest of the beam heads off into space. In the case of the Crab nebula, the Earth is in line with this beam and so one pulse is received at the same point during every rotation (see Figure 13.11). There must be many neutron stars whose magnetic axes never align with the Earth (estimated at 80%) and so can only be suspected as being present due to the glowing nebula that surrounds them. No pulsar has yet been detected at the centre of the Sanduleak supernova remnant.

The period of most pulsars is slowly increasing. For instance, that of the Crab pulsar increases by about 10 μs per year. This can be explained in terms of energy conservation. The neutron star loses rotational kinetic energy that is used to accelerate electrons. The electrons spiral in the magnetic field and lose their energy as synchrotron radiation. Some of this is beamed into space and some is used to excite fluorescence in the nebula (Figure 13.12).

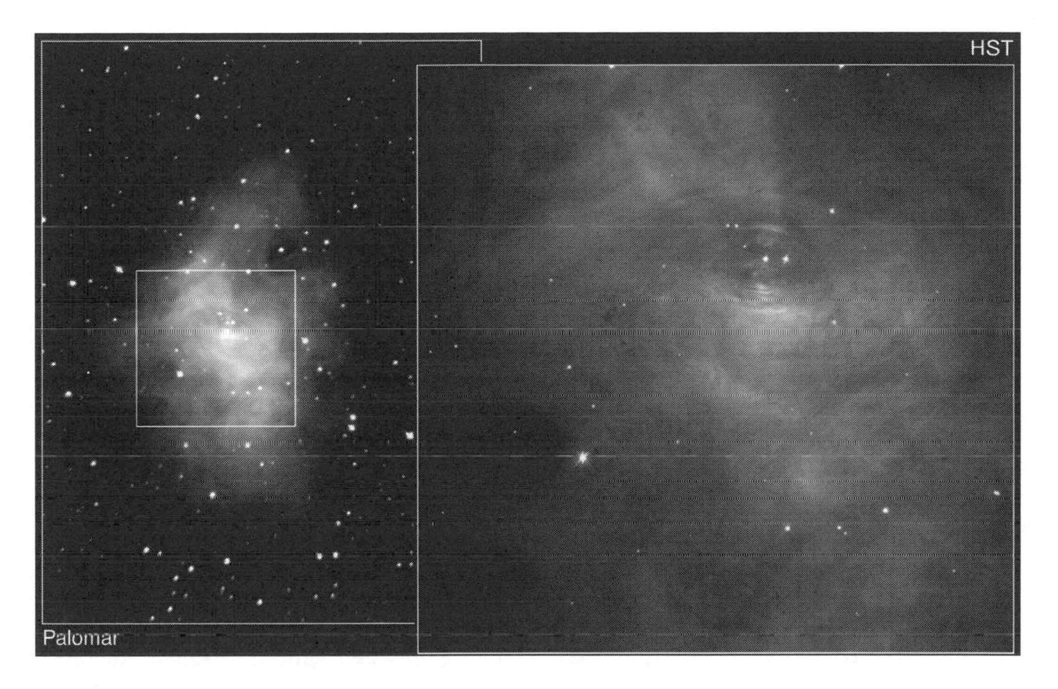

Figure 13.12 The Crab nebula as photographed from the Palomar Observatory (left) and an enlargement of the centre, including the pulsar, as taken by the Hubble Space Telescope. Much of the energy that illuminates the nebula derives from the pulsar. The matter, strewn over 3 parsecs, is the remnant of the supernova explosion observed on Earth in 1054. The neutron star that constitutes the pulsar is the collapsed core of the star that became the supernova (reproduced by permission of the Space Telescope Science Institute)

WORKED EXAMPLE 13.7

Q. How much energy is the Crab nebula pulsar losing per second due to the gradual decrease in its rotational rate given that the rotational kinetic energy of a sphere of mass, M, radius, R, and rotational period, T, is given by the expression below?

$$\text{Rotational kinetic energy} = \frac{4\pi^2 MR^2}{5T^2}$$

A. It is possible to solve this problem by differentiating the above expression with respect to time but a simpler technique will be used here. The kinetic energy will be calculated for two moments that are separated by one second, given that the pulsar is spinning with a period 10 μs greater every year. In one second the increase in period is therefore (10 μs/(365 × 24 × 3600) =) 3×10^{-13} s, symbolised by ΔT in what follows. The pulsar's period is 0.033 s and its mass and radius will be assumed to be about two Solar masses and 15 km respectively. The change in energy during 1 s is therefore

$$\frac{4\pi^2 MR^2}{5T^2} - \frac{4\pi^2 MR^2}{5(T+\Delta T)^2} = \frac{4\pi^2 MR^2}{5}\left\{\left(\frac{1}{T^2}\right) - \left(\frac{1}{T^2} - \frac{2\Delta T}{T^3} + \ldots\right)\right\}$$

The latter expression is a binomial expansion of the previous terms. Because the change in the period is proportionally very small, only the first two terms in the expansion are required and the final result is

$$\frac{8\pi^2 MR^2 \Delta T}{5T^3} = \frac{8\pi^2[4 \times 10^{30}(1.5 \times 10^4)^2 3 \times 10^{-13}]}{5(0.033)^3} = 10^{32}\text{ J}$$

So the decrease in the speed of rotation of the Crab nebula pulsar results in the loss of about 10^{32} J per second. In other words, the neutron star is losing power at the rate of 10^{32} W. The Crab nebula shines with an intensity of about 10^{31} W so about 10% of the rotational energy is being converted to fluorescence by the diffuse material that surrounds the pulsar. Again it is worth comparing the power output of the Sun at 4×10^{26} W, about 25 thousand times dimmer than the nebula and emitting about 250 thousand times less power than the pulsar is losing.

There are pulsars that do not appear to be surrounded by a nebula. This can simply be because the nebula has dispersed through old age or due to the source being too far away to see neighbouring fluorescence. The former explanation appears to be true of a suspected neutron star (no longer producing detectable pulsed radiation) that was discovered as an X-ray source by the Röntgen satellite in 1992 and observed in visible light by the Hubble Space Telescope in 1997. The star is about 25th magnitude (very dim) but has a time-invariant black body spectrum that shows its surface temperature to be a little above 1 million K. Its distance cannot be accurately determined but is known to be less than about 100 parsecs due to the presence of a molecular cloud of known distance behind it. Combining this data (see Figure 11.4) shows that the diameter of the star is no larger than 28 km. There are no other bodies that could have such a high temperature and small size and so the object has confidently been labelled a neutron star.

Trying to See Black Holes

The evidence for neutron stars is quite good but can it be possible to observe a black hole in which all the matter is concentrated into a single point? The Schwarzschild radius marks a position from which nothing can escape but does not represent an object that can be seen. It is essentially empty space, albeit rather unusual in its gravitational properties. The only way to see a black hole must be as something is falling into it. The intense gravitational field around the Schwarzschild radius can accelerate any passing matter to very high speeds thus heating it and causing ionisation. The resulting hot plasma may then emit a combination of blackbody and synchrotron radiation as it spirals into the black hole (see Figure 13.13). The conditions are so extreme that the peak wavelength of this emission is in the X-ray part of the electromagnetic spectrum, distinguishing it from any objects that have been discussed previously.

Hunts for X-ray sources have revealed several candidates for black holes. To be highly visible a good source of matter is required. A lone black hole, even in a molecular cloud, cannot suck in enough matter to be an intense enough object so as to be seen from Earth. It should be remembered that the gravitational field due to a ten Solar mass black hole is exactly the same as that due to a 10 solar mass supergiant at distances greater than the size of the supergiant's radius (see Figure 13.14). The gravitational field is only especially strong close to the black hole. Black holes cannot suck dry large regions of space as some imagine. A good source of matter is a partner star. If a black hole is part of a binary it is

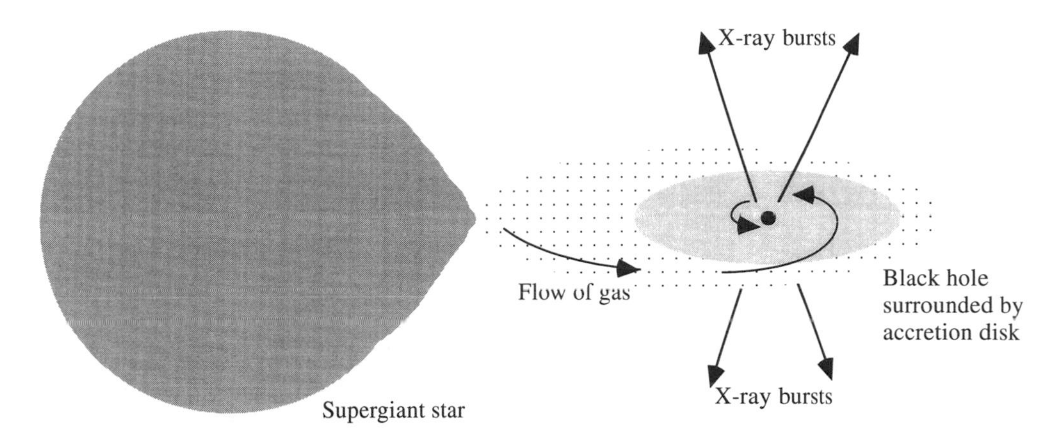

Figure 13.13 Binary system consisting of a supergiant and a black hole. As the supergiant increases in size its outer envelopes become more strongly attracted to the gravitational field of the black hole. Captured matter flows towards the black hole spiralling in through an accretion disk. As the matter moves towards the black hole it is warmed through the release of gravitational potential energy and releases much of this energy via electromagnetic radiation including X-ray bursts

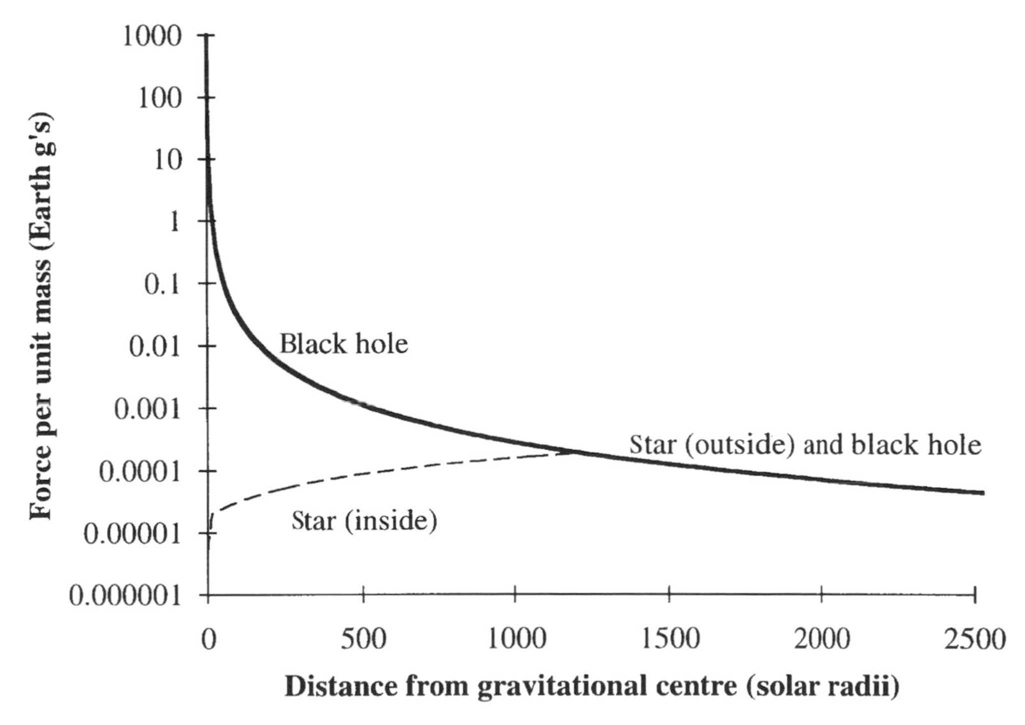

Figure 13.14 The gravitational field of a 10-solar-mass red supergiant such as Betelgeuse (1200 Solar radii) and a 10-solar-mass black hole. Close to the black hole the gravitational field becomes very intense (note logarithmic scale) whereas within the star the gravitational field decreases. Outside the body of the star, however, the gravitational fields are identical. Red supergiants have large radii compared to other stars but, nevertheless, are minute compared to the distance between stars. Thus a black hole has no more effect on large volumes of space than a star of the same mass

possible for material from the outer envelopes of the regular star to drift towards the black hole and then be accelerated inwards. The most famous candidate for such an arrangement is an O supergiant called HDE226868. The binary combination is known as Cygnus X-1 (number 1 X-ray source in the constellation Cygnus). It emits irregular X-ray bursts with powers of about 4×10^{30} W, around 10 thousand times more powerful than the Sun's entire electromagnetic output. The orbital period of the supergiant can be determined from its varying Doppler shift but, of course, the suspected black hole cannot be seen. As the system is not an eclipsing one the angle of the binary's plane of motion to the Earth's observation plane cannot definitely be determined. This means that, despite good estimates of the mass of the supergiant being available from theory, the black hole's precise mass cannot be calculated. Best guesses suggest about 9 solar masses with a very high probability that the mass is greater than the largest possible mass of a neutron star, the other possible contender for the invisible partner. There is a very good chance therefore that the X-ray emission from Cygnus X-1 is due to mass being accelerated into a black hole.

There are several X-ray sources that are black hole candidates. Unfortunately, all eclipsing X-ray binary sources for which precise masses can be determined have so far proved to have low mass invisible partners, showing them to be neutron stars. One, Centaurus X-3, is also a pulsar, adding further weight to the lighthouse model. Non-eclipsing binaries with strong X-ray emission do not all contain giants or supergiants. Some of the visible partners are stars of lower mass than the Sun. The search continues for new sources or new clues to known sources in the hope that a system can be labelled a black hole with 100% certainty.

Novae and Type I Supernovae

The idea of a black hole raises the question of how binary stars influence their partners' development. It might be expected that a supernova taking place only a few astronomical units away from a stable star might tend to make it, at best, unstable. Some partner stars are destroyed during such events but clearly some must survive and restabilise. The rule of thumb that is used is that at least half of the total mass of the pair must remain in the system after the supernova if the partner star is to survive.

During the early stages of stellar development there is no reason why partner stars cannot develop independently, a small exchange of stellar wind making little difference to the integrity of each. Under some circumstances, however, the presence of a partner star can result in spectacular changes. Consider two main sequence stars that have different masses both of which are close to or a little above a solar mass (Figure 13.15). The more massive star will evolve to become a white dwarf more quickly. During this process it will shed some mass to a planetary nebula, a little of which may be captured by the less massive star. This exchange of mass will shift the barycentre of the orbits towards the less massive star but not affect the evolution of either star markedly as the ejected matter of the more massive star would be mainly hydrogen and helium, the components of the less massive star. Eventually the less massive star will become a red giant. Both orbits remain largely unaltered at this stage as neither of the individual masses nor the centre of mass of either star has changed very much since the more massive star shed its outer envelopes. As the less massive star increases its size and moves across the red giant region this can change. If the stars are close enough together the outer envelopes of the red giant may begin to be more strongly attracted to the white dwarf and to shift towards this other star. The gravitational region of influence of an individual star is known as its Roche lobe. This is the case of a star attempting to outgrow its own Roche lobe.

Matter that leaves the Roche lobe of a red giant to enter the equivalent region of a white dwarf will no longer be in equilibrium and will begin to spiral in towards the white dwarf. A continual flow of matter will lead to the formation of an accretion disk around the white dwarf which will begin to accumulate a new envelope of hot hydrogen. Eventually, the base layer of this new material will reach a temperature high enough to restart nuclear fusion reactions. The electron degenerate state of the white dwarf means that, in a similar way to the helium flash, the onset of new nuclear reactions is rapid and explosive. In this case the flash is not well hidden inside a red giant and the results can be instantly seen as the star lights up and expands, sending external layers of matter off into space. The central star survives, however, dying back down to a white dwarf over a period of a few months. As observed from Earth this process seems to create a new star and so was originally named 'stella nova' in Latin. This is now shortened to nova.

Mass exchange between a red giant and a white dwarf has also been suggested as a possible mechanism to trigger supernovae (type I). In this case the white dwarf reaches the Chandrasekhar limit before causing a nova so that instead of only the outer envelopes of the star exploding the whole star collapses as electron degeneracy pressure is suddenly overcome. Type I supernovae spectra have a dearth of hydrogen lines showing that the exploding matter must have shed its outer envelope. This is contrary to type II supernovae discussed above in which hydrogen lines are easy to see. The explosion of a white dwarf is therefore a reasonable explanation for a type I supernova as much of a white dwarf is composed of carbon, but there must be some question as to what happens to the newly accumulated hydrogen.

A model for a type I supernova that does not rely on the presence of hydrogen involves a binary white dwarf system (Figure 13.16). In this scheme the two white dwarfs evolve independently but are quite closely separated so that they move at high speeds. General relativity then predicts that the stars will lose energy via gravity waves, analogous to the way in which electrically charged particles moving through magnetic fields lose energy through radiation of electromagnetic waves. Loss of energy from the system will cause the stars to slowly spiral in towards each other. At some stage one of the stars will be ripped apart by the tidal forces of the other thus forming an accretion disk of carbon (and oxygen) about the star that remains intact. The material will gradually spiral into the remaining white dwarf until its mass exceeds the Chandrasekhar limit and a supernova explosion ensues. Weight is added to this argument by the fact that most type I supernovae occur in regions populated mainly by old stars.

The theoretical ideas underlying the collapse of a binary system due to loss of energy through gravity waves have been proven through the study of a different binary system (known as PSR1913+16). Here one of the partners is a pulsar which is very convenient as its Doppler shift can be used to determine its motion and the gradual slowing of its period employed to measure orbital changes. The energy loss through gravity waves is proportionally tiny and could never be measured in an Earth-based experiment but such is the mass of a pulsar and the consistency of its pulses that it has been possible to make measurements that have confirmed Einstein's predictions.

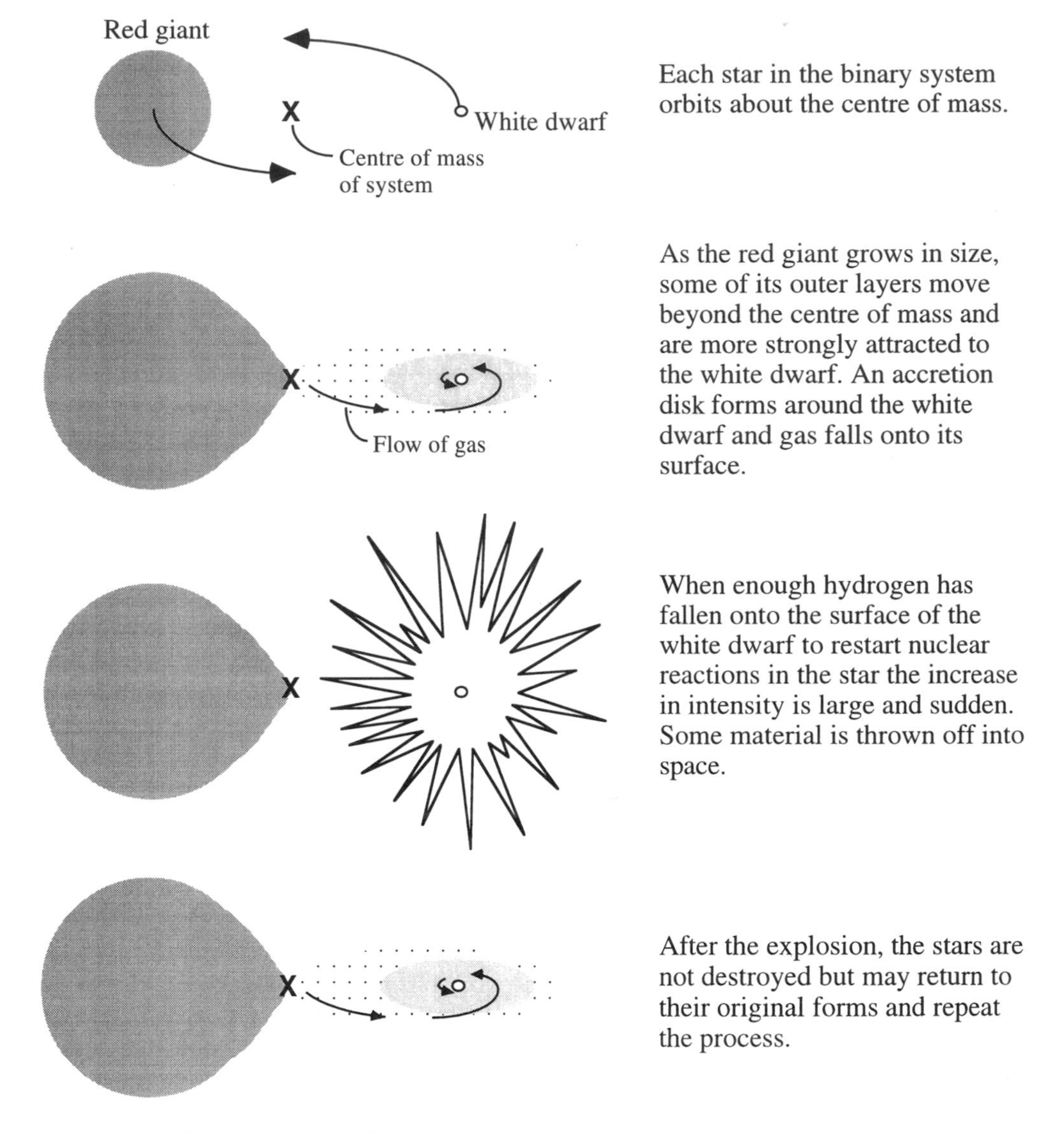

Figure 13.15 Processes involved in a nova

Variable Stars

Eclipsing binaries, whether they are telescopically separable or not, have a varying total light intensity to an observer on Earth. As each star eclipses its partner in turn a dip in the total intensity is registered. Such a system is known as an extrinsic variable. Intrinsic variables, single stars with unstable luminosities, also exist. Until now, this book has considered stars to be in a state of complete stability (e.g. small main sequence stars), slow change (e.g. stars moving off the main sequence) or critical instability (e.g. stars creating planetary nebulae or self-destructing in supernovae). In fact there are a large number of

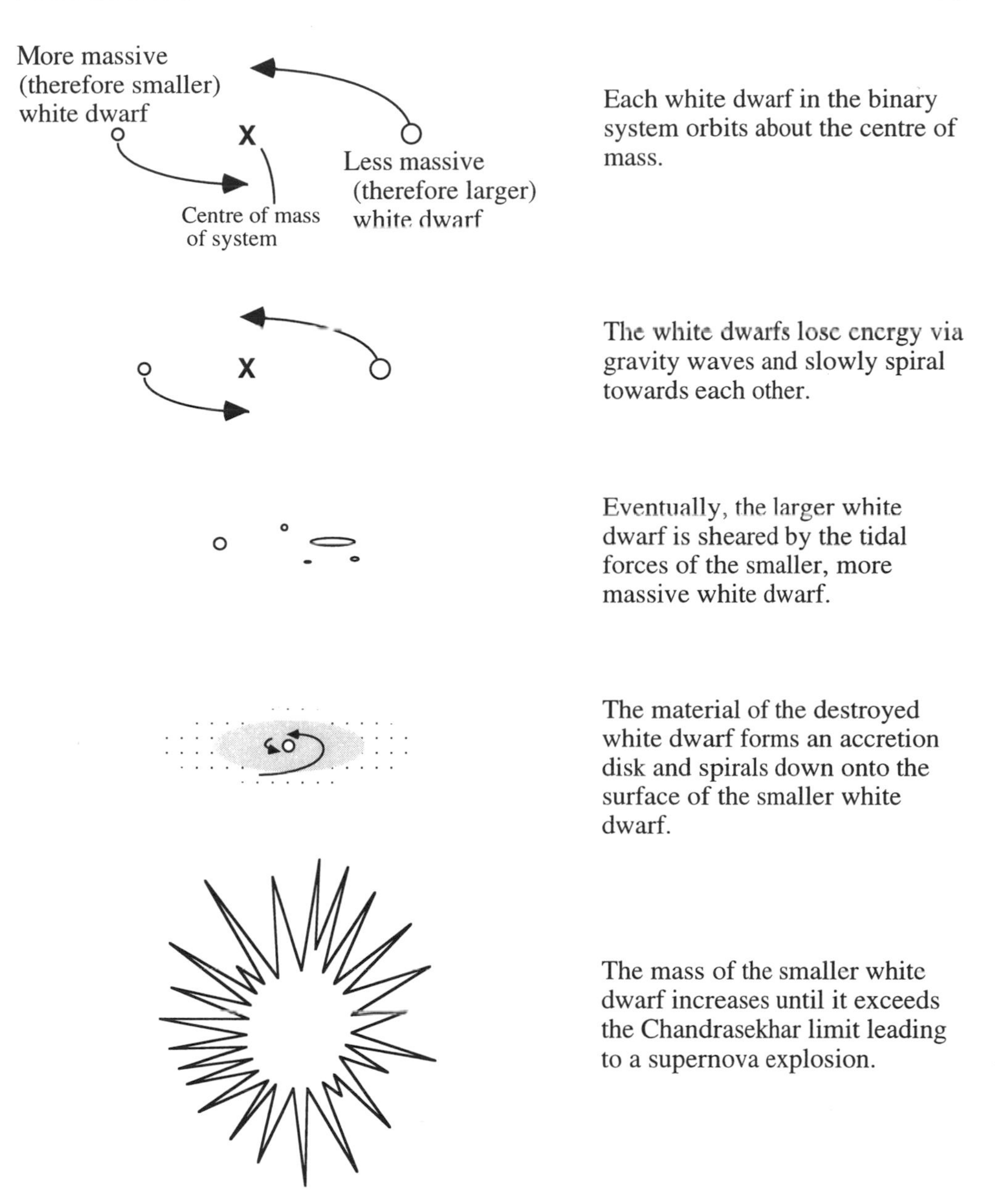

Figure 13.16 One model for the processes involved in a type I supernova

stars that can survive for considerable periods of time while their internal equilibrium oscillates between two extremes. A good analogy is to a spring loaded with a hanging mass. It has a stable equilibrium position but once displaced a little it will rock up and down as gravitational potential energy and the spring's reaction energy are periodically exchanged. For a star it is the balance between gravity and thermodynamic forces that can be disturbed to cause a cyclical energy exchange. The period of the oscillation is often a function of the star's size and density variation. In the same way that a guitar string plays a note determined by its length, tension and density then so too can a star oscillate at certain frequencies determined by its physical characteristics.

The most important type of intrinsic variable stars are known as Cepheids, after the first such star to be catalogued, δ Cephei. There are several subgroups of Cepheids but all pulsate according to specific physical rules. Classical Cepheids are bright giant stars that have evolved from high-mass stars. Calculations suggest that passage through a certain part of the Hertzsprung–Russell diagram, known as the instability strip, automatically imply that stars will become variables and classical Cepheids are found in this region. One pulsation lasts for a few days or weeks and may involve a change in luminosity of a factor of about two or three. During each pulsation the star's size and spectrum change accordingly. The star's position on the Hertzsprung–Russell diagram rocks up and down in a very regular way.

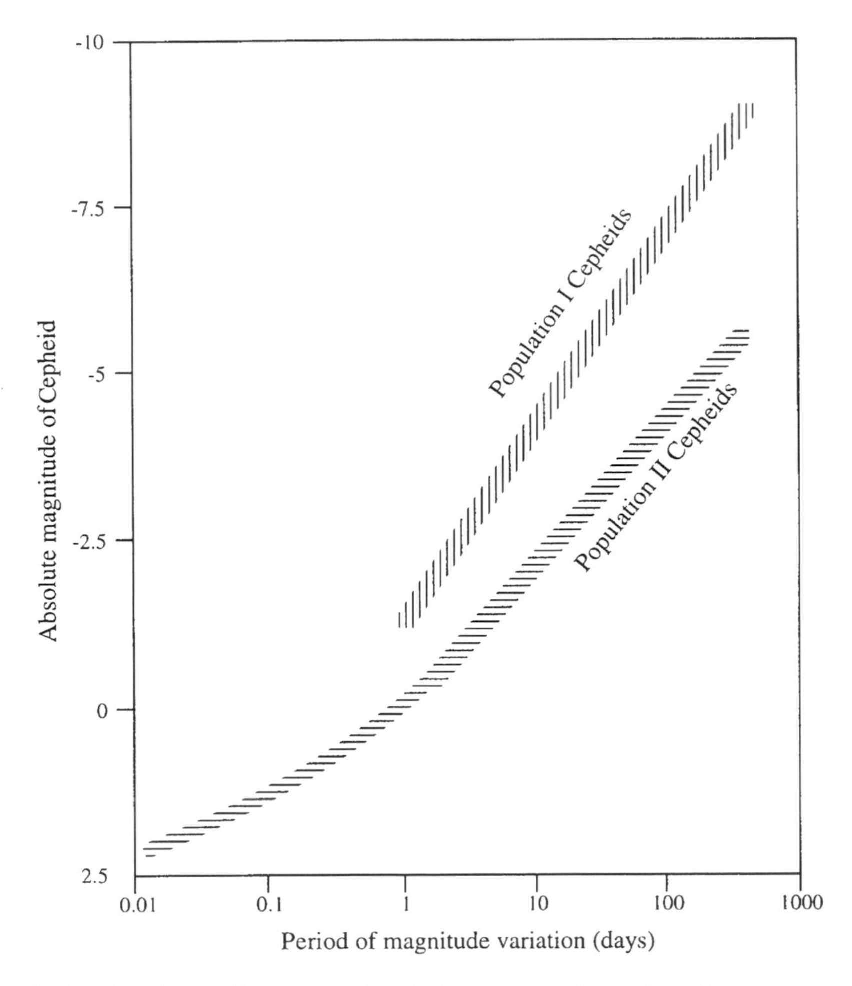

Figure 13.17 The relationship between the absolute magnitude and intensity variation period for Cepheids. There is a difference in the behaviour of population I and II Cepheids. Population I and II stars can be distinguished spectroscopically

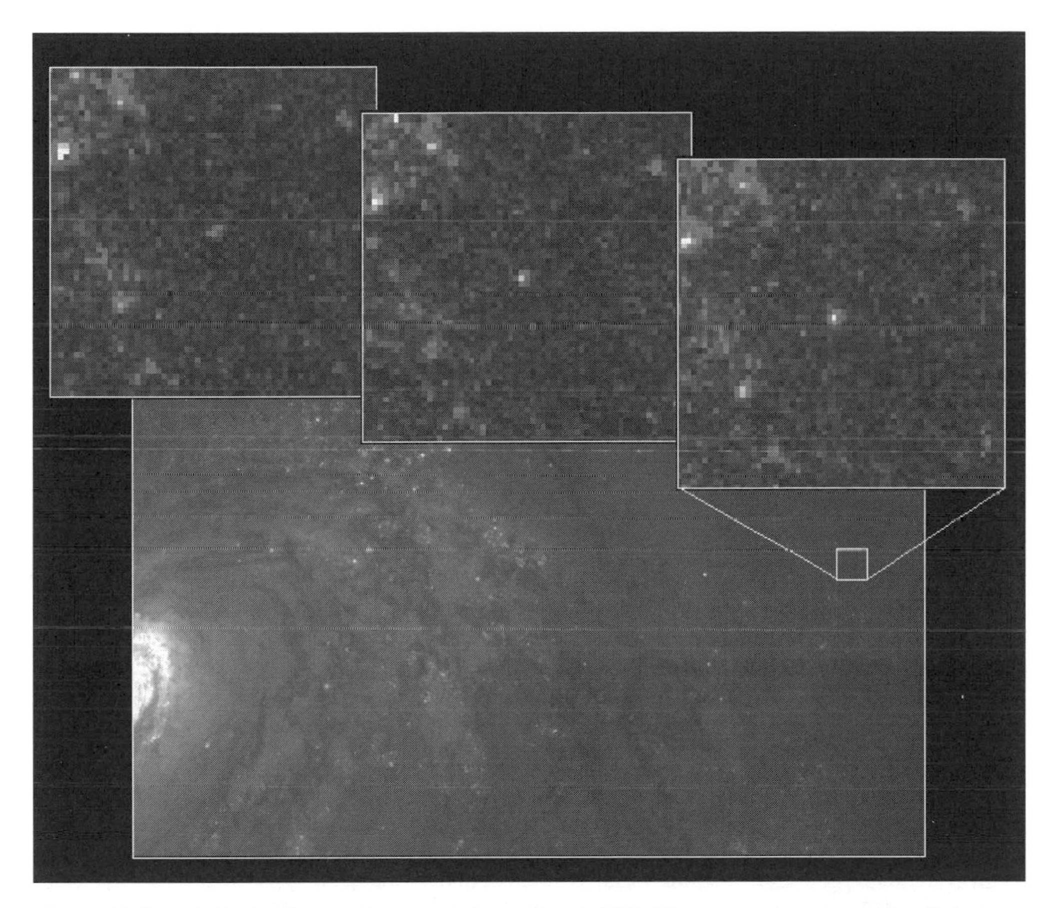

Figure 13.18 A Cepheid variable star in the galaxy M100. The apparent magnitude of the star varies between 24.5 and 25.3 over a period of 51 days. Combining this data leads to a distance for M100 of 17 million parsecs (reproduced by permission of the Space Telescope Science Institute)

The period of a Cepheid is determined by its mass, density and size. These parameters also relate directly to a star's luminosity and so it follows that there should be a period–luminosity relationship. Careful observation of known-distance Cepheids show that this is indeed the case (shown in Figure 13.17). Herein lies the importance of Cepheid stars. The period of a Cepheid can easily be measured, as can its apparent magnitude. From the Cepheid's period its absolute magnitude can be determined. Using equation (11.5) the star's distance can then be calculated as both its absolute and apparent magnitudes are known. Classical Cepheids are very bright and so can be detected at large distances (see Figure 13.18). Such stars thus serve as mileposts to the wider universe where simpler techniques for determination of distance are not applicable.

There is a whole range of variable star types, most of which do not have such well-defined behaviour as classical Cepheids. RR Lyrae variables are also old giant stars but are not so luminous as classical Cepheids. They are often found in globular clusters and as such are population II stars. The period of their intensity change is among the most rapid of variable stars, being measured in hours.

In contrast to RR Lyrae stars are Mira variables. These have very long periods, usually of more than a year, and, though almost as luminous as classical Cepheids, have much lower surface temperatures. Broadband spectra of these stars do not approximate well to blackbody radiation and line spectra are indicative of a range of temperatures. Coupled to the stars' enormous radii, several hundred times greater than the Sun's, this evidence points to the death of a star through pulsation. The pulses cause the outer envelopes of the star to be gradually expelled. Such is the distension of the star that the effective photosphere becomes quite deep, with different layers emitting light characteristic of varying temperatures. The broadband spectrum is therefore a composite of a range of different blackbody spectra. The pulsation process leads to the production of a planetary nebula and the death of the star in as little as a million years.

Not all variables are considerably evolved. At the other end of the stellar life to the variables discussed so far are pre-main sequence entities such as T Tauri stars, mentioned in the previous chapter. Their pulsations occur as the battle between gravity and thermodynamics first begins.

It is the simple contest between inward and outward forces that determines the whole of a star's life. At some stage most stars have temporary instabilities that cause them to become variables but for the most part stars sit statically or drift slowly on the Hertzsprung–Russell diagram. Eventually their lives will end either with a bang or a gradual fade. In both cases some matter will escape, enriched with heavy elements to create new population I stars, but some will remain trapped in a dark carcass, indefinitely.

Questions

Problems

1 (a) Calculate the Schwartzschild radius of a black hole of mass 2.0×10^{31} kg.
(b) What is the physical significance of the Schwartzschild radius (generally)?
(c) The supergiant that collapsed to the black hole in part (a) had an original absolute magnitude of -10. At this time its apparent magnitude, as viewed from Earth, was $+7$. If the death of the star was observed on Earth this year, when did it actually explode?

2 (a) How much energy is released in the fusion of a ^{12}C nucleus (12.0000 amu) with a ^{4}He nucleus (4.0026 amu) to produce a ^{16}O nucleus (15.9949 amu)?
(b) Compare the answer in part (a) with the energy released during a triple-alpha reaction.

3 Radio bursts are received from a pulsar 50 times per second but the period between pulses is measured to be slowing at a rate of 5 μs per year. Estimate the upper and lower limits on the power that the pulsar is losing.

4 A Cepheid variable is spectroscopically characterised as being a population I star. Its period of luminosity variation is 10 days and it has an apparent magnitude of $+11$. How far away is this star?

Teasers

5 Why is the path of a white dwarf across the Hertzsprung–Russell diagram very straight?

6 Why are planetary nebulae and supernova remnant nebulae so different in size and shape?

Exercises

7 Explain why the lifetime of a small star is longer than that of a large star.

8 Describe the processes that take place during the death of a red supergiant star. Include in the description the circumstances under which neutron stars and black holes are formed and how both may be detected.

9 Describe the main stages of evolution of a Sun-like star from birth until it becomes a red giant.

10 Are the following statements about a small star true or false?
(a) It starts its life as a white dwarf.
(b) Nuclear reactions cease suddenly and cause a supernova.
(c) It ends its life as a black hole.

14 Galaxies

Staring into the sky from Earth, early astronomers could divide the objects they viewed into those that moved on the celestial sphere and those that were more or less static. It seemed logical that the former group had to be nearby and they turned out to constitute the bodies of the Solar System. More distant objects were subdivided into stars and nebulae, distinguishing between point sources of light and those with measurable angular size. In earlier chapters several types of nebulae, all of which are composed of loosely scattered matter of different sorts and under different conditions, were discussed. It was only in the early part of the twentieth century that a further type of nebula had to be recategorised and given a new name—galaxies.

Galaxies are enormous collections of stars, all of which, except our own, are located at great distances from the Solar System. The closest are more than 50 thousand parsecs away. Even the smallest contain millions of stars and most contain many billions. The number of stars in a galaxy appears to vary over at least seven orders of magnitude and it seems that virtually all stars belong to such a collection. The Sun is no exception and our galaxy is known as the Milky Way. Almost everything that has been discussed in this book so far has been heavily based on observations of phenomena within a few thousand parsecs of ourselves, placing the action well and truly within the Milky Way. One exception to this is the Sanduleak supernova, discussed in the previous chapter, which took place in the nearest galaxy, the Large Magellanic Cloud.

The study of galaxies is fraught with technical difficulties. To examine the Milky Way it would be easier to nip outside its structure and take a few snaps looking back but such is the galactic scale that this is clearly not possible. Imagine the relative difficulty in mapping a maze while standing immobile in the middle of it compared to looking down upon it. Examining galaxies other than the Milky Way is therefore preferable in the first instance. The problem with this is that they are all so far away. Only a few are visible with the naked eye and large telescopes are required to make useful observations of a reasonable sample. To study these galaxies one big assumption is necessary. That is that the behaviour of stars and nebulae and indeed physics in general is the same throughout the universe. Actually, there is no choice other than to accept this but studies of individual stars, nebulae, supernovae and other phenomena in the closest galaxies, where good enough observations are possible, provide sufficient evidence to allow this assumption to be made.

Galactic Distances

The consistency of astrophysical phenomena between galaxies is essential for a confident calculation of galactic distance to be made. Clearly, trigonometrical parallax is of no use over such large distances and so phenomena such as the period–luminosity relationship for classical Cepheids becomes very important. Fortunately, classical Cepheids are fairly luminous and so can be identified in galaxies millions of parsecs away. They provide the first step in the process of galactic distance determination and perhaps the most reliable of all. Eventually, in looking deeper into space, these stars can no longer be distinguished. Brighter objects are required. Supergiant luminosities can be determined quite well from their spectra and this extends the range a little further, again relying on the relationship between apparent and absolute magnitude (equation (11.3)) to reveal distance. Occasional events such as novae and supernovae also have reasonably well-known light curves that can be used to determine even greater distances, now going beyond hundreds of millions of parsecs. Unfortunately, the science of such events is less well known and larger uncertainties are associated with distance determination through the measurement of the apparent magnitude of a nova or supernova. The Sandulcak supernova provided an example of the dangers of such calculations in that its behaviour was very much as expected except for its peak luminosity, which was unexpectedly low.

The data that is available from the comparison of apparent and estimated absolute magnitudes is quite considerable but it is worth bolstering using any other available technique. Two such methods involve studying the scattered matter present in a galaxy. Emission nebulae, such as the Orion nebula, have predictable sizes, considerably greater than the new stars at their centres that energise them. Measurement of their angular sizes thus provides a method of determining their distance and hence that of the galaxy that contains them. A more complex technique involves study of the neutral hydrogen in a galaxy. This species is very common throughout all (known) galaxies and emits 21-cm radio waves. By studying nearby galaxies it was ascertained that the linewidth (see Chapter 11) of the 21-cm emission increases in a well-characterised way with galactic luminosity. This is because larger, brighter galaxies have greater rotational speeds causing the emitting particles to take on a greater range of radial velocities relative to an observer on Earth. Different radial velocities relate to different Doppler shifts so that larger galaxies with a larger range of radial velocities have broader emission lines. The distance at which far-away galaxies lie can thus be found by measuring the width of the radio emission centred at 21 cm, translating this to an absolute magnitude and, as usual, comparing this to the apparent magnitude to ascertain range.

Hubble's Law

The importance of galactic distance determination may seem low, almost esoteric, but it is the most important clue to help in solving perhaps the biggest remaining mystery in science. This question will be addressed in the next, final, chapter of the book. For now, some more connections must be made. The velocity at which a galaxy is either receding or approaching can easily be determined by examining the Doppler shift of any spectral line. In this case it is the shift of the centre of the line (see Figure 14.1) that is important in determining the overall galactic speed rather than the line broadening as for determina-

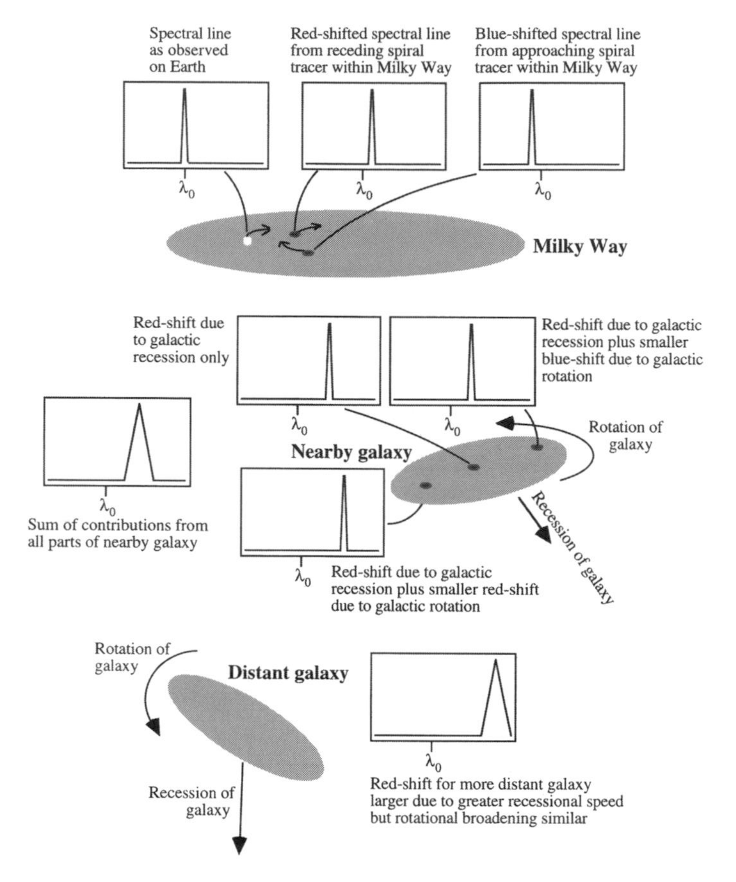

Figure 14.1 How the Doppler shift can be used to determine the relative motion of bodies within and outside the Milky Way

tion of rotational speed. It is always worth looking for correlations between different physical parameters, even when no obvious connection exists. It has been found that there is a linear relationship between a galaxy's distance and its speed of recession. The word 'recession' is used as almost all Doppler shifts are to lower energy (red-shifts) implying movement away. In other words, all galaxies are moving away from us and the further away they are, the faster they are retreating. The universe is expanding!

The relationship between galactic distance, d, and recession speed, v_r, is known as Hubble's law and is given by

$$v_r = Hd \tag{14.1}$$

where H is Hubble's constant. The idea is that Hubble's constant can be determined by making measurements of galaxies in our corner of the universe so that the range of very far-distant galaxies can be determined from their red-shifts. Unfortunately, an unambiguous value for Hubble's constant remains elusive to date. Different techniques for distance determination give different values. All that can be stated with certainty is that Hubble's constant is $65 \pm 20\,\mathrm{km\,s^{-1}\,Mp^{-1}}$ (1 Mp = one million parsecs).

WORKED EXAMPLE 14.1

Q. One of the prominent Lyman emission lines of hydrogen has a rest wavelength of 121.6 nm and a very narrow linewidth. If this line is observed to have a wavelength spread of between 129.6 nm and 129.8 nm in the spectrum of a certain galaxy what can be said about the galaxy?

A. The shift of the centre of the line can be used to calculate the recessional speed of the galaxy and from that its distance can be estimated using Hubble's law. The Doppler shift of the line (equation (1.5)) allows the recessional speed to be found:

$$\Delta\lambda \approx \frac{v}{c}\lambda \quad \therefore v_r \approx \frac{\Delta\lambda}{\lambda}c = \left(\frac{129.7 - 121.6}{121.6}\right)3 \times 10^8 \approx 2 \times 10^7\,\mathrm{m\,s^{-1}}$$

Inserting this velocity into equation (14.1) (with H expressed in $\mathrm{m\,s^{-1}\,Mp^{-1}}$) gives

$$d = \frac{v_r}{H} = \frac{2 \times 10^7}{6.5 \times 10^4} = 300\,\mathrm{Mp}$$

The galaxy is therefore 300 million parsecs away if the value for Hubble's constant of $65\,\mathrm{km\,s^{-1}\,Mp^{-1}}$ is accepted. The limits of uncertainty in the value of H mean that the distance can only be said with certainty to be between about 200–400 Mp.

The width of the emission line is broadened by the rotation of the galaxy. The corresponding recessional speeds for each side of the line are calculated as for the centre:

$$v_{\mathrm{min}} \approx \frac{\Delta\lambda}{\lambda}c = \left(\frac{129.6 - 121.6}{121.6}\right)3 \times 10^8 = 1.97 \times 10^7\,\mathrm{m\,s^{-1}}$$

$$v_{\mathrm{max}} \approx \frac{\Delta\lambda}{\lambda}c = \left(\frac{129.8 - 121.6}{121.6}\right)3 \times 10^8 = 2.02 \times 10^7\,\mathrm{m\,s^{-1}}$$

This means that some parts of the galaxy are receding faster than others and must be due to the galaxy's rotation (one edge is rotating away and the other towards the observer). The difference in recessional speeds is $5 \times 10^5\,\mathrm{m\,s^{-1}}$ or $500\,\mathrm{km\,s^{-1}}$. The maximum speed of rotation of galaxial components is therefore about $250\,\mathrm{km\,s^{-1}}$, assuming that this rotation is observed in a direct line. If the galactic rotation is at an angle to the observer's line of sight then the actual rotation speeds are larger. This induces some uncertainty because the angle of observation of galaxies cannot always be determined. However, uncertainty in Hubble's constant is not important for this measurement. From the galaxy's rotational behaviour estimates of its mass and thence absolute magnitude can be made.

Types of Galaxies

It would be considerably easier to determine galactic distances if their luminosities and general behaviour were as well understood as those of stars. Though this is not the case a great deal is known and much progress has been made in classification and categorisation. Most galaxies fall into one of three classes: elliptical, spiral and irregular. Spiral galaxies are further split into two varieties; normal and barred (see Figure 14.2). Within this classification further distinctions can be drawn as to the tightness and number of spirals or the shape of the defining envelope for ellipticals. Details are not important for this discussion. Irregular galaxies have a generally patchy appearance, making this class

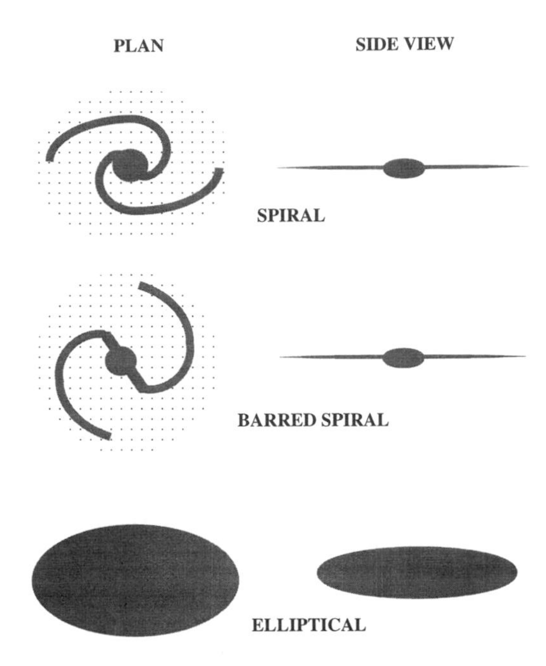

Figure 14.2 Shapes of the most common types of galaxy. The tightness and number of spirals may vary in the upper two classes while the eccentricity (proximity to a circle, see Appendix 3) of the outline of elliptical galaxies may also vary

difficult to subdivide. Beyond these, there are a few galaxies that have inexplicable shapes, such as rings, and these remain beyond classification for now.

Elliptical Galaxies

Elliptical galaxies vary over enormous size ranges. Dwarf ellipticals may contain only a few million stars spread throughout an ellipsoid with a largest dimension of only a few hundred parsecs. Supergiant ellipticals can have radii as large as a million parsecs. The stars that comprise ellipticals tend to be older, redder stars, regardless of the galactic size. There is a dearth of free gas and dust out of which new stars form and consequently very few young stars. As large, blue stars have the most rapid rate of recycling then they are largely absent from elliptical galaxies. Ellipticals therefore appear red as viewed by telescope, their oval shape brightening towards the centre but otherwise lacking internal structure.

Spiral Galaxies

Spiral galaxies are more interesting than ellipticals. As their name suggests, and as illustrated in Figures 14.2 and 14.3, they contain variations in their luminous matter distribution that give rise to spiral arms, sometimes including a straighter, central bar. When viewed edgeways, they are relatively flat with a central bulge. Of course, any

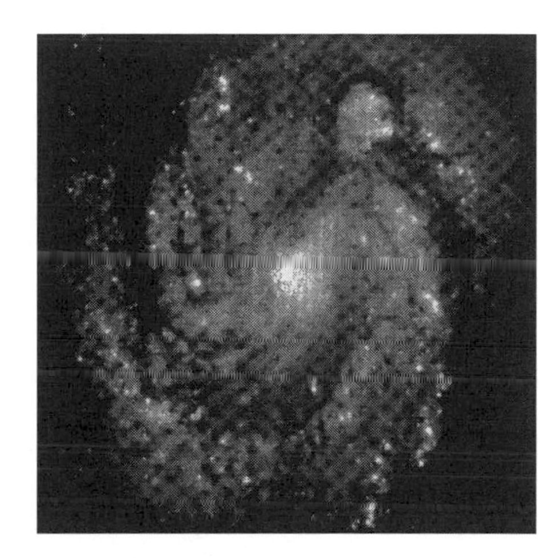

Figure 14.3 The spiral galaxy M100 as photographed by the Hubble Space Telescope (reproduced by permission of the Space Telescope Science Institute)

individual galaxy can only be viewed from the angle at which it happens to be aligned with the Earth. Fortunately, there is a large enough observable sample so that similar galaxies (of all types) have been observed from all possible angles thus enabling a three-dimensional picture to be built up. Spiral galaxies have three main components: the nucleus which causes the spherical, central bulge, the spiral arms that are principally responsible for giving these galaxies their interesting shapes and the halo which is a much more diffuse region containing some gas and dust but also a large number of globular clusters orbiting the galactic centre on long, elliptical orbits. As previously discussed, globular clusters contain mainly old, population II stars. The nucleus of spiral galaxies also has older, redder stars. The spiral arms contain the regions of star formation, each curl appearing considerably bluer then the nucleus. It is the plane of the arms, the disk of the galaxy, that must therefore also contain the bulk of the neutral gas and dust required for star formation. This is best seen when viewing spiral galaxies edgeways on. A dark band can often be observed slicing the galaxy into two slivers along the thin edge. This represents the strong optical absorption associated with gas and dust that is gravitationally gathered in the plane of the galaxy. A similar phenomenon has already been encountered in discussing the formation of the Solar System. All the matter in the disk of the solar nebula fell into the orbital plane and condensed there before accreting solid bodies. Galactic scales are enormously larger but similar processes take place with gas and dust gathering in a plane where it is able to contract to create stars.

The study of gravitational interactions in spiral galaxies has been very rewarding as it has resulted in wide-ranging conclusions. There are two techniques that can lead to the determination of a galaxy's mass. As stated previously, gravitational interactions are the only definite indicators of the mass of distant objects. Galaxies occasionally exist in pairs and, though orbits are highly complex with extremely long periods, masses can be inferred from observations of such binaries. While galactic binaries may reveal less information than stellar binaries, compensation comes from the fact that galaxies are composed of many components, each moving within the galaxy. By measuring the rotational speed of stars as a function of their separation from the nucleus information on the total mass can be obtained as well as a mass distribution relationship. Again, the mathematical relationships are more complex than the simple dynamics required to explain the motions of planets about the Sun. In the case of the dynamics of a single galaxy this is because there is a distribution that is not completely dominated by a central mass. For instance, a star in a spiral arm is influenced both by stars closer to the nucleus than it and those further distant. Though the problem is a multi-body one it is possible to model and data that is obtained from Doppler shift measurements (Figure 14.1) can be used to calculate galactic masses and dynamics.

An important conclusion derived from the dynamics of spiral galaxies is that spirals are unstable and must break up after a few turns! Complex mathematics is not required to see this. The radial velocity distributions (rotation curves) show that spiral arm components do not move at the required speeds to keep the structure in place. Some stars are moving too fast and some too slow to maintain the spiral shape. Though one turn of a galaxy does take a time typically measured in hundreds of millions of years, it is reasonable to assume that all galaxies were formed many billions of years ago so that spiral structures should have dissipated a long time ago. While star formation and death continues cyclically within most galaxies, the packaging of matter into galaxies was complete in the early days of the universe. Observations from Earth suggest that the majority of galaxies are spirals though the significant presence of bright, blue stars in their constitution does cause their luminosities to be generally greater than ellipticals that are dominated by duller, red stars. Spirals therefore appear to be more numerous than they really are. Nevertheless, accounting for this optical illusion, there remain more than one third of galaxies with a spiral form. The most plausible explanation for the longevity of spiral structure, accepting that all galaxies are genuinely ancient, is that the bright bands represent stable density waves.

A density wave is best understood in terms of a familiar example from everyday life, the sound wave. It is a very different type of wave from light. Sound requires the presence of matter to propagate. A disturbance, maybe the vibrations of a human vocal chord, causes matter, say air for instance, to be alternately compressed and rarefied. This disturbance then moves through the air with a speed determined by ambient conditions. At any given position through which the wave passes, the air density is caused to oscillate between high and low. Individual air molecules move over short distances accordingly, rocking backwards and forwards towards positions that require higher densities, but do not move with the wave as a whole. The wave is passed by the matter but does not carry matter with it. Musical instruments produce waves with well-defined frequencies that are determined by the geometry of the instrument. For instance, longer organ pipes produce deeper-pitched sounds than do short ones. Guitar strings operate in an analogous way. The types of vibrations set up in musical instruments are known as standing waves. Because the waves bounce backwards and forwards within the instrument certain frequencies interfere constructively and remain static in space, setting up permanent regions of high- and low-density air. While the waves in galaxies are not sound waves it is thought that a pattern of low and high densities is established in the matter of the galaxy. High-density zones are more likely to create star-forming regions and are therefore illuminated by bright, blue stars. As the matter in the galaxy rotates, passing through high- and low-density regions, the density wave maintains its shape and the spiral galaxy remains a spiral galaxy.

The idea of density waves in galaxies is well established but the cause of these waves is not. Galactic nuclear cores are tremendously active places and may hold the key to understanding such phenomena. When these regions are better understood the model of galactic density waves may become more complete. It is worth mentioning that the energy source for density waves appears to have a lifetime less than that of galaxies, that is, that the spiral does eventually disappear. By comparing the proportion of galaxies within a few million parsecs of the Milky Way with those more than, for instance, a billion parsecs away it is found that more spiral galaxies are found at greater distances. One billion parsecs is the equivalent of about 3 billion light years and so these very far-distant galaxies bring light emitted at a time much earlier in the universe's history. There must have been more spiral galaxies in the past than there are now and so the indications are that spirals evolve into elliptical or irregular galaxies.

The Structure of the Milky Way

The Milky Way remains a spiral galaxy. Despite the difficulties involved with observing from within, the structure of our galaxy is well understood and artists' impressions have been constructed. The Milky Way is a medium-sized spiral galaxy containing about 100 billion stars. Its disk has a radius of around 20 000 parsecs and the Sun is a tiny dot in one of the spiral arms about half-way between the nucleus and the outer reaches. The nucleus is a slightly flattened sphere with a radius of about 2000 parsecs. It therefore bulges out from the disk which has a thickness of only about 300 parsecs. The halo provides a diffuse spherical surround to the whole of the galaxy but is denser in central regions.

In order to map from the inside, special detective work is required. Observations in the plane of the disk are particularly difficult due to the optical density of the gas and dust that must be peered through. This also manifests itself in extragalactic observations. The distribution of the galaxies on the celestial sphere is quite homogeneous except for a band, known as the zone of avoidance, which appears to be deficient in numbers. This represents the effect of the gas and dust in the plane of the Milky Way. Light approaching the Earth from outside the galaxy must travel through the whole of the thickness of the scattering and absorbing gas when coming from the zone of avoidance. Only very bright galaxies have sufficient intensity to be able to penetrate the haze. The relative thinness of the disk means that light from other galaxies has only a few hundred parsecs to penetrate before reaching Earth. Though this effect slightly restricts a complete view of the universe it does provide a simple clue as to the relative flatness of the Milky Way. Actually, there is a much easier way to see that the collection of stars around us has a flat shape. A simple glance at the night sky when viewing conditions are reasonable reveals a diffuse but easily visible swathe of white light. The number of distinguishable stars in this region may not be considerably greater than elsewhere but the general brightness of this part of the sky bears testimony to the tremendous number of stars in that direction and to the increased density of gas and dust that scatters the light. The brightest stars seem evenly distributed about the celestial sphere which is reasonable as the galaxy has a real thickness. The brightest stars are very likely to be within a radius of half the thickness of the Milky Way meaning that they are homogeneously distributed in the night sky (see Figure 14.4). Indeed, 17 of the 20 brightest stars are within 160 parsecs of the Sun. The spherical homogeneity of the night sky implies that the Sun is close to the centre of the disk's thickness.

To determine our position in the plane of the galaxy is more difficult than pinpointing our place in its thickness. One technique involves assuming that the globular clusters, that make up the most tangible part of the galactic halo, orbit about the centre of the galaxy

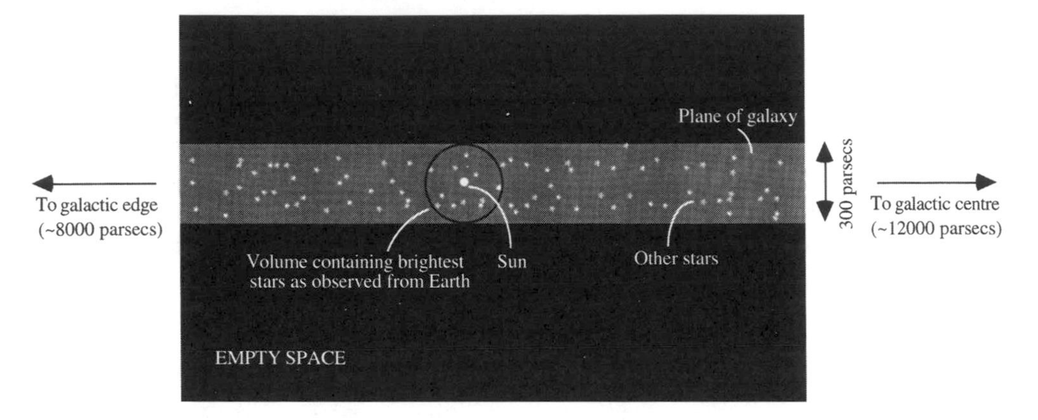

Figure 14.4 The position of the Sun in the Milky Way. Though the radius of the plane of the galaxy is much larger than its thickness, bright stars are uniformly distributed in the Earth's night sky because the galaxy is sufficiently thick so that a large number of stars are contained within the spherical volume indicated. Looking in the plane of the galaxy reveals many more stars but most are much further away and therefore fainter. To the eye the diffuse light from the plane of the galaxy can be seen arcing across the night sky

and that their distribution is homogeneous. By measuring the positions and Doppler shifts of globular clusters, the centre of the galaxy can be found relative to them. As we know our position relative to the globular clusters, the centre of the galaxy can be found relative to ourselves. It turns out that the Sun is located on a spiral arm about half-way between the nucleus and the galactic edge. This same technique can be used to find the Sun's rotational speed about the galaxy, by measuring our speed relative to the globular clusters. This speed is about 250 km s^{-1} and is almost ten times greater than the Earth's rotational speed about the Sun. As discussed above, the galaxy's mass can be calculated from rotation curves in which speed is plotted as a function of separation from the galactic centre. The Milky Way shows little variation in rotation speed outside the nucleus to a distance of 20 000 parsecs. Note how different this is from the manner in which planetary speeds decrease with separation from a dominant stellar mass. The distribution of mass in a galaxy is much more even and so the rotation curves show very different characteristics. The Milky Way's mass is calculated in this way to be around a few hundred billion solar masses.

The Sun's position in the plane of the galaxy makes mapping the Milky Way difficult. Whichever part of the galaxy is investigated, the light must travel through the relatively thick molecular clouds that occupy the same plane. To a distance of perhaps 5000 parsecs the visible part of the electromagnetic spectrum can be used to distinguish regions of low and high stellar densities. The stars that are used to do this are known as spiral tracers and include ever-reliable classical Cepheids as well as the populous O and B stars. The irony of utilising classical Cepheids as spiral tracers is that their absolute magnitudes are, in a sense, better known than their apparent magnitudes. The formula that relates absolute and apparent magnitudes (equation (11.3)) relies on light intensities decreasing with the square of separation from the source for purely geometrical reasons. In the plane of the galaxy, however, light is also scattered by gas and dust, quite strongly when the separation is large. The principal difficulty is that, starting from a position in which the galaxy is uncharted, it is not possible to say precisely how strong the dissipation of light is in any given direction. Only when the galaxy is mapped does this become possible!

To add information on the whereabouts of molecular and neutral hydrogen clouds radio astronomy is employed. Radio waves have much deeper penetration through diffuse matter and extend the range through which meaningful observations of spiral tracers can be made. Unfortunately, the spiral tracers that are the most useful emitters in the long-wavelength part of the electromagnetic spectrum do not have such characterisable behaviour as, for instance, classical Cepheids. Again the Doppler effect has to be employed to determine the relative speeds of clouds orbiting the galactic core. From this their separation from the core can be inferred. The 21 cm emission of hydrogen is employed and carbon monoxide spectral lines also prove very useful. More conclusive results have been obtained while observing the outer reaches of the galaxy where observations can be made without the background of the central bulk of the Milky Way. In observing the outer parts of the Milky Way, unusually, blue-shifts of spectra are obtained. Though the rotational speed of most bodies in the spiral arms is similar, the Sun is always catching up with objects that are further from the nucleus than it because the Sun has a smaller orbit. For these objects a blue-shift is obtained while objects left behind are red-shifted. From these Doppler shifts, the positions of different regions of diffuse matter can be determined. The rotational periods involved are very long at around a quarter of a billion years in the Sun's case.

There are obviously difficulties with a system that uses Doppler shifts to determine rotation curves and rotation curves to determine distances, especially as the data derived must then feed back into itself a second time to calculate absorption characteristics. No single set of data can provide definite answers about the structure of the galaxy. Nevertheless, all the data simultaneously represents a single system so that careful mathematical analysis allows consistent maps to be produced. Such maps suggest that the Milky Way has four main, fairly tight spiral arms surrounding the nucleus.

The Centre of the Milky Way

When radio detectors are aimed towards the centre of the galaxy very different data is obtained. A very strong source of radio energy appears to mark the precise centre of the galaxy. Around it, covering a size of a few parsecs, are weaker radio sources with irregular shapes. The radio emission is composed of a mixture of line, blackbody and synchrotron radiation. This small region is contained in a cloud of neutral gas about 10 parsecs across, consisting of two lobes centred on the radio point source. This region is also a strong X-ray emitter and has been extensively studied in this part of the spectrum. X-rays are emitted from a much larger region than the radio-producing nucleus but the central few parsecs do contain regions of irregular and X-ray burst emitters. More familiar emission also appears from the nucleus deriving from a tremendous population of mainly old stars that orbit the nucleus with a very high density. These stars are best observed through transparency windows in the infrared. Very large areas of star formation also exist in the vicinity of the galactic centre.

The data available from the galactic nucleus is complex and potentially confusing. There is a lot of everything! To move towards a possible model it is worth considering one particular piece of information. The speed of rotation of the large number of stars at the centre of the galaxy is greater than that of the Sun. Analysis is complicated but the rapid rotation of inner stars implies a large central mass. One candidate for this large central mass is a supermassive black hole of some millions of solar masses. Radio emission, in particular synchrotron emission, as well as X-ray bursts can thus be explained in terms of the infall of material on to the black hole in a similar way to that discussed for X-ray sources such as Cygnus X-1 in the previous chapter. It should be noted that the mass of the suggested galactic black hole is considerably greater than that of supernovae cores and that the origin of such objects is still highly unclear. In fact much work remains before the existence of a black hole at the centre of the Milky Way can be confirmed.

Active Galaxies

The Milky Way's core appears to exhibit another type of structure not so far discussed. Observed at radio wavelengths, it seems that clumps of ionised gas are being expelled by the central object (be it a black hole or not). These clumps move in specific directions, perpendicular to the plane of the galaxy, both up and down. In other galaxies these structures can be much more prominent, to the extent that they come to dominate the galactic emission characteristics. A normal galaxy shows a spectrum that is a sum of the spectra of a large collection of normal stars that emit as blackbodies. So-called active galaxies show considerably distorted spectra. Though normal stars make a contribution, synchrotron emission and its effects dominate. Synchrotron emission can be observed throughout the spectrum but is dominant at radio wavelengths in active galaxies. Its influence on the gas and dust of the galaxy is to heat it to an average of a few hundred kelvin. The dust particles then reradiate the energy they receive as blackbodies. The spectral peak is shifted considerably to the infrared compared to the emission of stars as dictated by Wien's displacement law (equation (1.4)). Spectra of active galaxies are therefore quite different from those of ordinary galaxies (see Figure 14.5). Of course, the

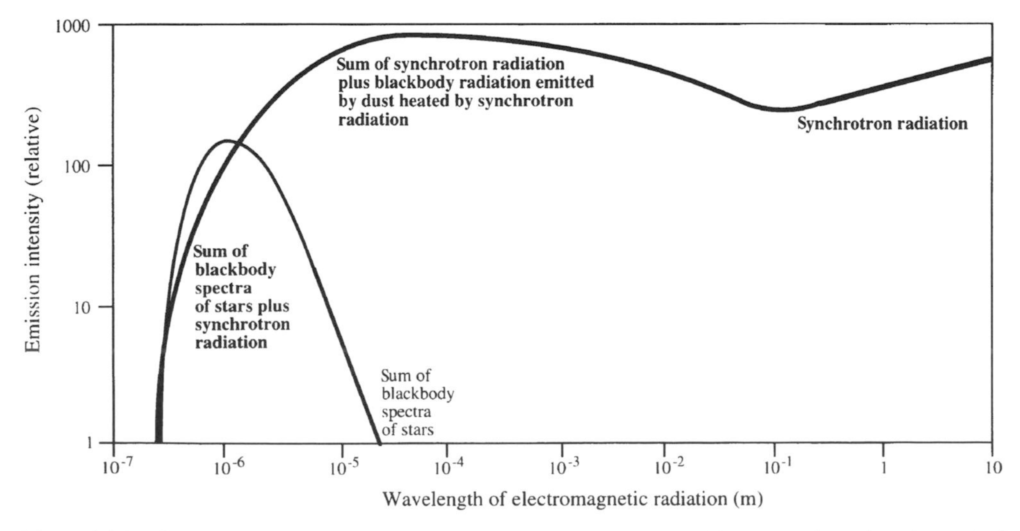

Figure 14.5 Comparison of the spectra of a typical normal galaxy (thin line and type) and a typical active galaxy (thick line and type). The contributions to each part of the spectrum are indicated for each curve

gas may also give rise to sharp line emission spectra which prove useful for accurate Doppler shift measurements when required.

As with seemingly all groups of astronomical bodies there is a large variety in the appearance of active galaxies. Details of this variation are beyond the scope of this book and so only a generalised model of a 'typical' active galaxy will be given. The most common features are jets and lobes. Typically a pair of narrow jets spew charged matter in tight beams from the galaxy. The jets are always directed in diametrically opposite directions from the parent galaxy though sometimes only one of the jets can be detected, probably for observational reasons. The lengths of the jets may be around the same size as the galaxy itself, extending over hundreds of thousands of parsecs until they arrive at the lobes. The lobes are enormous regions with filamentary structure where the charged particles of the jets end their linear motion to produce massive eddying clouds with large, swirling magnetic fields that must inevitably result from structures composed of charged particles (see Figures 14.6 and 14.7). An enormously miniaturised analogy for a jet and lobe system is a furiously boiling kettle. From the spout a laminar flow of steam appears

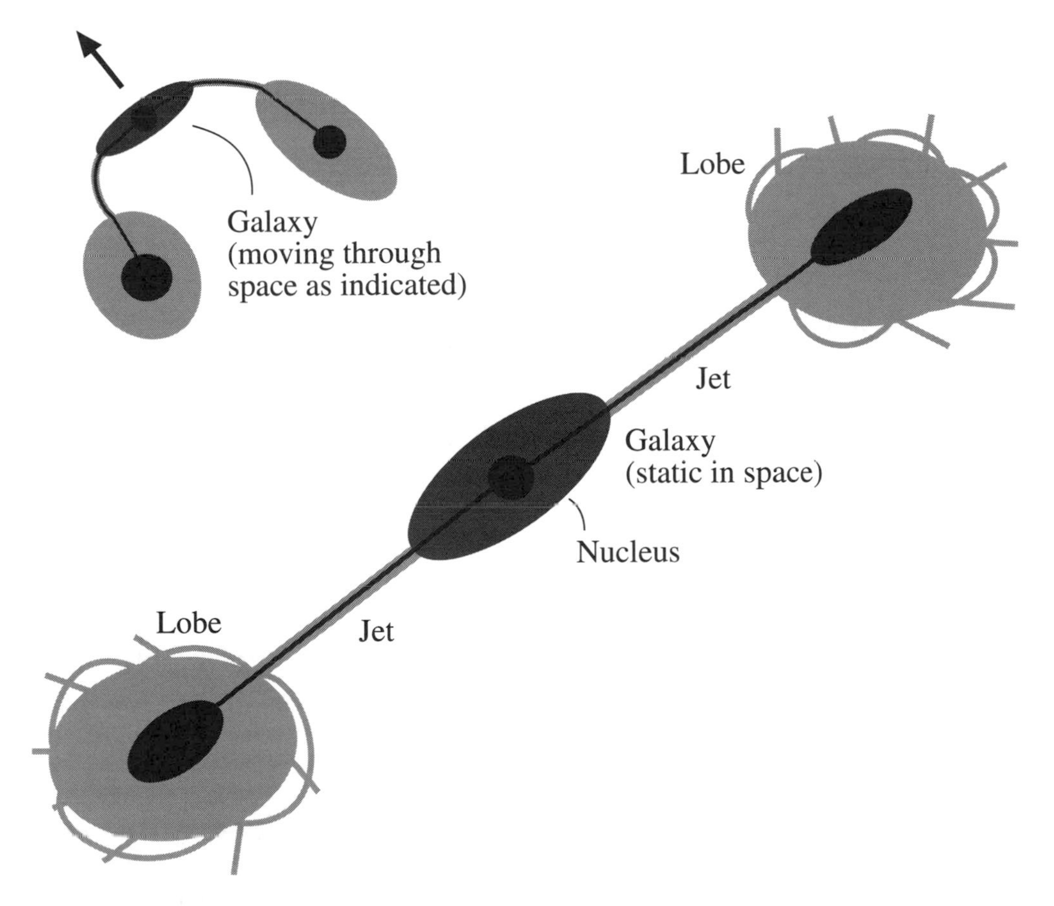

Figure 14.6 Typical structure of active galaxies. The main diagram shows a galaxy that is static in space and a moving galaxy is shown at the top left. Matter is indicated in grey while the main regions of radio emission are indicated in black

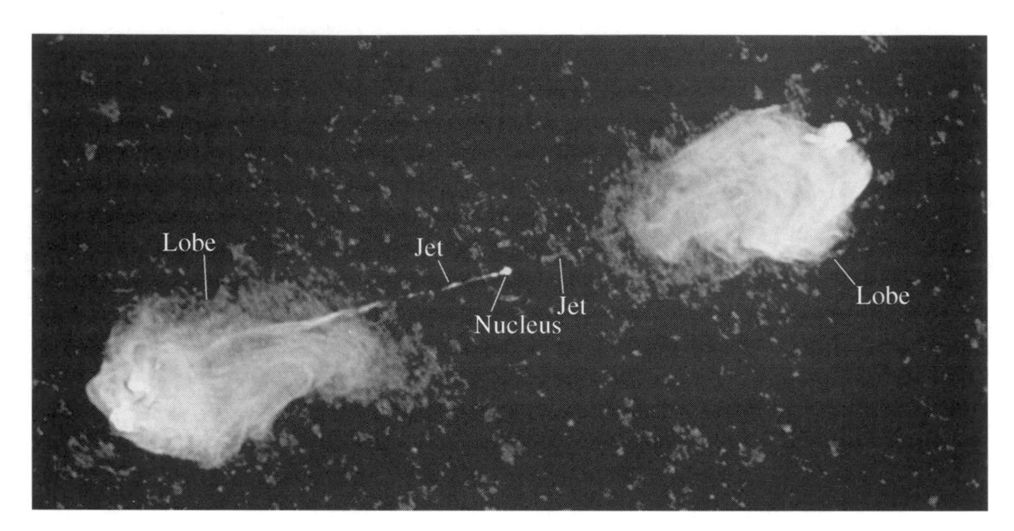

Figure 14.7 The jets and lobes of Cygnus A as observed by the Very Large Array radio interferometer (reproduced by permission of NRAO/AUI)

and forces its way through the air until it is slowed to produce a billowing, chaotic ball of water vapour. Some galaxies produce bent jets. The Earth-based analogy can be extended to a steam train. As the train moves the steam jet and lobe are sequentially left behind thus bending the overall structure. The same is thought to occur for active galaxies in motion. One flaw in the analogy is the relative length of the jet. How galactic jets remain stable over such large distances remains a mystery. The sparseness of matter outside the galaxy and the magnetic field that the charged particles bring with them are likely to be important. As discussed for the Sun (Chapter 10), magnetic fields in plasmas can be both rope-like and self-sustaining.

The model that is used to explain the tremendous structures of active galaxies is precisely that already discussed for the Milky Way though on a much larger scale. A supermassive black hole attracts matter to it which swirls around in the plane of the galaxy as it falls towards the Schwarzschild radius. The accelerating matter is ripped into its constituent atomic components to produce a plasma that radiates strong synchrotron emission. Resulting radio and X-ray sources are usually found between the jets at the centre of the galaxy. The orbital motion of the plasma about the nucleus creates a strong magnetic field and it is this that is thought to blast some of the charged particles deep into space as jets. Eventually the jets, still emitting synchrotron radiation, become turbulent and produce the lobes.

It should be remembered that as science probes deeper and deeper into outer space models gradually become more tenuous until they are eventually merely suggestions. Nevertheless, the Hubble Space Telescope has been able to observe the centres of active galaxies in detail . For instance, gas with a temperature of about 10 000 K has been observed whirling about the centre of a galaxy known as M87. Doppler shift measurements on gas at about 20 parsecs from the galactic centre show its speed to be about 550 km s^{-1}. The magnitude of the central mass must therefore be about 3 billion solar masses. The resolution of the observation shows that this central mass is contained within a volume smaller than the Solar System. This object is a strong radio source and jets emanate from it. Observations of other active galaxies reveal similar results, for instance M84 appears to have a central black hole with a mass of about 300 million solar masses. The evidence for the presence of supermassive black holes at the centres of active galaxies is thus becoming compelling.

WORKED EXAMPLE 14.2

Q. In the model for the core of an active galaxy an accretion disk forms around the supergiant black hole. This is composed of former stellar material ripped apart by the tidal forces of the black hole and has a radius of about 10^{14} m. Estimate the upper limit for the power emitted by the system if about one solar mass moves from the accretion disk to the black hole (of 10^8 solar masses) per year.
A. The Schwarzschild radius of the black hole is given by equation (13.1):

$$r_S = \frac{2GM}{c^2} = \frac{2 \cdot 6.7 \times 10^{-11} \cdot 10^8 \times 2 \times 10^{30}}{(3 \times 10^8)^2} = 3 \times 10^{11}\ \mathrm{m}$$

Energy released after matter has entered within the Schwarzschild radius will not be observable and so, in calculating the release of gravitational potential energy, this represents the 'end point'. The rate at which matter enters the Schwarzschild radius is 2×10^{30} kg per year or ($2 \times 10^{30}/(365 \times 24 \times 3600) =$) $6 \times 10^{22}\ \mathrm{kg\,s^{-1}}$. The gravitational potential energy released is given by the gravitational potential energy of the matter while in the accretion disk (radius r_a) minus that at the Schwarzschild radius (see Worked Example 13.6), given by equation (6.13);

$$\text{Energy released, } E = \frac{GMm}{r_S} - \frac{GMm}{r_a} = GMm\left(\frac{1}{r_S} - \frac{1}{r_a}\right)$$

$$= (6.7 \times 10^{-11})(10^8 \times 2 \times 10^{30})(6 \times 10^{22})\left(\frac{1}{3 \times 10^{11}} - \frac{1}{10^{14}}\right)$$

$$= 3 \times 10^{39}\ \mathrm{J}$$

This is equal to the amount of energy released per second, in other words, the active galaxy has a potential output of more than 10^{39} W.

Some of the objects that occupy deepest space are active galaxies known as quasars (QUAsi-StellAR objects). These galaxies at first appeared to be stars with strong radio emission and very unusual optical line spectra. The mystery was partly solved when it became clear they are very distant galaxies (see Figure 14.8) with enormously red-shifted spectra. Rather than spectral line shifts of a few per cent the lines appear in altogether different parts of the electromagnetic spectrum after shifts by factors often greater than 100% and up to 450%. The red-shift is normally stated as the proportional change for such large effects, e.g. $450\% \equiv 4.5$.

WORKED EXAMPLE 14.3

Q. If a quasar has a red shift of 4.5 how fast is it moving away and how far away is it?
A. A red shift of 4.5 implies that $\Delta\lambda/\lambda$ is 4.5 so that, according to equation (1.5) (see also Worked Example 14.1) v/c is also 4.5. This implies that the quasar is receding at a speed of more than four times the speed of light. This contravenes the laws of relativity which show that velocities cannot exceed the speed of light. A relativistic expression for Doppler shift is given below (not derived here)

Figure 14.8 A quasar at a distance of 3 billion parsecs (centre). Above and to the right is an elliptical galaxy at a distance of only 2 billion parsecs (reproduced by permission of the Space Telescope Science Institute)

and must be used when v becomes very large ($\sim$c/10):

$$\Delta\lambda = \left\{ \left(\frac{1 + v/c}{1 - v/c} \right)^{\frac{1}{2}} - 1 \right\} \lambda \tag{14.2}$$

If $\Delta\lambda/\lambda$ is 4.5 it is possible to solve for v:

$$\frac{\Delta\lambda}{\lambda} = 4.5 = \left(\frac{1 + v/c}{1 - v/c} \right)^{\frac{1}{2}} - 1$$

$$\therefore\ 5.5^2 = \frac{1 + v/c}{1 - v/c} \Rightarrow v/c = 0.94$$

The quasar is receding at a speed of 94% of the speed of light or $2.8 \times 10^8\,\mathrm{m\,s^{-1}}$. Substituting this value into equation (14.1) gives

$$\therefore\ d = \frac{v_r}{H} = \frac{2.8 \times 10^8}{6.5 \times 10^4} = 4300\,\mathrm{Mp}$$

Again the uncertainty in Hubble's constant leads to an actual distance for the quasar of between 3 billion and 6 billion parsecs.

The very large red shift of quasars implies that they are a very long way away according to Hubble's law. The strength of the electromagnetic energy that arrives so far away at the Earth implies that quasars must be extremely intense objects. Their spectra contain large contributions from synchrotron radiation and jets have been detected emanating from the small, central source of radio emission. Very high resolution imaging, for instance using the Hubble Space Telescope, has managed to spatially resolve the outer part of the galaxy that surrounds the much more intense nucleus. Precise models of quasars have not yet been agreed upon by all astronomers but the evidence suggests that these galaxies, the furthest of which are being viewed in the early stages of the universe (see Chapter 1), are quite similar to active galaxies closer to home except for their much greater energy output. Quasar models therefore mimic those for active galaxies and merely require the presence of a more massive central black hole, of many billions of Solar masses.

Observations of galaxies of all types are beginning to suggest the presence of central massive black holes as being ubiquitous. There is strong evidence for this from the Milky Way as well as other nearby normal galaxies for which stellar motions can be observed, such as the Andromeda galaxy. Active galaxies have core-centred energy outputs as much as a million times greater than that of the Milky Way. Quasars are typically a thousand times more intense still with a power output of around 10^{42} W.

Galactic Distribution

A final topic for this chapter concerns the distribution of galaxies in space. Are galaxies spread homogeneously or is there a tendency for them to clump? Careful mapping has shown that the distribution is far from homogeneous. Galaxies bunch together in clusters (see Figure 14.9 for instance). The Milky Way is part of a cluster known as the Local Group that contains almost twenty galaxies. Clusters rarely have well-defined shapes and the Local Group has a loose dumbbell structure with the Milky Way and the Andromeda galaxy dominating each end, separated by about 1 million parsecs. The Andromeda galaxy is the furthest object that can be seen with the naked eye. The closest companions

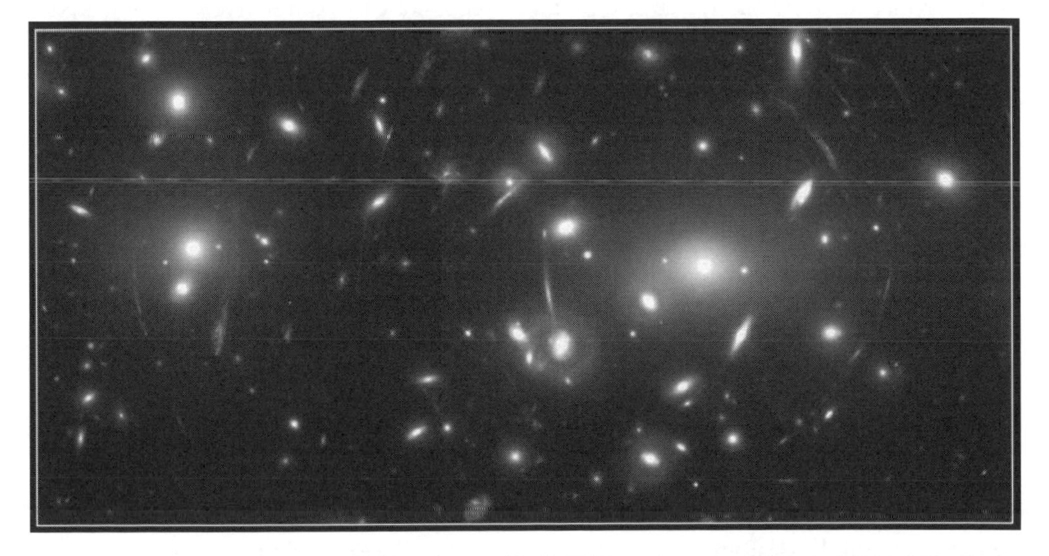

Figure 14.9 A cluster of galaxies known as Abell 2218 as photographed by the Hubble Space Telescope. Some of the objects are actually more distant galaxies that have been gravitationally lensed (and distorted) by the cluster (see Chapter 15) (reproduced by permission of the Space Telescope Science Institute)

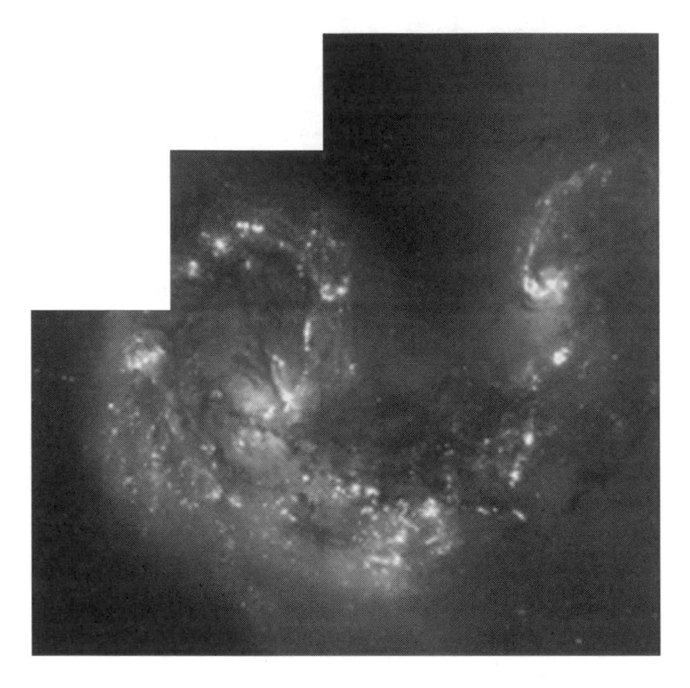

Figure 14.10 A collision between two nearby galaxies. Though the universe is expanding, local inhomogeneities can cause galaxies to approach each other and collide. The ensuing disruption leads to rapid star formation in molecular cloud regions (reproduced by permission of the Space Telescope Science Institute)

to the Milky Way are the Large and Small Magellanic Clouds, relatively small galaxies that appear quite large in the sky due to their proximity.

Clusters are held together by gravity. Though separations between galaxies are huge compared to the gravitationally tied systems discussed previously, the masses involved are also extremely large so that galaxies do interact with each other. In some cases galaxies may collide, not in the same way that a meteorite crashes into the Earth but in a gradual way that untangles the structure of each galaxy and forms a new compound galaxy (see Figure 14.10). Close encounters may also disrupt galactic structure through tidal interactions. These processes are very slow so that, like a star's evolution, changes in structure cannot be followed in real time. Nevertheless, pairs of galaxies in the process of mutual deformation can be seen.

Some galactic clusters may contain as few as only two galaxies but others are much more numerous, having thousands of members. The Virgo cluster is just such a highly populated group and consequently has a very high mass. As it is relatively close to the Milky Way, at less than 20 megaparsecs, the Local Group is being gravitationally tugged towards the Virgo cluster. This is truly an example of action at a distance. That a pair of galactic clusters are moving together may seem to contravene Hubble's law but it should be remembered that Hubble's law is a consequence of the expansion of the universe. The motion of component elements may vary randomly on a local scale. Hubble's law only really applies on a very large scale where local inhomogeneities become unimportant.

If clusters of galaxies interact with other clusters of galaxies then perhaps there are clusters of clusters of galaxies? There are. For instance, the Local Group is a member of the Local Supercluster that contains the Virgo Cluster among others. The Local Supercluster is about 30 megaparsecs across and contains hundreds of clusters. Between superclusters are regions of apparently empty space known as voids. As usual, the structure of superclusters and voids can be highly variable. The Local Supercluster appears to be of average size though making maps of deep space remains prone to error and uncertainty.

It seems that there are no clusters of clusters of clusters of galaxies! Once the network of superclusters and voids has been established space becomes homogeneous, a little like a sponge with regions of high density separated by large pores. So this book has reached the stage where no more can be said about the structure of the universe at a local, intermediate or full-scale level. All that remains is to explain why the universe is the way it is and make predictions for what will become of it in the future.

Questions

Problems

1 Deneb has an apparent magnitude of +1.2 and an absolute magnitude of −7.1. To which galaxy must it therefore belong?

2 Neutral hydrogen atoms emit radio waves at 21.11 cm with a very narrow linewidth. If this emission is observed to have a wavelength spread of between 21.80 cm and 21.82 cm in the spectrum of a certain galaxy what can be said about the galaxy?

3 How fast are galaxies moving away that have the following red-shifts and how far away are they, assuming Hubble's constant to be 65 $\mathrm{km\,s^{-1}\,Mp^{-1}}$?
(a) 1 (b) 2 (c) 3 (d) 4

Teaser

4 How might it be possible to predict whether a distant galaxy is spiral or elliptical when it is being observed by a telescope that is unable to produce a clear image of it?

Exercises

5 What does the distribution of spiral and elliptical galaxies imply about galactic evolution?

6 Describe techniques for the determination of galactic distances.

7 What is the principal observational difference between normal and active galaxies?

8 Describe how stars are distributed throughout the universe.

9 What does the distribution of active galaxies imply about galactic evolution?

10 Why, if the Milky Way has a very flat structure in the vicinity of the Sun, are the bright stars of the night sky homogeneously spread?

15 Cosmology

This book has discussed the nature of the universe, both by describing its appearance and by explaining the physical processes that cause it to be just the way it is. The collections of matter that are the universe's most obvious components, planets, stars and galaxies, have been dealt with in turn. Finally, the biggest problems of all have to be confronted, those of creation and fate. This study of Everything is known as cosmology.

The Expansion of the Universe

The biggest clue to unravelling the history of time has already been discussed. Hubble's law shows that the universe is expanding. The more distant a galaxy is, the faster its recessional speed. The relationship is linear which means that, if this relationship has been constant throughout the history of the universe, all the matter in the universe must have been at the same place at the same time on some occasion in the past. To see this, consider two galaxies, one of which is twice as far away from an observer as the other. According to Hubble's law, the more distant galaxy must be moving away at a rate twice that of the near galaxy. The galaxies must therefore have set off from the same place at the same time (see Figure 15.1). Such is the consistency of this observation that the conclusion that all the universe's matter was once concentrated in space is almost irrefutable. It is important to realise that the universe's expansion is not geocentric. Wherever the universe is observed from, Hubble's law will hold as all galaxies are moving away from all others (ignoring the effects of local inhomogeneities). Whichever viewing position is chosen all objects recede with a speed that increases with separation. Such a phenomenon is an inevitable consequence of any expansion. It is therefore not possible to point to the centre of the universe.

Hubble's constant gives a measurement of the period of time since the matter of the universe began to expand. There remains considerable uncertainty as to the value of H, as discussed in Chapter 14, but at least an order of magnitude calculation can be made. If H is taken to be 65 km s^{-1} Mp^{-1} then a galaxy that is 1 Mp from the Milky Way is moving away at a rate of 65 km s^{-1}. The time at which the two were in the same place must therefore be equal to the time required to move apart 1 Mp at a rate of 65 km s^{-1}. There are about 3×10^{19} km in 1 Mp which gives the age of the universe to be ($3 \times 10^{19}/65 =$) 4.5×10^{17} s or 15 billion years. Apart from the uncertainty in the current value of H there is the further complicating factor that H has changed as the matter density of the universe

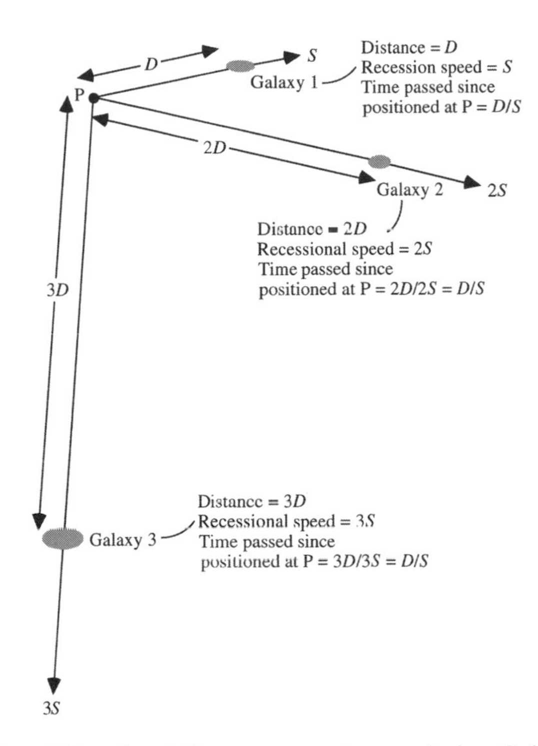

Figure 15.1 The results of Doppler shift measurements on galaxies. Galaxies recede at a rate proportional to their distance. Hence it would appear that all galaxies started their motion from the observer at P. However, in such an expansion, observers positioned anywhere in the universe would observe the same phenomenon

has decreased. The contracting influence of gravity is gradually slowing the universe's expansion but at an ever-decreasing rate as matter continues to fly apart. In other words, gravity's action is always to contract but as the mass density decreases the overall force becomes smaller. Calculations can be made as to how H has varied throughout time but its accuracy depends on many unknowns that will be discussed below. Nevertheless, the universe's age since its matter was scattered can be said to be between 10 and 20 billion years with some certainty. The event that marked the start of this era of the cosmos is generally known as the Big Bang.

The Cosmic Blackbody Microwave Radiation

In most astronomical bodies that have been considered in this book so far, gravity has acted to draw matter together. The result of this is always to heat the matter through the release of gravitational potential energy. The Big Bang scenario involves the opposite occurrence. The matter of the universe has been scattered over an enormous distance and must therefore be enormously cooler than it once was. At some time the temperature must have been too hot for atoms and molecules to hold together so that the matter existed in a plasma state similar to that in the core of a star. The density would also have been high and so the characteristic electromagnetic radiation would have been blackbody in nature. As in the core of a star, scattering events would be frequent so that the charged particles and mainly high-energy photons that made up the universe at that time could be considered to be coupled. That is, the matter and light particles interacted together in a mutually dependent way.

The difference in the behaviour between plasmas and neutral, diffuse matter is profound and so the stage of the universe's evolution at which radiation became decoupled from matter was an important one. As the universe expanded and cooled, protons and other nuclei were able to trap electrons to form atoms or weakly charged ions. Of course, matter and light continued to interact, in particular through absorption and re-emission, but the matter particles and photons became essentially independent. The temperature would then have been about 3000 K, just that at which plasmas and uncharged matter change state in stars or in experiments on Earth. What has happened to the enormous amount of electromagnetic energy that was trapped in the plasma now that it is free to wander the universe? The answer is that nothing has happened to it, it can still be observed. Very demanding experiments have been performed that map the sky's radiation after the removal of sources associated with matter (planets, stars, galaxies, etc.). The experiment maps not only the intensity of this radiation but also its spectrum. The findings of such experiments have been conclusive. First, this radiation is isotropic to within a few parts in a million. Second, its spectrum is that of a perfect blackbody with a temperature of 2.73 K. A quick look at Wien's displacement law (equation (1.4)) shows that the peak wavelength of this radiation curve is at about 1 mm, in the microwave part of the electromagnetic spectrum.

There are two ways to explain why the background temperature is so low and the radiation centred so far to the low photon energy part of the spectrum. Most simply, as described in the previous paragraph, the universe has expanded enormously since the decoupling of radiation and matter so that the effective temperature must have reduced significantly. The second approach is to consider the early universe's blackbody radiation to represent a source which is very far distant. The justification is simply the converse of the argument given in Chapter 1 that looking at very far-distant objects is the equivalent to looking into the past. The difference is that the microwave echoes of the Big Bang must be isotropic because the electromagnetic radiation of the early universe filled space and must therefore still do so. However, as the source is considered to be very far distant it must be moving away very rapidly according to Hubble's law. The original spectrum, peaking in the high energy photon part of the spectrum, is thus very strongly red-shifted to the microwave region. This isotropic source is referred to as the cosmic blackbody microwave radiation and provides further strong evidence for the Big Bang model.

Mass and the Universe

The conceptual difficulty associated with the isotropy of the cosmic blackbody microwave radiation is worth discussing further. Why doesn't it come from the 'middle of the universe'? In considering the evolution of the universe it is important to remember that this entity contains Everything. This applied at the first instant and will always be the case. The plasma that held the radiation early in the universe's history was isotropic and so, as space has expanded, the radiation field must have remained isotropic. A true understanding of these concepts requires knowledge of Einstein's theory of general relativity. The mass of the universe curves space so that, though light travels locally in straight lines through space, it cannot escape the region of the universe that contains the matter. In other words, the light and matter are locked in a region that defines the universe. Evidence for the bending of space by mass has already been mentioned in discussing the shift in position of stars adjacent to the Sun in the sky during an eclipse (see Chapter 5). Further evidence can be seen in photographs of certain distant galaxies which appear as multiple images due to the light that carries the image being bent in various ways by an intervening galaxy. In other words the light may take more than one path to an observer and thus appears to be coming from more than one place (see Figures 15.2–15.4).

There is no doubt that Einstein's theory of general relativity is sound and that the mass of the universe controls the motion of photons within it. An important question to ask, though it may initially seem esoteric, is whether there is a direction in which a laser beam can be pointed such that it orbits the universe and returns to hit the sender on the back of

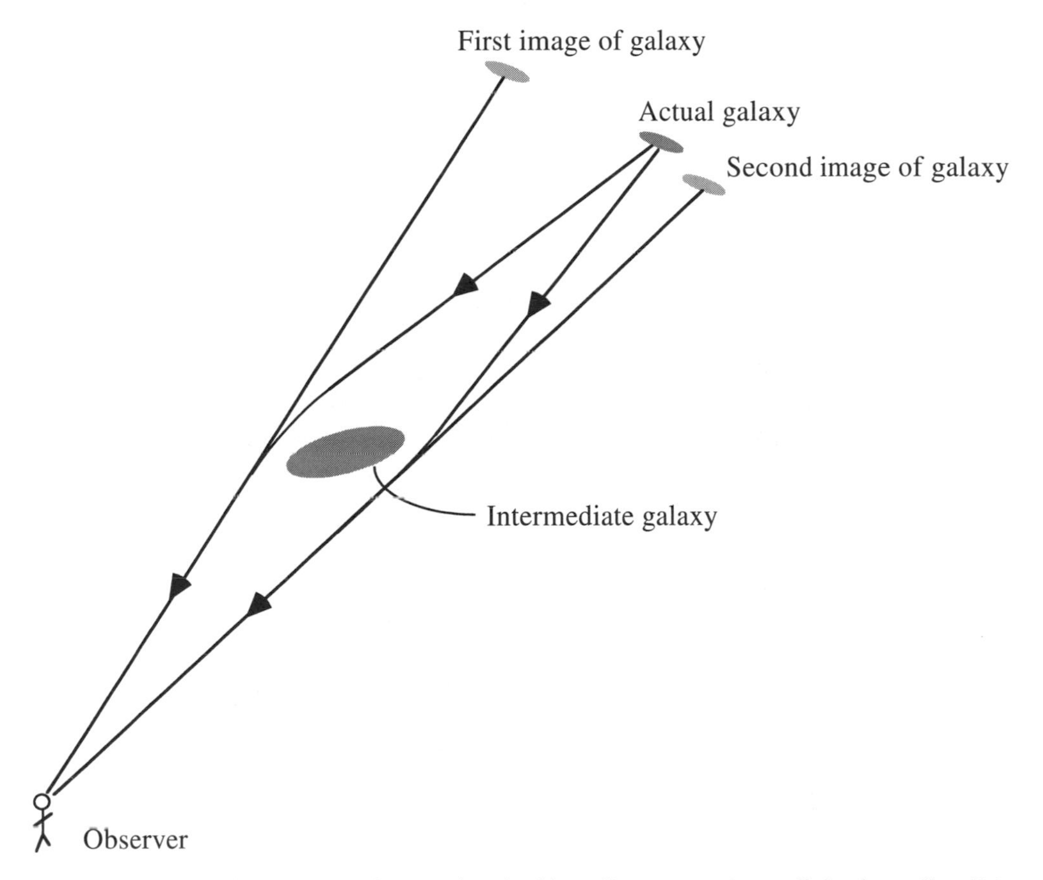

Figure 15.2 How a large mass such as a galaxy is able to distort spacetime sufficiently to allow light leaving a distant galaxy to take two different routes to the same observer. The observer knows only the direction that the light arrives from and interprets the observation as the existence of two galaxies

Figure 15.3 A cluster of galaxies at a distance of 5 billion light years acts as a gravitational lens to produce a multiple image of a more distant galaxy (indicated by rectangle). The light from the more distant galaxy set off when the universe was just 7% of its current age of about 15 billion years. This galaxy is therefore regarded as being the most distant object ever seen, at 14 billion light years (reproduced by permission of the Space Telescope Science Institute)

Figure 15.4 Four images of the same galaxy. The light from a distant galaxy takes four different paths around the intermediate galaxy on its way to the Hubble Space Telescope (reproduced by permission of the Space Telescope Science Institute)

the head (many billions of years later). This question is equivalent to asking whether the universe will continue to expand indefinitely, will slow to a steady state or will stop expanding and begin contracting again. Both answers depend on the density of the universe. In the first case, photon paths will be bent sufficiently to return to the sender if the density is above a critical value. This critical density corresponds to that required to allow a recontraction of the universe, the so-called closed universe scenario. If the universe has a density less than critical it will continue expanding indefinitely and the beam of light will spiral out through the universe, never returning to its starting position. When the density is precisely critical the beam will still not return but the universe asymptotically approaches a steady state.

Clearly, sending a laser beam off into space and waiting to see if it comes back is not a practical way of determining the geometry of space and, by association, the future of the universe. The critical density must be calculated from general relativity theory and the actual density must be measured. The critical density is a seemingly tiny 5×10^{-27} kg m^{-3}. The mass of a hydrogen atom is almost 2×10^{-27} kg. Such is the emptiness of space, however, that visible matter, mainly in the form of stars, can muster only about 1% of the required density. Dark matter, in the form of black holes, neutron stars, black dwarfs and brown dwarfs, may be able to add as much as 4% to the total. The existence of dark matter can be verified by its influence on nearby visible matter as discussed in Chapter 13 for the case of supernova remnants in binaries and in Chapter 14 for the rotation of galaxies. To reach critical density, however, an entirely different source of mass is required. There is no definite evidence for other entities with mass as yet but two possibilities remain. One is that neutrinos have a very small mass. As there are so many of them, their total mass could add up to a significant amount. The second is that other exotic particles could have been created in the early stages of the universe that have not yet been detected.

The Big Bang and Just After

The first few seconds of the universe are of great importance to what follows, in particular with reference to the critical density. The theories that explain those first few moments will therefore be (briefly) examined and their consequences considered. The word 'moment' is something of an exaggeration as the crucial period of time does not even amount to a billionth of a billionth of a billionth of a second. After that period had passed the universe was on a one-way journey.

During the first 10^{-35} of a second after the universe began to expand from its unknown origin, the laws of physics are thought to have operated in a very different way from those familiar even in the strange and wonderful scenarios already investigated in this book. There are four fundamental forces of nature: gravity, electromagnetism, the strong nuclear force (that holds nuclei together) and the weak nuclear force (important in radioactive decay processes and the proton–proton chain in stellar fusion). Throughout the whole of the present known universe (with the possible exception of black hole interiors) these four forces have their own domains and quite different effects, each on their own set of particles at characteristic ranges. During the first few instants of the universe, such were the conditions that the forces are thought to have been coupled, in other words to have acted together. The laws of nature were not different at this time but manifested themselves in a way that cannot be seen under normal circumstances. The ideas that describe such phenomena are known as grand unification theories. At present there is no grand unification theory that is entirely self-consistent but that is what makes this area of study so exciting.

The most important occurrence of the first femtosecond of the universe starts to take place after about 10^{-35} seconds and lasts for about 10^{-32} seconds. At this time the strong nuclear force decoupled from the electroweak force (the name given to the two remaining coupled forces, electromagnetic and weak nuclear, gravity having decoupled earlier) to release a tremendous quantity of energy. The size of the universe suddenly increased at a stupendous rate, a phenomenon known as inflation. Precisely how extensive this expansion was is still not fully agreed upon but a good guess takes the universe's total size from much smaller than a proton to something more imaginable on a human scale, maybe the size of a football. This represents growth by a factor of about 10^{24}. Some estimates are much greater, closer to 10^{50}. It may seem strange to say that variation in estimates by a factor of 10^{26} is secondary but in this case a few tens of zeroes can be bandied about without affecting the main consequences of the inflationary period. Regardless of specific quantification, the two most important results of inflation are both related to the extraordinary stretch that the universe undertook at that time and the incredible rate at which it took place. First, a homogeneous temperature was ensured. In changing from a very small entity, across which temperature fluctuations are certain to be tiny, to something much larger in a vanishingly short period of time, the temperature is guaranteed to be spread evenly. The consequences of this have already been discussed. The cosmic blackbody microwave radiation is almost perfectly isotropic, as it has been since the inflationary period. The second important consequence of inflation is to imply that the universe's density is close to critical. Without a good understanding of general relativity, rare among professors of physics let alone introductory astronomy students, the statement 'spacetime is flattened', means little. The underlying idea is that the expansion is so fast that gravitational knots, likely to cause the hypothetical laser beam discussed above to take sudden curves, are untied. Like the radiation in the universe, the matter becomes evenly distributed and at a specific density. Now those rows of zeroes do become important. The extent of the expansion during the inflationary period (the number of zeroes in the expansion factor) determines whether the universe is closed in a cycle of big bangs and big crunches or whether it will continue to expand indefinitely. The number of zeroes remains elusive to experimentalists and theoreticians alike at the moment. Nevertheless, the density of the universe immediately after inflation had to be very close to critical. After 15 billion years of expansion any deviation from critical density would have been magnified by many orders of magnitude. That the actual density is within one order magnitude of critical is strong evidence that the inflation process really did take place.

After the tiny period of time during which inflation occurred the universe remained a very hot place, filled with high-energy photons such as gamma-rays. Under such conditions matter is created. The conversion of mass to energy that occurs, for instance, during nuclear fusion reactions in stellar cores may take place in reverse. Particles with mass can be produced when two photons 'collide'. As for the conversion of mass to energy, the reverse process is governed by Einstein's mass/energy equivalence formula (equation (10.2)). When gamma-ray photons are annihilated to produce mass the product is in the form of matter/antimatter particle pairs. The very high energy photons that would have been present just after the inflationary period created relatively massive particles but such was the density of energy and matter that the reverse reactions could also take place with matter and antimatter particles colliding and annihilating each other to recreate photons. The relative numbers of photons and matter particles was thus a delicate balance determined by the rates of backward and forward reactions. The same might be expected to be true of the balance between matter and antimatter particles though the equilibrium was clearly broken at some stage as now only matter remains. The imbalance between matter and antimatter first set in during the very early universe when the temperature was greater than 10^{28} K. At this time particles known as X particles (not to be confused with X-rays) were present but, upon decay, produced electrons, positrons, quarks (the component particles of neutrons and protons) and antiquarks. There is an imbalance in the symmetry of the decay of X particles so that more matter quarks are produced than antimatter quarks. As subsequent annihilation and creation events proceeded, the balance in favour of matter remained. As creation events became more rare, all antimatter particles were annihilated leaving only the excess matter to remain. The conditions of the early universe just after the inflationary period can be recreated on Earth (on an enormously smaller scale) using very powerful particle accelerators and such experiments are a great aid to theoretical work that might otherwise have become almost unfettered by reality.

WORKED EXAMPLE 15.1

Q. What is the maximum wavelength of electromagnetic radiation required to produce electrons and protons?

A. The energy of a photon is given by equation (1.6) whereas the amount of energy required to produce a mass, m, is given by equation (10.2):

$$E_P = \frac{hc}{\lambda} \tag{1.6}$$

$$E = mc^2 \tag{10.2}$$

Two photons are converted to a matter/antimatter pair so that the energy per photon required is simply given by mc^2. Hence,

$$\frac{hc}{\lambda} = mc^2 \Rightarrow \lambda = \frac{h}{mc} = \frac{2.2 \times 10^{-42}}{m}$$

The mass of an electron is 9.1×10^{-31} kg meaning that a pair of photons of wavelength 2.4×10^{-12} m are required. A proton's mass is 1.67×10^{-27} kg so that photons of wavelength 1.3×10^{-15} m are needed. Both of these wavelengths correspond to γ-rays. As an indication of the conditions under which such photons are present recall Wien's displacement law:

$$\lambda_P T = 2.9 \times 10^{-3} \tag{1.4}$$

It is not necessary for the required photons to be the most common but equation (1.4) shows the temperatures at which the peak blackbody radiation is at 2.4×10^{-12} m and 1.3×10^{-15} m are about 1 billion kelvin and 2 trillion (10^{12}) K respectively. Such conditions are well beyond those in the core of a star but may be achieved in very small spaces in the target chambers of particle accelerators on Earth!

Isotopic Abundances

Eventually (after a whole microsecond!), the temperature of the universe dropped sufficiently so that the rate of heavy particle (protons and neutrons) production became much less than that of lighter particles (electrons). The number of protons and neutrons (and their antimatter counterparts) thus decreased due to annihilation while the electron/positron population continued to increase. In this soup of fundamental particles a whole series of reactions took place. Protons and electrons combined to produce neutrons while neutrons and positrons combined to produce protons. As the temperature dropped by a further order of magnitude or so, these combinatorial reactions decreased in frequency and neutrons, being less stable than protons in free space, tended to decay back into protons and electrons more quickly than the reverse process could restore their numbers. The result of this series of reactions was that the universe ended up containing more protons than neutrons.

The consequence of the universe containing more protons than neutrons can be seen by examining the next stage of its evolution. Below a billion kelvin or so, matter production stops completely and nuclear combination starts to take place. When this temperature was reached the universe's age had become imaginable, being measured in seconds. The imbalance of protons and neutrons meant that most protons could not find a partner neutron to combine with and so remained single. Those that did combine produced deuterium nuclei. Further fusion reactions followed to produce helium and very small quantities of lithium and beryllium. These processes were similar to those that occur in stars except that the time scale available was drastically reduced. As the universe expanded the temperature declined so that these nuclear fusion reactions soon ended and the make-up of the matter of the universe became locked. Only nucleosynthesis in stellar cores has changed it since the first few minutes of the universe. The unmatched protons became hydrogen nuclei when the universe had cooled enough for electrons to be captured. Most deuterium nuclei combined to become helium nuclei. The proportions of deuterium, helium and other nuclei that remain at the present time are the final pieces of evidence that support the Big Bang model. Theory predicts that helium should be the initial resting place for about 25% of the created matter with hydrogen making up almost all of the rest. Since then nucleosynthesis in stars has changed these proportions but stellar evolution theory gives good predictions for the scale of these changes. The observed proportions now are in good agreement with Big Bang theory. The much smaller proportion of deuterium that remains can also be predicted from Big Bang models. The deuterium fraction at formation turns out to be very sensitive to the density of the universe at that time. This means that a measurement of the present proportion of deuterium in the universe can be used to predict the density of the early universe. Projecting this density forward to the present day, using measured expansion rates, predicts a matter density close to that observed. That is, a second piece of evidence predicts the universe to have less than critical density implying that it is expanding indefinitely. Again, it must be remembered that considerable uncertainties surround both the theoretical models and experimental measurements so that the question of the universe's eventual fate remains unconvincingly answered.

Inconsistencies: Explained and Still to be Explained

One final question remains. Though there are still doubts about the details of the Big Bang model, there is definite qualitative and reasonable quantitative evidence to support all aspects. However, the matter in the universe is not isotropically distributed. This book has spent fourteen chapters explaining the properties of the lumps of matter that make up the universe. In the final chapter the concept of inflation has been described and evidence of the isotropic distribution of the cosmic blackbody microwave radiation has been given. If the radiation and particle baths were stretched into homogeneity during the inflationary period, how did the present uneven distribution of matter come about? There must have been some inhomogeneity in the early universe. Very careful observations by an orbiting satellite (the Cosmic Background Explorer) have found these ripples in the early universe. The cosmic blackbody microwave radiation is not perfectly isotropic after all. The blackbody temperature of radiation is slightly different when measured in different directions. These differences are measured in millikelvin and correspond to density fluctuations in the early universe of just a few parts in a million but their existence gave gravity something to grab hold of during the formation of galaxy clusters and the galaxies themselves. Observation of deep space suggests that galaxy formation occurred quite quickly after matter decoupled from radiation. Though galaxies appear to evolve they do not seem to self-destruct or reform in the way that their main components, stars, do. The large-scale structure of the universe seems to be set before matter and radiation decouple and remains broadly the same indefinitely, notwithstanding continuous expansion.

The current model of the universe is based on three main pieces of evidence: the galactic red-shift observations that led to Hubble's law, the cosmic blackbody microwave radiation and isotopic abundances. A neat model of a big bang that incorporates inflation, matter creation from photons and subsequent nuclear reactions has become generally accepted. If it survives the onslaught of new data that will pour in over the next century it will become accepted as fact. Before then, a lot more questions remain to be answered. Some unsolved mysteries survive within our own Solar System, many surround the behaviour of stars and galaxies but the quantification of the universe's density and therefore a determination of its eventual fate is perhaps the biggest problem that still faces science.

Questions

Problem

1 In a particle accelerator experiment on Earth, is it easier to create conditions for the formation of electron/positron pairs or proton/antiproton pairs?

Exercises

2 Explain how Hubble's law leads directly to the conclusion that all matter in the universe was once concentrated in a small volume.

3 Why is the characteristic temperature of the cosmic blackbody microwave radiation so low?

4 Why is it impossible to point to the centre of the universe?

5 What is the fundamental importance of the mean density of the universe?

6 How can the presence of dark matter be ascertained?

7 Give a brief history of the universe so far!

8 How do observed nuclear abundances in the universe give credence to the Big Bang theory?

9 Why are variations in the cosmic blackbody microwave radiation of importance in explaining the present structure of the universe?

10 What are the most important pieces of observational evidence that support the Big Bang theory?

A1 Measurement and Units

To understand quantitative aspects of astronomy it is necessary to comprehend some simple physics. Most of this is outlined in the main body of the book when it becomes necessary but the narrative assumes that the reader is already familiar with ideas of measurements and units. For those that are not, this short passage provides an introduction.

To describe, for instance, how large a planet is or how much power a star is emitting, a comparator is required. Sometimes it is convenient to say that a planet has a size of three Earth radii or that a star emits twelve times as much power as the Sun. Such units are not helpful in providing a systematic method of measurement that can usefully compare the size of a football and a planet or the power emitted by a star and a light bulb. Such a system exists and relies on a few precise definitions. By combining these standards a whole series of other units can be defined. The so-called SI system (Système International d'unités) relies on seven base units, only five of which are used in this book:

metre	the SI unit of length (symbol, m)
second	the SI unit of time (symbol, s)
kilogram	the SI unit of mass (symbol, kg)
kelvin	the SI unit of temperature (symbol, K)
ampere	the SI unit of electrical current (symbol, A)

Each of these units is defined according to some measurable phenomenon, for instance, the boiling and freezing temperature of pure water (under precisely defined conditions) in the case of the kelvin. Almost all equations are written in such a way that data is entered as SI units. If one quantity in an equation is an unknown then entry of all data into the equation yields the unknown parameter in SI units.

To make the units more convenient, prefixes may be added to the symbols to indicate multiplication or division by powers of a thousand. The SI unit for mass already includes such a prefix, k for kilo, indicating that 1 kilogram is equal to 1000 grams. Other prefixes are:

E	exa (a factor of 10^{18} or 1 000 000 000 000 000 000 or one billion billion)
P	peta (a factor of 10^{15} or 1 000 000 000 000 000 or one million billion)
T	tera (a factor of 10^{12} or 1 000 000 000 000 or one thousand billion or one trillion)
G	giga (a factor of 10^{9} or 1 000 000 000 or one billion)
M	mega (a factor of 10^{6} or 1 000 000 or one million)
k	kilo (a factor of 10^{3} or 1000 or one thousand)
m	milli (a factor of 10^{-3} or 0.001 or one thousandth)
μ	micro (a factor of 10^{-6} or 0.000001 or one millionth)
n	nano (a factor of 10^{-9} or 0.000000001 or one billionth)
p	pico (a factor of 10^{-12} or 0.000000000001 or one thousand billionth or one trillionth)
f	femto (a factor of 10^{-15} or 0.000000000000001 or one million billionth)
a	atto (a factor of 10^{-18} or 0.000000000000000001 or one billion billionth)

In addition, there are a number of multiplying prefixes that are not preferred SI units but which have common everyday use. The most common of these is c for centi, indicating a factor of one hundredth, for instance 1 cm is one hundredth of a metre. It is important to remember that when entering data in an equation the raw SI unit must be used, for instance 5×10^{-6} m rather than 5 μm and 12 kg rather than 1.2×10^4 g.

To combine units it is necessary to have a physical understanding of the measurable of interest. A simple example is speed. The equation that describes the average speed of an object is

$$\text{Speed} = \frac{\text{Distance travelled}}{\text{Time taken}}$$

The units of speed are obtained from the units of the variables that are combined in the equation. Hence, in this case, the units are m/s (metres per second), or, as used in this book, m s^{-1}. Units can always be found in this way if the expression that quantifies the measurable is known.

Often particular combinations of units are more conveniently given their own special name. For instance, Newton showed that a force applied to a mass causes it to accelerate in the direction of the force in proportion to the magnitude of the force, expressed mathematically as

$$\text{Force} = \text{mass} \times \text{acceleration}$$

The unit of force is therefore the product of the units of mass and acceleration. The latter measurable refers to the rate of change of speed and therefore has units of metres per second per second or m s^{-2}. The unit of force is thus kg m s^{-2}. This is rather a clumsy unit and so it is replaced by a unit named after Newton, symbolised by N, so that one newton is exactly equal to 1 kg m s^{-2}. Multiplying factors can be used as before, so that, for instance, 1 mN = 0.001 N.

A second example of unit replacement uses the newton in the original expression. The work done (or energy required) to move an object through a certain distance is given by the following expression:

$$\text{Work done} = \text{force used} \times \text{distance moved}$$

so that the unit for work, or energy, is N m. This unit is replaced by the joule, J. Energy can take on many different forms but the same unit is always used. When a star shines it radiates energy that is quantified in joules though this energy cannot easily be measured in terms of forces and distances moved. More normally, the rate at which energy is radiated is required, in other words how much energy is radiated per second. The replacement unit for J s^{-1} is the watt, W.

The examples given so far are actually used to define units. This is not always the case. For instance, the Stefan–Boltzmann law states that the amount of power radiated by a black body is proportional to the fourth power of the absolute temperature of the body. A constant of proportionality, σ, is required, however:

$$P_A = \sigma T^4$$

Using SI units, σ is measured to be numerically equal to 5.67×10^{-8}. The units of σ must cause the units on both sides of the equation to be the same, hence, as the units of P_A are W m^{-2} and of T^4 are K^4 then σ is given by 5.67×10^{-8} W m^{-2} K^{-4}. Though this is a bulky unit it simply describes a constant and not a variable and so no alternative unit is used.

This simple system can be applied to any physical situation and is used by scientists and engineers of all types throughout the world.

A2 Atoms, Ions and Molecules

The building blocks of matter; atoms, ions and molecules, are referred to throughout this book. Relevant properties are described when they arise. Here a brief introduction is given, allowing all the simple ideas to be written together in one place.

Much astronomical study is of very large objects. Planets, stars and galaxies behave according to the gravitational forces that act upon them, both at a distance and within their own bodies. Nevertheless, a large amount of information that is obtained about the processes occurring within astronomical bodies is extracted using a knowledge of the physics of their component particles. The behaviour of matter on this much smaller scale is dominated not by gravitational forces but by electrical forces.

The concept of the atom was originally conceived to represent the smallest component of matter, 'That which can be cut no more.' However, scientists of about one century ago realised that atoms were in fact composed of still smaller particles. It turned out that atoms comprise mainly empty space. The mass of an atom is dominated by a small, positively charged nucleus at its centre while considerably less massive, negatively charged electrons orbit at high speeds. The diameter of an atom, in other words the extent of electronic orbits, varies between about 0.1 nm and 0.3 nm whereas the nucleus is about 100 000 times smaller, varying between about 1 fm and 7 fm.

Electrons all have the same (negative) charge, defined in terms of the SI unit for current, the ampere. The following equation defines charge:

$$\text{Charge} = \text{current flowing} \times \text{time current flows}$$

so that the unit for charge is As, renamed coulomb, C. A useful analogy is that of water flowing out of a pipe. The amount of water that flows is simply the product of the flow rate and the time that the water flows for. One coulomb is the amount of charge that flows through a wire carrying one ampere in one second. The charge on an electron is just (negative) 1.6×10^{-19} C so that a current of 1 A passes $(1/(1.6 \times 10^{-19}) =)$ 6.2×10^{18} electrons per second. In the same way that gravitational forces are proportional to the masses of the interacting bodies then electrical forces are proportional to the charges of the interacting particles (see Chapter 10). Gravity is always an attractive force but electrical force may be attractive or repulsive. It is attractive when the charges involved are of different signs and repulsive when they are the same.

The overall charge of an atom is always zero. The positive charge of the nucleus is always balanced by the negative charge of the electrons. This implies that the nuclear charge must also vary in steps of (positive) 1.6×10^{-19} C. The nuclear particles that have this charge are known as protons. Thus atoms contain equal numbers of protons and electrons. The electrical attraction between the protons in the nucleus and the orbiting electrons holds the atom together as the gravitational force holds together the Solar System, for instance.

The number of protons or electrons in an atom is known as the atomic number and defines the type of atom. For instance, hydrogen has just one proton and one electron while helium has two of each, carbon six of each, oxygen eight of each, iron 26 of each, lead 82 of each and uranium, the largest naturally occurring atom, has 92 of each. All combinations up to 92 are possible though some are unstable. Some larger atoms have also been produced artificially.

The mass of a proton is 1.6726×10^{-27} kg and of an electron 1836 times less at 9.1094×10^{-31} kg. When the mass of most atoms are measured, however, they cannot be wholly accounted for by these two particles. A third particle is required to make up the mass. This particle is electrically neutral but has a mass almost equal to that of the proton at 1.6749×10^{-27} kg. It is known as a neutron. Neutrons are also particles that reside in the nucleus with the protons. Most nuclei contain similar numbers of protons and neutrons with larger nuclei (atomic number $> \sim 20$) often containing a few more neutrons than protons. A given atom need not always have the same number of neutrons. For instance, copper atoms all have 29 protons but 69% have 34 neutrons while 31% have 36 neutrons. None have 35 neutrons. Atoms that have different numbers of neutrons but the same number of protons are known as isotopes. The next atom to copper in the Periodic Table, nickel, has five different isotopes whereas the atom next to it, cobalt, has only one. Nuclear physics beyond the scope of this book is required to explain the reasons for the relative stability of different isotopes.

Electrons in atoms may only occupy certain specific orbitals. This is a consequence of quantum mechanics that it is not necessary to understand here. A rule known as Pauli's exclusion principle states that only two electrons may occupy a single orbital (and each of these two electrons must have different spin orientations). An electron in a given orbital has a precisely defined energy so that multi-electron atoms contain electrons that have a series of different energies (though some orbitals have equivalent energies). Low-energy orbitals fill up with electrons first but there remains a large number of unfilled orbitals at higher energies. Occasionally, an electron is able to obtain precisely the correct amount of energy to enable it to hop to a higher orbital. Usually it will stay there for a short period and then hop back again, sometimes via intermediate energy orbitals. The quantities of energy gained and lost in such orbital hops is often in the region of 10^{-18} to 10^{-19} J, in the same energy range as visible photons of electromagnetic radiation. Much astronomical investigation relies on measuring the wavelengths of light that particles in different regions of space absorb as their electrons are excited to higher levels and the wavelengths of light emitted as they relax back down again (see Chapter 11). This is a particularly important technique as each type of atom has its own unique series of energy levels so that particles can be identified by the wavelengths of light that they absorb and emit. In combination with Doppler effect studies (see Chapter 1) and other physical phenomena, astronomical spectroscopy is a very powerful tool.

In the latter part of the description of spectroscopy above, the word 'atom' was replaced with the word 'particle'. In many regions of space the temperature is very hot. Under these circumstances, one or more atomic electrons may often be able to gain sufficient energy not only to hop to a higher orbital but to escape the atom entirely. The remaining particle therefore has an overall positive charge. Atoms with unequal numbers of protons and electrons are known as ions. Ions also have different allowed electronic energies so that they can be spectroscopically identified. For instance, an oxygen atom, an oxygen ion with one less electron than proton and an oxygen ion with two less electrons than protons all have different sets of energy levels.

Under some circumstances, usually when the temperature is cool ($< \sim 1000$ K) atoms may join together to form conglomerates known as molecules. A molecule contains a number of atomic nuclei that share some of their electrons. For instance, an oxygen molecule contains two nuclei, composed of eight protons and eight neutrons each, and sixteen electrons. Inner electrons orbit their nearest nucleus whereas outer electrons form orbits about both nuclei. Once again, the energy levels of molecules are distinct from other particles. Molecules may also exist in an ionic form when they lose electrons. However, when temperatures are hot enough for thermal ionisation the intramolecular forces that hold the atoms together also tend to be overcome. Highly ionised molecules are therefore rarely stable but small molecules with perhaps one or two missing electrons do exist.

For simple studies of stars little more information about atoms, ions and molecules is needed though nuclear reactions are discussed in Chapters 10, 11 and 12 and the internal structure of protons and neutrons is mentioned in Chapter 15. For planetary studies, temperatures tend to drop considerably and atoms and molecules coagulate into more substantial forms. When a gas is cooled it will eventually condense into a liquid. The small forces between independent molecules at low temperatures are greater than their thermal energies and they roll together with no particular structure. When the temperature becomes lower still a certain order freezes into the liquid and it becomes a solid. Sometimes, the material can be thought of as being an enormous molecule with all the nuclei sharing their outer electrons (for instance, in metals) but sometimes the solid is better described as being a regular array of ions whose mutual attraction binds them together (for instance, in minerals). At very high temperatures gases become so hot that all constituent particles are ionised (sometimes so that all nuclei and electrons become separated). Such a soup of electrically charged particles is called a plasma and this state of matter constitutes stars.

The visible part of the universe consists almost entirely of matter that is in solid, liquid, gas or plasma form and so an understanding of the structure of such matter is clearly essential to explain astronomical phenomena. The passage above barely scratches the surface of the current understanding of matter (and is intentionally simplified) but should provide a useful reference for the main body of the narrative.

A3 Ellipses

Gravitationally coupled pairs such as star/planet, planet/moon or star/star combinations move about their barycentre in elliptical orbits. Most of the mathematical analysis in this book approximates ellipses as being circles. A circle is actually a specific type of ellipse as will be shown below. Kepler's laws have been given in the text in terms of both circles and ellipses. This appendix is a short introduction to ellipses so that the interested reader can apply the information given here to the examples and equations that are given in the text.

It is instructive to begin by drawing an ellipse symmetrically on xy-coordinates as :

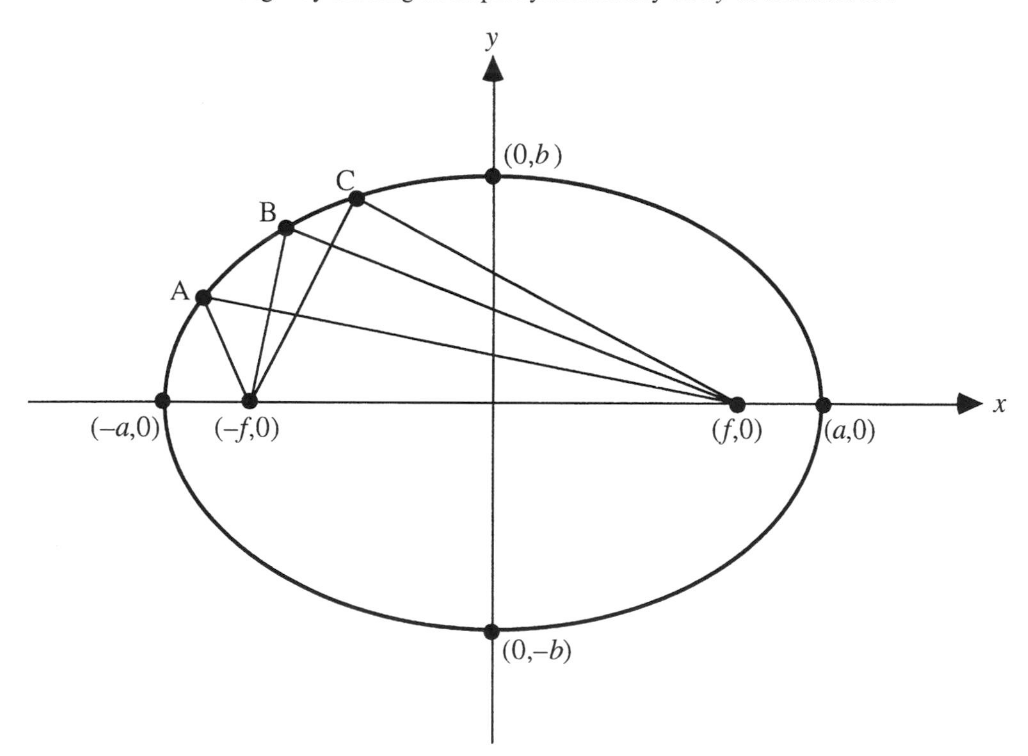

This ellipse is described by the following equation:

$$\frac{x^2}{a^2} + \frac{y^2}{b^2} = 1$$

The distance from the centre of the ellipse to its furthest points is a, known as the semi-major axis (being half of the length of the longest axis). The distance from the centre of the ellipse to the closest points is b, known as the semi-minor axis (being half of the length of the shortest axis). Two points on the major axis are defined as being the foci, both of which are at a distance f from the centre of the ellipse at coordinates $(f, 0)$ and $(-f, 0)$. The sum of the distance from both foci to any point on the ellipse is a constant. This is illustrated for three points, A, B and C, on the diagram. The sum of the lengths of the two lines emanating from each of the three points is a constant. The distance of the focus from the centre can be simply calculated, for instance by considering that the sum of the distances from both foci to $(a, 0)$ and to $(0, b)$ are the same:

$$(a-f)+(f+a)=(f^2+b^2)^{1/2}+(f^2+b^2)^{1/2}$$

$$\Rightarrow f=(a^2-b^2)^{1/2}$$

This leads to a definition of the eccentricity of an ellipse. The eccentricity, e, is given by the ratio of the focal distance to the semi-major axis, that is,

$$e=f/a=\frac{(a^2-b^2)^{1/2}}{a}=\left(1-\frac{b^2}{a^2}\right)^{1/2}$$

As the eccentricity increases (towards a maximum of 1) the ellipse becomes flatter and flatter and the foci move toward the edge of the ellipse. When the eccentricity is 0, $a=b$, which leads to the following equation for this particular ellipse, putting, $r=a=b$:

$$\frac{x^2}{r^2}+\frac{y^2}{r^2}=1$$

$$\Rightarrow x^2+y^2=r^2$$

which is the equation of a circle of radius, r. Hence, a circle is a special type of ellipse and obeys the same mathematical rules.

For planetary orbits about the Sun, the Sun is effectively static while the planets orbit in ellipses about it. Under these circumstances, the Sun's position is at one of the foci of each orbital ellipse. Note that the ellipse in the figure is not particularly elongated though its eccentricity is about 0.73. The largest eccentricities for planetary orbits are for Mercury and Pluto at 0.21 and 0.25 with all other planetary eccentricities being under 0.1. Thus planetary orbits can be quite well approximated as circles. If the eccentricity and the semi-major axes of an orbit are known then the distance of perihelion (position of closest approach to the Sun) is simply given as being

$$a-f=a-(e\times a)=a(1-e)$$

and the distance of aphelion (furthest distance from Sun) is given by

$$a+f=a+(e\times a)=a(1+e)$$

The proportional variation in distance from the Sun for Pluto is thus $(a(1+e)/a(1-e)=1.25/0.75=)$ 1.7. For Earth, the same calculation with $e=0.017$ gives a proportional change of 1.035 so that the Earth's distance from the Sun varies by only 3.5%. Substituting, $a=149.6$ million kilometres gives perihelion and aphelion distances of 152.1 million km and 147.1 million km.

The approximation to circular orbits does not work as well for binary stars. As the mass distribution is also not usually strongly skewed, it is necessary to consider each star as moving in its own elliptical orbit about the barycentre. Though mathematical proofs become more complex, the resulting physical laws operate precisely as for circles with $a/2$ substituting for r.

Finally, bodies that orbit with highly eccentric orbits should be mentioned. The best examples of such are comets. Halley's Comet has an eccentricity of 0.967. Its path takes it to closer than 90 million km from the Sun before sweeping out to more than 5 billion kilometres, beyond the orbit of Neptune. Note that Kepler's third law still applies. The cube of the semimajor axis of the orbit of Halley's Comet (~ half of 5 billion km) is in the same proportion to the square of its period (76 years) as the cube of the Earth's distance from the Sun (150 million km) to the square of its period (1 year). Some comets have eccentricities of precisely 1. This implies that they are following non-returning (hyperbolic) paths, having been perturbed from their slow motion in the outer Solar System (probably in the Oort cloud) and travelled towards the Sun only to be sling-shot out of the Solar System forever.

A4 Historical Milestones in Astronomy

BC

1600 (and onwards for more than a thousand years) Babylonians become the first civilisation to complete star catalogues, track planetary positions and make detailed observations of eclipses.

520 Pythagoras postulates a universe with a spherical Earth at its centre.

340 Aristotle adds to Pythagoras' model by including spherical motions.

250 Aristarchus proposes a Sun-centred model of the universe with the Earth spinning on its axis once per day.

200 Eratosthenes calculates the size of the Earth using shadow lengths.

AD

130 Ptolemy produces a thorough model of the universe with the Earth at the centre.

1054 Chinese and Japanese astronomers note supernova visible during daytime for 23 days.

1543 Copernicus publishes *De Revolutionibus orbium coelestium* (On the Revolution of the Celestial Orbs) on his deathbed. A model of the universe with the Sun at the centre finally begins to become established.

1572 Brahe makes observations of a supernova and is rewarded with the opportunity to build an observatory (Uraniborg, Castle of the Heavens). Here (and later in Prague) he amasses data on planetary motions and develops a new (incorrect) theory of the universe.

1590 Galileo shows that bodies of different masses fall to the Earth at the same rate.

1594 Kepler publishes *Mysterium Cosmographicum* (The Cosmic Mystery) supporting the Copernican model with new evidence.

1608 Lippershey invents the refracting telescope.

1609 Kepler (now Brahe's successor in Prague) publishes his first two laws of planetary motion in *Astronomia nova* (New Astronomy) based on the observations of Brahe.

1610 Galileo discovers Io, Europa, Ganymede and Callisto (the Galilean moons) and observes mountains on the moon using a rudimentary telescope.

1618 Kepler publishes his third law of planetary motion.

1632 Galileo publishes *Dialogue on the Two Chief World Systems*, a discussion of the geo- and heliocentric models and is placed under house arrest by the Inquisition.

1655 Titan discovered.

1668 Newton invents the reflecting telescope.

1671 Newton recognises that white light is composed of a continuum of component colours, each of which interacts with matter in a different way.

1675 Roemer measures the speed of light.

1687 Newton publishes *Principia Mathematica* (Principles of Mathematics) in which his theory of gravity and laws of motion are included.

1705 Halley predicts return of a comet (Halley's comet) 76 years after he previously observed it in 1682.

1781 Herschel discovers Uranus.
1795 Herschel discovers the first binary stars.
1800 Herschel observes the heating effect on a thermometer of the portion of the spectrum of white light beyond the visible thereby discovering infrared radiation.
1803 Dalton proposes an atomic theory of matter.
1826 Niepce produces the first permanent photographs.
1835 *Dialogue on the Two Chief World Systems* is taken off the Catholic church's forbidden book list.
1839 Schönbein discovers ozone.
1843 Schwabe points out variation in sunspot cycle.
1845 Adams calculates position of Neptune from orbital deviations of Uranus and transmits his results to Airy who is sceptical and takes no action.
1846 Leverrier makes similar calculation as Adams (1845) and transmits his result to Galle who discovers Neptune immediately.
1859 Kirchhoff expounds ideas explaining differences between atomic and condensed matter spectra.
1862 Clark observes (what is later to be seen to be) the first white dwarf, Sirius B.
1864 Maxwell publishes a theory that suggests light to be an electromagnetic wave.
1868 Janssen and Lockyer spectroscopically discover helium in the Sun.
1885 Balmer devises empirical formula to describe regular patterns of lines in stellar spectra.
1888 Hertz discovers a new type of electromagnetic wave, the radio wave.
1895 Röntgen discovers a further type of electromagnetic wave, X-rays.
1896 Becquerel discovers radioactivity.
1897 Thomson discovers the electron and measures its properties.
1900 Planck publishes blackbody radiation theory and indicates that light energy is quantised.
1900 Villard discovers the final type of electromagnetic radiation, gamma-rays.
1903 Tsiolkovsky publishes theories on rocket propulsion.
1905 Einstein proves the quantum nature of light energy via the photoelectric effect.
1905 Einstein publishes the special theory of relativity ($E = mc^2$, etc.)
1908 Hale discovers strong magnetic fields of sunspots.
1911 Rutherford discovers the atomic nucleus.
1913 Bohr publishes a theory of the atom in which energy levels are quantised to explain the Balmer series, etc.
1916 Einstein publishes the general theory of relativity (relating to gravitational curvature of spacetime, etc.).
1917 Schwarzschild calculates size of the event horizon of a black hole.
1919 Rutherford proves that protons exist in atomic nuclei.
1920 Rutherford proposes the existence of neutrons.
1924 Hubble proves that the Andomeda galaxy is an entity not connected to the group of stars to which the Sun belongs thereby proving the existence of galaxies.
1926 Goddard launches the first liquid-fuelled rocket.
1926 Hubble proposes scheme to classify galaxies.
1929 Hubble promulgates his observation that galaxies are receding in proportion to their distance from us (Hubble's law).
1930 Tombaugh discovers Pluto.
1931 Pauli predicts the positron.
1932 Anderson discovers the positron.
1932 Urey discovers deuterium.
1932 Chadwick discovers the neutron.
1932 Jansky publishes the first observation of a radio source in outer space (which later turns out to be at the centre of the Milky Way).
1935 Watson-Watt invents radar.
1935 Chandrasekhar applies theories of electron degeneracy to stars.
1948 Gamow (and others) announce the Big Bang theory.
1957 The first artificial satellite (Sputnik 1) is launched.

1959 First probe goes to Moon (Luna 1) and following probe (Luna 3) returns first pictures of the far side of the Moon.
1960 Sandage and Matthews discover the first quasar.
1961 Gagarin becomes the first human in space.
1965 Penzias and Wilson discover the cosmic blackbody microwave radiation.
1966 Surveyor 1 soft lands on the Moon and transmits pictures from the surface.
1967 Burnell and Hewish discover pulsars.
1968 Experimental evidence for quarks from several sources.
1968 First manned lunar orbiter (Apollo 8).
1969 Armstrong and Aldrin become the first people to walk on the Moon as part of Apollo 11 mission.
1970 Uhuru X-ray observatory launched.
1971 Mariner 9 orbits Mars and returns pictures.
1972 Apollo 17 is most recent mission to send people to the Moon.
1972 Pioneer 10 travels to Jupiter and returns data.
1973 Mariner 10 flies by Mercury (three times) and returns images.
1975 Venera 9 becomes first successful probe to return images from Venusian surface.
1976 Viking 1 sends first images from Martian surface.
1977 Voyager 1 and 2 launched.
1978 International Ultraviolet Explorer satellite launched.
1978 Einstein Observatory X-ray telescope launched (and produces images of the galatic centre, etc.).
1978 Charon discovered.
1979 Voyager 1 and 2 flyby of Jupiter, return images, data and discover new satellites.
1980 Very Large Array radio interferometer completed.
1980 Voyager 1 flyby of Saturn returns images, data and discovers new satellites.
1981 Voyager 2 flyby of Saturn returns images and data.
1981 First flight of the Space Shuttle.
1986 The first permanently manned space station, Mir, is launched.
1986 Voyager 2 flyby of Uranus returns images and discovers new satellites.
1986 Giotto encounters Halley's Comet, returning images and chemical analysis.
1987 Shelton detects supernova 1987A.
1989 Launch of Cosmic Background Explorer that goes on to detect inhomogeneities in the cosmic microwave blackbody radiation.
1989 Voyager 1 flyby of Neptune returns images and discovers new satellites.
1989 Launch of Magellan probe which spends three years making detailed radar images of Venus.
1990 The Röntgen satellite launched to study X-ray sources.
1990 The Hubble Space Telescope launched from the space shuttle.
1991 Galileo probe takes close-up images of an asteroid (Gaspra).
1992 Galileo cleared of his official condemnation of 1632 by the Catholic church.
1993 The Hubble Space Telescope's flawed mirror corrected for during space walk from space shuttle.
1996 Galileo space probe orbits Jupiter, returning pictures of Jupiter and its satellites and sending entry probe into Jupiter's atmosphere.
1997 Cassini space probe to Saturn and Titan is launched to fly through Saturn's rings and land on Titan.
1997 Mars pathfinder lands on Mars with surface rover and returns pictures plus chemical analysis.
1998 Lunar prospector launched to produce high-resolution maps of the Moon and detects frozen polar water-ice deposits.

A5 Compendium of Astronomical Data

The Planets of the Solar System

	Mercury	Venus	Earth	Mars	Jupiter	Saturn	Uranus	Neptune	Pluto
Distance from Sun (semi-major axis, in millions of km)	57.9	108.2	149.6	227.9	778.4	1424	2871	4499	5906
Sidereal period (Earth years)	0.241	0.615	1.00	1.88	11.9	29.5	84.0	165	248
Orbital eccentricity (see Appendix 3)	0.21	0.01	0.02	0.09	0.05	0.06	0.05	0.01	0.25
Direction of revolution (relative to direction of Sun's rotation)	Same	Same	Same	Same	Same	Same	Same	Same	Same
Angle of plane of revolution to ecliptic (Earth by definition)	7.0°	3.4°	0	1.8°	1.3°	2.5°	0.8°	1.8°	17.1°
Angle of plane of revolution to spin axis	0.1°	178°	23.5°	25.2°	3.1°	26.7°	97.9°	29.6°	122°
Rotation period (Earth days)	58.7	243	1.00	1.03	0.41	0.43	0.72	0.67	6.4
Volumetric mean radius (km)	2440	6050	6370	3390	69 900	58 200	25 400	24 600	1140
Surface temperature (K)	100–700	730	260–310	190–240	110–150	97	58	58	50
Mass($\times 10^{24}$ kg)	0.33	4.87	5.98	0.64	1900	569	86.8	102	0.013
Mean density (kg m^{-3})	5430	5200	5520	3930	1330	690	1320	1640	2050
Number of known moons	0	0	1	2	16	18	17	8	1
Albedo	0.06	0.72	0.39	0.16	0.70	0.75	0.90	0.82	0.15
Synodic period (Earth days)	116	584	n/a	780	399	378	370	368	367
Main colour	Grey	White	Blue	Red	Yellow	Yellow	Blue	Blue	Red
Ring system	No	No	No	No	Yes	Yes	Yes	Yes	No

The Moons of the Solar System

Parent planet	Name of moon	Distance to planet (thousands of km)	Orbital period (Earth days)	Volumetric mean radius (km)	Mass where known (kg)	Mass as percentage of parent planet (if >0.001%)	Year of discovery
Earth	(the) MOON	384	27.32	1740	7.35×10^{22}	1.2	
Mars	Phobos[a]	9	0.32	~12	1.1×10^{16}		1877
	Deimos[a]	23	1.26	~7	1.8×10^{15}		1877
Jupiter	Metis	128	0.29	20	9×10^{16}		1979
	Andrastea[a]	129	0.30	~10	2×10^{16}		1979
	Amalthea[a]	181	0.50	~100	7×10^{18}		1892
	Thebe[a]	222	0.67	~50	8×10^{17}		1979
	IO	422	1.77	1820	8.9×10^{22}	0.0047	1610
	EUROPA	671	3.55	1570	4.8×10^{22}	0.0025	1610
	GANYMEDE	1070	7.15	2630	1.5×10^{23}	0.0078	1610
	CALLISTO	1880	16.7	2400	1.1×10^{23}	0.0057	1610
	Leda	11 100	239	5	6×10^{15}		1974
	Himalia	11 500	251	85	9×10^{18}		1904
	Lysithea	11 700	259	12	8×10^{16}		1938
	Elara	11 700	260	40	8×10^{17}		1905
	Ananke[b]	21 200	631	10	4×10^{16}		1951
	Carme[b]	22 600	692	15	9×10^{16}		1938
	Pasiphae[b]	23 500	735	18	2×10^{17}		1908
	Sinope[b]	23 700	758	14	8×10^{16}		1914
Saturn	Pan	134	0.56	10			1990
	Atlas[a]	138	0.60	~15			1980
	Prometheus[a]	139	0.61	~50	1.4×10^{17}		1980
	Pandora[a]	142	0.63	~50	1.3×10^{17}		1980
	Epimetheus[a]	151	0.69	~60	5.5×10^{17}		1980
	Janus[a]	151	0.69	~100	2×10^{18}		1980
	Mimas	186	0.94	199	3.8×10^{19}		1789
	Enceladus	238	1.37	249	7.3×10^{19}		1789
	TETHYS	295	1.89	530	6.2×10^{20}		1684
	Telesto[a]	295	1.89	~12			1980
	Calypso[a]	295	1.89	~12			1980
	DIONE	377	2.74	560	1.1×10^{21}		1684
	Helene[a]	377	2.74	~16			1982
	RHEA	527	4.51	764	2.3×10^{21}		1672
	TITAN	1220	15.9	2580	1.4×10^{23}	0.024	1655
	Hyperion[a]	1480	21.3	~150	2×10^{19}		1848
	IAPETUS	3560	79.3	718	1.6×10^{21}		1671
	Phoebe[b]	13 000	550	110	4×10^{19}		1898
Uranus	Cordelia	50	0.34	13			1986
	Ophelia	54	0.38	16			1986
	Bianca	59	0.43	22			1986
	Cressida	62	0.46	33			1986
	Desdemona	63	0.47	29			1986
	Juliet	64	0.49	42			1986
	Portia	66	0.51	55			1986
	Rosalind	69	0.56	29			1986
	Belinda	75	0.62	34			1986
	Puck	86	0.76	77			1986
	Miranda	130	1.41	235	6.6×10^{19}		1948
	ARIEL	191	2.52	579	1.4×10^{21}	0.0018	1851
	UMBRIEL	266	4.14	585	1.2×10^{21}	0.0012	1851
	TITANIA	436	8.71	789	3.5×10^{21}	0.0068	1787
	OBERON	583	13.5	761	3.0×10^{21}	0.0069	1787
	S/1997 U2	5600	400	80			1997
	S/1997 U1	5700	410	40			1997

The Moons of the Solar System *(continued)*

Parent planet	Name of moon	Distance to planet (thousands of km)	Orbital period (Earth days)	Volumetric mean radius (km)	Mass where known (kg)	Mass as percentage of parent planet (if >0.001%)	Year of discovery
Neptune	Naiad	48	0.29	29			1989
	Thalassa	50	0.31	40			1989
	Despina	53	0.33	74			1989
	Galatea	62	0.43	79			1989
	Larissa	74	0.55	96			1989
	Proteus	118	1.12	208	4×10^{19}		1989
	TRITON[b]	355	5.88	1350	2.1×10^{22}	0.021	1846
	Nereid	5510	360	170	2×10^{19}		1949
Pluto	CHARON	19	6.39	586	1.7×10^{21}	7.9	1978

[a] Moons that are known to be irregular in shape (diameters given as approximate average dimension). Other moons that have not yet been well imaged may also be irregular.
[b] Moons that orbit their planet in a retrograde direction.
The most significant moons (mass > 10^{20} kg) are capitalised.

The Planetary Ring Systems of the Solar System

Planet	Distances from planetary centre (thousands of km)					Size of typical constituent bodies
	Planetary surface	Inner ring	Inner ring (main)	Outer ring (main)	Outer ring	
Jupiter	71.5	72	122	129	180	10 μm
Saturn	60.3	67	71	140	480	1 cm to 10 m
Uranus	25.6		42	51		Uncertain
Neptune	24.8	42	53	63		Uncertain

The Four Largest Asteroids

Asteroid	Average distance to Sun (millions of km)	Sidereal period (Earth years)	Orbital eccentricity	Angle of plane of revolution to ecliptic	Equatorial radius (km)	Approximate mass ($\times 10^{20}$ kg)
Ceres	414	4.60	0.08	10.6°	487	10
Pallas	415	4.61	0.24	34.8°	269	2.5
Vesta	353	3.63	0.09	7.1°	263	3
Hygiea	470	5.59	0.14	3.8°	215	1.5

Data for Selected Stars (including 20 brightest stars and 5 closest stars)

Name of star	Distance (parsecs)	Apparent visual magnitude	Absolute visual magnitude	Type of star	Spectral class	Constellation	Comments
Sun	4.85×10^{-6}	−26.7	+4.9	MS	G2		Only special feature is proximity.
Proxima Centauri	1.3	+11.0	+15.5	MS	M6	Centaur (*Centaurus*)	Closest star beyond Sun.
α Centauri A	1.3	0.0	+4.4	MS	G2	Centaur (*Centaurus*)	2nd closest star. 4th brightest star.
α Centauri B	1.3	+1.3	+5.7	MS	G2	Centaur (*Centaurus*)	3rd member of triple star
Barnard's star	1.8	+9.5	+13.2	MS	M4	Ophiuchus the serpent bearer (*Ophiuchus*)	Largest proper motion (10.3″/year).
Sirius A	2.6	−1.5	+1.4	MS	A1	Great Dog (*Canis Major*)	Brightest star of night sky.
Sirius B	2.6	+8.3	+11.2	WD	n/a	Great Dog (*Canis Major*)	Binary partner of A. 1st WD discovered
Procyon A	3.5	+0.37	+2.6	sG	F5	Little Dog (*Canis Minor*)	
Procyon B	3.5	+10.7	+13.0	WD	n/a	Little Dog (*Canis Minor*)	Binary partner of A.
Altair	5.1	+0.77	+2.3	MS	A7	Eagle (*Aquila*)	
Fomalhaut	6.9	+1.2	+2.0	MS	A3	Southern Fish (*Piscis Austrinus*)	
Vega	7.7	0.0	+0.6	MS	A0	Lyre (*Lyra*)	Used to define magnitude scale.
Arcturus	10	0.0	+0.2	G	K2	Herdsman (Boötes)	3rd brightest star.
Pollux	11	+1.1	+0.7	G	K0	Twins (*Gemini*)	
Capella	13	+0.1	−0.4	G	G6	Charioteer (*Auriga*)	
Aldebaran	18	+0.9	−0.3	G	K5	Bull (*Taurus*)	
Achernar	21	+0.5	−1.3	MS	B3	River Eridanus (*Eridanus*)	
Regulus	21	+1.4	−0.3	MS	B7	Lion(*Leo*)	
Canopus	23	−0.7	−2.5	G	A9	Keel (of Argo) (*Carina*)	2nd brightest star.
Spica	67	+1.0	−3.2	MS	B1	(Virgin) (*Virgo*)	
Hadar	98	+0.6	−4.4	G	B1	Centaur (*Centaurus*)	
Becrux	140	+1.3	−4.7	G	B1	Southern Cross (*Crux*)	
Antares	160	+0.9	−5.2	SG	M2	Scorpion (*Scorpius*)	
Rigel	430	+0.1	−8.1	SG	B8	Orion the hunter (*Orion*)	
Betelgeuse	430	+0.5	−7.2	SG	M2	Orion the hunter (*Orion*)	1st star to be imaged as real size object.
Deneb	460	+1.3	−7.2	SG	A2	Swan (*Cygnus*)	

MS = main sequence, G = giant, SG = supergiant, sG = subgiant (between MS and G), WD = white dwarf

A6 Some Fundamental Physical Constants

Constant	Value	Units	Symbol
Speed of light (in vacuum)	2.9979×10^{8}	$m\,s^{-1}$	c
Planck's constant	6.6262×10^{-34}	J s	h
Stefan–Boltzmann constant	5.6696×10^{-8}	$W\,m^{-2}\,K^{-4}$	σ
Wien's constant	2.8979×10^{-3}	m K	
Gravitational constant	6.6682×10^{-11}	$N\,m^{2}\,kg^{-2}$	G
Permittivity of free space	8.8542×10^{-12}	$F\,m^{-1}$	ε_0
Proton mass	1.6726×10^{-27}	kg	m_p
Neutron mass	1.6749×10^{-27}	kg	m_n
Electron mass	9.1095×10^{-31}	kg	m_e
Electron charge	1.6022×10^{-19}	C	e

A7 Multiple-Choice Quiz

Learnt anything? Find out . . .

1 A technique developed to overcome scintillation is:

A Adaptive optics.
B Chromatic aberration.
C The Coudé system.
D Retrograde motion.

2 The advantage of building telescopes with larger apertures is:

A Improved angular resolution only.
B Improved angular magnification only.
C Improved angular resolution and light grasp.
D Improved angular resolution and angular magnification.

3 The declination of Polaris as viewed from Glasgow (latitude $= 55^\circ\ 50'$N) is:

A $+55^\circ\ 50'$.
B $+34^\circ\ 10'$.
C $-34^\circ\ 10'$.
D $+90^\circ$.

4 How often does the vernal equinox cross the observer's meridian?

A Always once every 24 sidereal hours.
B Always once every 24 solar hours.
C It varies, being more frequent in summer than in winter.
D It varies, being less frequent in summer than in winter.

5 What is the approximate surface temperature on Pluto?

A 4 K.
B 40 K.
C 400 K.
D 4000 K

6 The main reason that the surface of Venus is difficult to observe is:

A Venus is always close in the sky to the Sun.
B Venus is covered by thick clouds of sulphuric acid.
C Venus has a very low albedo.
D Venus is a very long way from Earth.

7 Where is the Great Red Spot?

A The Sun.
B Mars.
C Jupiter.
D Sirius.

8 The main process that creates energy in the Solar System is:

A Nuclear fusion.
B Nuclear fission.
C Gravitational contraction.
D Gravitational differentiation.

9 Are the surfaces of white stars hotter than red stars?

A Yes, always.
B No, always.
C Yes, except if the white star is a white dwarf.
D Yes, except if the red star is a red giant.

10 A very large star is likely to end its life as:

A A neutron star.
B A black hole.
C A white dwarf.
D A red giant.

11 Details of the light output of eclipsing binary stars can be used to determine which of the following stellar properties?

A Masses only.
B Masses and surface temperatures only.
C Masses and diameters only.
D Masses, surface temperatures and diameters.

12 The Milky Way is an example of which of these types of galaxy?

A Elliptical.
B Quasar.
C Active.
D Spiral.

13 The Earth's sidereal period is 365 days. The Moon's sidereal period is 27.3 days. The system's synodic period is 29.5 days. How often do spring tides take place?

A Once every 365 days.
B Once every 13.7 days.
C Once every 14.8 days.
D Once every 12.4 days.

14 How many of the following provide evidence for the Big Bang theory?

(1) Doppler shift of galactic spectra; (2) The microwave background; (3) Chromatic aberration.

A Only one.
B Only two.
C All three.
D None.

15 How many solar days are equal to 183.0 sidereal days?

A 182.5.
B 183.0.
C 183.5.
D None of the above.

16 What is the surface temperature of the Sun?

A 580 K.
B 5800 K.
C 58 000 K.
D 580 000 K.

17 Which of the following is not a part of the Solar System?

A Vega.
B Neptune.
C Halley's Comet.
D The Sun.

18 One parsec is a distance of:

A 500 nm.
B 3.26 km.
C 150 000 km.
D 3×10^{13} km.

19 The coordinate that always has a value of $+90°$ for Polaris is:

A Declination.
B Latitude.
C Altitude.
D Azimuth.

20 How many times does the Earth spin on its axis during one complete orbit around the Sun?

A 28.5.
B 365.
C 365.24.
D 366.26.

21 Which of the following is not associated with the solar wind?.

A Aurorae.
B Bursts of noise in radio communications.
C Comet tail direction.
D Volcanic activity.

22 Proxima Centauri has a parallax of 0.77″. How far away from Earth is this star?

A 0.77 parsec.
B 1.0 parsec.
C 1.3 parsec.
D 1.5 parsec.

23 Large, young stars are which one of the following in comparison to small, young stars?

A Hotter.
B Redder.
C Dimmer.
D None of the above.

24 Which one of the following statements is true?

A Mercury has a dense atmosphere.
B Venus has an increased surface temperature due to the greenhouse effect.
C The Moon's upper atmosphere contains a lot of ozone.
D Mars has no moons.

25 Which of the following statements is true?

A Jupiter is composed mainly of hydrogen and helium.
B Jupiter is a perfect sphere.
C Jupiter is smaller than the Earth.
D Jupiter has no ring system.

26 Consider the following statements about a very large star:

(1) Upon initial stabilisation it is red; (2) Nuclear reactions cease suddenly and cause a supernova; (3) It ends its life as a black hole.

Which of the three statements are true?

A All three statements.
B Only (1) and (2).
C Only (2) and (3).
D Only (1).

27 Consider the following statements about Mars:

(1) The surface is covered in dried-up canals that are generally believed to be evidence of former civilisation; (2) The most prominent feature is the Great Red Spot; (3) It is a large, gassy planet of the 'failed star' type.

How many of the three statements are true?

A None of them.
B One of them.
C Two of them.
D All of them.

28 Which of the following is a type of active galaxy?

A Pulsar.
B Quasar.
C Radar.
D RR Lyrae.

29 What is the characteristic temperature of the cosmic blackbody microwave radiation?

A 2.7 K.
B 27 K.
C 270 K.
D 2700 K.

30 How many of the following processes were involved in the formation of the Solar System?

(1) Accretion; (2) Gravitational contraction; (3) Condensation; (4) Scintillation.

A One of them.
B Two of them.
C Three of them.
D All of them

A8 Selected Answers

Chapter 1

1 $\gamma \leftrightarrow x$, 10^{-13} J; $x \leftrightarrow$ uv, 10^{-16} J; uv $\leftrightarrow$ vis, 6×10^{-19} J; vis $\leftrightarrow$ ir, 2×10^{-19} J; ir $\leftrightarrow \mu$, 10^{-21} J; $\mu \leftrightarrow$ rad, 10^{-24} J

2 (a) 6 km s^{-1}; (b) 600 km s^{-1}

3 (a) 2 min, 17 s; (b) 4.2 years; (c) 1400 years; (d) 25 years; (e) 100 000 years; (f) 35 years; (g) 170 000 years; (h) 2.2 million years; (i) 10 billion years

4 violet, 360 nm, 8.3×10^{14} Hz; indigo, 400 nm, 7.5×10^{14} Hz; blue, 470 nm, 6.4×10^{14} Hz; green, 530 nm, 5.7×10^{14} Hz; yellow, 580 nm, 5.2×10^{14} Hz; orange, 610 nm, 4.9×10^{14} Hz; red, 700 nm, 4.3×10^{14} Hz

5 (a) 150 nm; (b) 290 nm; (c) 800 nm

6 sixty times larger

7 (a) $\sim 4 \times 10^{16}$; (b) $\sim 10^{10}$

9 (a) B; (b) A

Chapter 2

1 (a) $R = \dfrac{2.52 \times 10^{-4}\lambda}{D}$; (b) 1″

2 (a) 1.2×10^{-7} radians or 0.025″; (b) 12 million km (about 8% of the Earth–Sun separation); (c) 36 km

3 0.12 m

4 (a) 0.16 m; (b) 39

5 (a) 1.2×10^{-3} radians or 250″

Chapter 3

1 (a) 72°; (b) 6°

3 (a) −79° to +90°; (b) −52° to +90°; (c) −26° to +90°; (d) −90° to +67°; (e) −90° to +38°

4 (a) 42°; (b) circumpolar, 69°, 7° (c) circumpolar, 85°, 33°; (d) 8°; (e) never visible

5 Glasgow, 180°, 74°; Abidjan, 0°, 55°

6 30° 20′

7 15h
8 $-70°$
10 (a) $-25°$; (b) $+72°$

Chapter 4

1 (a) 12h 46m; (b) 8h 18m; (c) 2h 25m
2 $46° 0'$
3 23h
4 (a) $+90°$ or $-90°$ only; (b) $+66.5°$ to $+90°$; (c) $+90°$ or $-90°$ only; (d) $-66.5°$ to $-90°$
5 (a) 3h 12m; (b) 12h 31m; (c) 1h 32m, yes, no; (d) 1h 36m, yes, no
6 Hint: effective solar constant – 1400 W $\times \cos \phi$
7 365.2425 days (actual value: 365.2422 days so further adjustments required)
8 England, hint: consider angle of motion of Sun relative to horizon

Chapter 5

1 $46.3°$
2 Mercury : Earth : Pluto—10 400 : 1600 : 1
4 0.52 (→117 Earth Solar days per Venusian Solar day)
5 (a) Equator of Jupiter, 45 000 km h^{-1}; (b) Mercury, 170 000 km h^{-1}
6 (b) 29.5 days

Chapter 6

1 (a) 3.6×10^{22} N; (b) 3.6×10^{22} N; (c) 1.2×10^{18} N, invisible or very thin crescent with large angular size; (d) 3.0×10^{16} N, small and circular (full) though close in the sky to the Sun; (e) 1.7×10^{15} N, small and circular (full) though close in the sky to the Sun; (f) 3.0×10^{16} N, large and circular (full), particularly easy to see as in the sky during the darkest part of the night
2 2.2×10^{17} N, 1.1×10^{16} N
3 (a) 58 million km; (b) 790 million km; (c) 6.0 billion km
4 (a) 6.0; (b) 1.3; (c) 16
5 2.40
6 71% of the distance between the Earth's centre and its surface (in the direction of the Moon) or 4600 km from the Earth's centre
7 5.4×10^{26} kg
8 9.1 m s^{-2}
9 Every 14.8 days, half Moon (alternating between first quarter and third quarter)

Chapter 7

1 (a) 2×10^{29} J; (b) 700 K
2 (a) 2×10^{-3}; (b) 4×10^{30} J
3 15%
4 6 million years

Chapter 8

1 250 K

Chapter 9

1 (a) 65 K; (b) 80%

Chapter 10

1 (a) 5800 K; (b) 4×10^{26} W; (c) 5×10^{38} s^{-1}
2 6.7×10^{-31} kg, 1.0×10^{-29} kg, 2.2×10^{-29} kg (b) 4.3×10^{-29} kg
3 $\sim 10^{12}$
5 1.2×10^{36}

Chapter 11

1 440 parsecs
2 2.65 parsecs
3 1.6
4 (a) 64; (b) +0.4
5 (b) +6.9; (d) $\sim$ 70 000 km; (e) $\sim 7 \times 10^{25}$ W
6 2.04×10^{5}
7 −18.7

Chapter 12

1 (a) 1.1×10^{27} W
2 1.7×10^{30} kg, 6.9×10^{30} kg
3 4.3×10^{29} kg, 2.1×10^{29} kg (very small stars)
4 (a) 16 times; (b) 2.1 times
6 3×10^{30} kg, 6×10^{30} kg

Chapter 13

1 (a) 30 km; (b) 77 000 BC

2 (a) 1.2×10^{-12} J; (b) almost the same

3 3×10^{32} W (little variation if 5 solar mass neutron star has radius of 10 km and 1.4 solar mass neutron star has radius of 20 km)

4 10 kp

Chapter 14

1 Milky Way

2 Distance, between 100 and 200 Mp; rotational speed $\geqslant 140\ \mathrm{km\,s^{-1}}$

3 (a) $1.8 \times 10^{8}\ \mathrm{m\,s^{-1}}$; 2.8 Gp; (b) $2.4 \times 10^{8}\ \mathrm{m\,s^{-1}}$, 3.7 Gp; (c) $2.6 \times 10^{8}\ \mathrm{m\,s^{-1}}$, 4.0 Gp; (d) $2.8 \times 10^{8}\ \mathrm{m\,s^{-1}}$, 4.3 Gp

Appendix 7

1	A	**11**	D	**21**	D
2	C	**12**	D	**22**	C
3	D	**13**	C	**23**	A
4	A	**14**	B	**24**	C
5	B	**15**	A	**25**	A
6	B	**16**	B	**26**	C
7	C	**17**	A	**27**	A
8	A	**18**	D	**28**	B
9	A	**19**	A	**29**	A
10	B	**20**	D	**30**	C

Index

Page numbers in italic signify reference to worked example
Page numbers in bold signify reference to figure